Designing Successful Products with Plastics

PLASTICS DESIGN LIBRARY (PDL)
PDL HANDBOOK SERIES
Series Editor: Sina Ebnesajjad, PhD (sina@FluoroConsultants.com)
President, FluoroConsultants Group, LLC
Chadds Ford, PA, USA
www.FluoroConsultants.com

The **PDL Handbook Series** is aimed at a wide range of engineers and other professionals working in the plastics industry, and related sectors using plastics and adhesives.

PDL is a series of data books, reference works and practical guides covering plastics engineering, applications, processing, and manufacturing, and applied aspects of polymer science, elastomers and adhesives.

Recent titles in the series
Biopolymers: Processing and Products, Michael Niaounakis (ISBN: 9780323266987)
Biopolymers: Reuse, Recycling, and Disposal, Michael Niaounakis (ISBN: 9781455731459)
Carbon Nanotube Reinforced Composites, Marcio Loos (ISBN: 9781455731954)
Extrusion, 2e, John Wagner & Eldridge Mount (ISBN: 9781437734812)
Fluoroplastics, Volume 1, 2e, Sina Ebnesajjad (ISBN: 9781455731992)
Handbook of Biopolymers and Biodegradable Plastics, Sina Ebnesajjad (ISBN: 9781455728343)
Handbook of Molded Part Shrinkage and Warpage, Jerry Fischer (ISBN: 9781455725977)
Handbook of Polymer Applications in Medicine and Medical Devices, Kayvon Modjarrad & Sina Ebnesajjad (ISBN: 9780323228053)
Handbook of Thermoplastic Elastomers, Jiri G Drobny (ISBN: 9780323221368)
Handbook of Thermoset Plastics, 2e, Hanna Dodiuk & Sidney Goodman (ISBN: 9781455731077)
High Performance Polymers, 2e, Johannes Karl Fink (ISBN: 9780323312226)
Introduction to Fluoropolymers, Sina Ebnesajjad (ISBN: 9781455774425)
Ionizing Radiation and Polymers, Jiri G Drobny (ISBN: 9781455778812)
Manufacturing Flexible Packaging, Thomas Dunn (ISBN: 9780323264365)
Plastic Films in Food Packaging, Sina Ebnesajjad (ISBN: 9781455731121)
Plastics in Medical Devices, 2e, Vinny Sastri (ISBN: 9781455732012)
Polylactic Acid, Rahmat et. al. (ISBN: 9781437744590)
Polyvinyl Fluoride, Sina Ebnesajjad (ISBN: 9781455778850)
Reactive Polymers, 2e, Johannes Karl Fink (ISBN: 9781455731497)
The Effect of Creep and Other Time Related Factors on Plastics and Elastomers, 3e, Laurence McKeen (ISBN: 9780323353137)
The Effect of Long Term Thermal Exposure on Plastics and Elastomers, Laurence McKeen (ISBN: 9780323221085)
The Effect of Sterilization on Plastics and Elastomers, 3e, Laurence McKeen (ISBN: 9781455725984)
The Effect of Temperature and Other Factors on Plastics and Elastomers, 3e, Laurence McKeen (ISBN: 9780323310161)
The Effect of UV Light and Weather on Plastics and Elastomers, 3e, Laurence McKeen (ISBN: 9781455728510)
Thermoforming of Single and Multilayer Laminates, Ali Ashter (ISBN: 9781455731725)
Thermoplastics and Thermoplastic Composites, 2e, Michel Biron (ISBN: 9781455778980)
Thermosets and Composites, 2e, Michel Biron (ISBN: 9781455731244)

To submit a new book proposal for the series, or place an order, please contact Edward Payne, Acquisitions Editor at edward.payne@elsevier.com

Designing Successful Products with Plastics

Fundamentals of Plastic Part Design

Mark T. MacLean-Blevins

ELSEVIER

William Andrew
Applied Science Publishers

William Andrew is an imprint of Elsevier
The Boulevard, Langford Lane, Kidlington, Oxford, OX5 1GB, United Kingdom
50 Hampshire Street, 5th Floor, Cambridge, MA 02139, United States

Library of Congress Cataloging-in-Publication Data
A catalog record for this book is available from the Library of Congress

British Library Cataloguing-in-Publication Data
A catalogue record for this book is available from the British Library

ISBN: 978-0-323-44501-6

Publisher: Matthew Deans
Acquisition Editor: Edward Payne
Editorial Project Manager: Ana Claudia Garcia
Production Project Manager: Poulouse Joseph
Designer: Mark Rogers

Typeset by Thomson Digital

In honor of my family and in memory
of my son, Benjamin.

Contents

Foreword		**xi**
Preface		**xv**
Acknowledgments		**xvii**
Section I	**Background Knowledge**	**1**
1	**Introduction**	**3**
	1.1 Plastic part design versus product design	4
	1.2 Why choose plastics?	6
	1.3 The *Four Pillars* of plastic part design	11
	1.4 Integration of the *Four Pillars* in every decision during the design process	13
	1.5 Conclusions—forward looking lessons	15
	1.6 Scope of this text	16
	References	17
2	**Material selection—which plastic to use?**	**19**
	2.1 Selection by material properties	21
	2.2 Selection by aesthetic properties	42
	2.3 Selection by economic properties	45
	2.4 Final "How To" select a material	48
	References	49
3	**Process selection—which plastics process to use?**	**51**
	3.1 Selection by functional requirements	51
	3.2 Selection by capacity (parts per year)	58
	3.3 Selection by cost economics	72
	3.4 Selection considering secondary processes	72
	3.5 Selection by other criteria	75
	Reference	76
4	**What tooling will be required?**	**79**
	4.1 Process dependent	79
	4.2 Capacity dependent	88
	4.3 Material dependent	90
	4.4 Manufacturing location dependent	90
	4.5 Part design dependent	91

4.6 Tool quality relationship to part quality 91
4.7 Specifying the tool within the part design specification 92
4.8 Sourcing the tooling 93
Reference 96

Section II Designing With Plastics 99

5 Conceptual design—how do we begin the design process? 101
5.1 Understand the "Need" and the "User" for the product, part, or assembly 101
5.2 Solution ideation 109
5.3 Selection (choices, choices, choices) 112
5.4 Conceptual design review 118
5.5 Functional specifications 120
5.6 Action items upon departing the conceptual stage 123
References 124

6 Preliminary detailed design 127
6.1 Seeing the part(s), the assembly, and the process 127
6.2 Initial computer-aided design (CAD) layout in 3D 130
6.3 Design rules review 160
6.4 Prepare a preliminary bill of materials 182
6.5 Prototyping during active design 183
6.6 Review of the preliminary design solution 186
6.7 Action items upon departing the preliminary detailed design stage 192
Reference 194

7 Final detailed design 197
7.1 Iterative updates to the CAD 3D layout and BOM 197
7.2 Breakout for detailed design of features and parts 198
7.3 Integration of all parts back into a new 3D CAD master checking assembly layout 201
7.4 Detailed part drawings and specifications 203
7.5 Detailed part design checking 209
7.6 Review and update FMEA 216
7.7 Prepare for patent review 216
7.8 Final design review 217
7.9 Preparing for implementation 221

Section III Implementation of the Design 225

8 Gathering production estimates 229
8.1 First cost estimates 229
8.2 Nonrecurring cost estimates 235

8.3 Production unit cost estimates 237
8.4 Total cost evaluation 241
8.5 Preparation for project verification and implementation 243

9 Project review and verification—risk management 245
9.1 Gather the verification team 245
9.2 Verification of the design 246
9.3 Agreement to move to production implementation planning 249
9.4 Preparation for transitioning into production 250

10 Production implementation 253
10.1 Placing capital tooling orders 253
10.2 Review the preliminary tooling design layout 255
10.3 Quality planning for production (again management of risk) 263
10.4 First tooling trials 266
10.5 Prepare and file patent applications 271
10.6 Prepare for product qualification 272

11 Production qualification and hand-off 275
11.1 Product performance verification 275
11.2 Quality plan review 278
11.3 Hand-off to primary manufacturing and assembly 279
11.4 Periodic production monitoring and review 282

Section IV Lessons 285

12 Creating value through good design practices 287
12.1 The *Four Pillars*—complete integration is a requirement 287
12.2 Basis for design decisions rests on the functional specification 290
12.3 Verification and qualification is mandatory 290
12.4 Final part or product safety is our responsibility 290
12.5 Big picture thinking 291
12.6 Thanks and do your part! 292

Index 295

Foreword

The Plastics Industry has recovered from the shock of off shoring. Reshoring is picking up momentum. The market place is expanding. The industry's ability to keep pace is being hampered by its inability to find and hire technically competent people. The industry and academia have failed to produce enough, technically trained people to support the growth of the industry. This very real problem also contains an opportunity for anyone searching for a rewarding and secure career.

The plastics industry in the United States employees close to 1 million people. On average, the plastics industry is growing at a faster rate than the gross domestic product.

The industry employs unskilled labor, chief executive officers, and everything in-between. An increasing number of these jobs require special skills and/or technical knowledge. That brings me to the justification for the book—"Designing Successful Products with Plastics."

The plastics industry is concerned with producing plastic parts that can be sold at a profit. The author has restricted the focus of this book to parts that are structural in nature. This can be interpreted to refer to rigid three-dimensional parts that will be formed or molded. Even the simplest of three-dimensional parts require designer consideration to establish the size and shape of that part.

The author has chosen to concentrate on the design of products and parts made of plastic materials. However, as explained in Chapter 1, design is only one of the four main integrated disciplines that combines to produce a plastic part. These four disciplines are: (1) the design of the product or part, (2) the selection of the optimal plastic material, (3) the design and construction of the tooling or mold, and (4) the forming or molding of the actual part.

Other disciplines, such as decorating, trimming or machining, assembly, quality assurance, and so on, may or may not be required. However, the four main disciplines will always be there. The author envisions these integrated disciplines as *Four Pillars* supporting the optimum part.

All four of these disciplines must be handled correctly in order to produce an optimum part. All four disciplines are of equal value. Design is no more important than the other three disciplines. However, everything starts with design. If there is no design, there is no need for a plastic material, a mold or a molding machine. There is nothing to mold, nothing to sell and no opportunity to make a profit. No matter how elegant a product may be it will not be successful unless it can attract a buyer who will purchase it for a profit.

Each of the four disciplines represents a well-defined group of technologies. All of these subtechnologies represented by the four main integrated disciplines combine to create a mind boggling number of details that must be considered and dealt with in order to produce an optimum plastic part. Over the years each of these disciplines have been perfected. Today, the technology has become so complex that it is doubtful that any one person can be an expert in all parts of the four integrated disciplines.

The emphasis is on design and this book makes no attempt to teach the reader how to master all of the other technologies. Detailed information of that type is available from other sources.

This book is intended to teach the reader enough about each of the four integrated disciplines to allow the design of optimum plastic products and parts using the different plastic materials, molds, and processes that are available.

A further objective of this book is to present a proven process for organizing and managing all of the complex details that must be considered while designing a successful plastic product. All of these details will not apply to every new part. All of the details have however, been included to avoid the possibility of a busy designer overlooking an obscure but important detail.

The growth of this industry is such that it will attract employees, with no prior experience with plastics technology. Sadly the majority of students pursuing technical or scientific degrees graduate without being taught plastics technology. If these students majored in engineering they will leave the University with a thorough understanding of all of the latest computer aided engineering technologies. They will not however, have any experience on how to apply these technologies to real life plastics industry problems and opportunities.

Those wishing to pursue a plastic product design career without prior training or experience will find this book was written for them. The detailed procedure for organizing and managing a product design project will teach the novice what needs to be done. The later chapters that go into the details of the four integrated disciplines will explain how to achieve each step in the design project management process. If there are questions regarding the next step it is time to ask for clarification and help from more experienced colleagues.

I am nearer the end of my career than the beginning. However, I can still vividly remember how difficult it was for me as a new college graduate to admit to not knowing everything. Admitting what I did not know only encouraged others to not only teach me what I needed to know but also to help me through those first few formative years.

As you delve into this book you will discover that it is different from the other product design texts. It is not a dry dissertation of facts and figures. It is written in a very personal and refreshing style. After a few pages it will seem as though you are listening to a senior product designer telling stories about his successes and failures and what he learned from each of those experiences. The author started his design career in 1978. In 1993 he formed his own design-consulting firm. In more recent years his work has become international in scope. Through this book he will be giving you access to almost 40 years of real life experience that the beginning designer does not have. Experienced designers will smile at being reminded of their own similar successes and failures as they wonder if their solution to a problem was better or worse than the authors.

There will be instances when you would like to raise your hand and ask a question. You will not hear an answer to that question. However, there is a high probability that you will find the answer somewhere in the following chapters of this book.

I met Mark MacLean-Blevins in 2006 when he was elected to the Board of Directors of the Society of Plastics Engineers Product Design and Development Division. He still serves on that Board today. As a new Board member he was eager to contribute and volunteered for tasks that needed doing. His eyes twinkled and he had the creative person's ability to see things from more than one perspective. I have watched Mark for over 10 years and have learned that he is a capable, award winning product designer and a generally a nice gentlemen. He also has a serious side. It is that part of Mark that took time from his business and made the effort to write this book and record his experiences for posterity.

I read the early drafts of the chapters in this book. I am familiar with the book and the author. I have no hesitation in recommending this book to anyone wanting to know how to design plastic parts and manage a new product design project. Experienced designers wishing to benchmark and improve on what they are already doing will benefit from the author's experiences.

Glenn L. Beall
Libertyville, IL
2017

Preface

Every written work is composed for some distinct purpose, or so I am told. This work came from outside my peripheral vision, sort of sideways (not a Pinot Noir reference) and without much fanfare. Out of the blue, the email came and said, "Hey, we are looking for someone to write about plastics design …" and so, here it is. The timing was just about correct, and we all know that timing is everything. As an experienced engineer and designer of hundreds of plastic products made up of thousands of custom designed parts, I knew that there was a large void, which needed to be filled. There are some large numbers of engineers who graduate from colleges and universities all over the globe each year, perfectly prepared and ready for that first employment opportunity. Those who land in manufacturing companies and are blessed to fall into the product design group or team are suddenly faced with the realization—they never taught me this stuff about plastic design or processing in school! What do I do now? Hence, the efforts invested for this book.

The path to product design was for me somewhat out of the ordinary, to say the least. I worked a full-time job as a mechanical design draftsman, pencils and vellum, over a drafting board, 8 hours a day; then attended night classes, 4 nights per week, for 8 years, in the engineering program at Johns Hopkins University. The day job taught me how things were made, how things were designed, how things were tested, why validation was important, and a host of other design process skills. The night classes taught me the science and math needed to explain and prove the design creations that I worked on during the day. Together, the day job and the night classes taught me discipline and perseverance.

In 1983, I worked on my first real plastic design project; it was a large housing for a lottery ticket machine molded from structural foam. The industrial designer on the project was extremely patient with me as I asked hundreds of questions, one right after the other, most he couldn't even answer. He did teach me about plastic design and a tremendous amount of general design wisdom as well. Structural and mechanical parts from plastics came soon thereafter and I soaked up everything I could learn. I subscribed to this "new" trade magazine called *Plastic Design Forum*, from which I collected articles from this guy named Glenn who taught us how to design parts correctly. Then in the early 1990s, I pulled the corporate plug and became an independent product design services provider, specializing in, you guessed it, products and parts manufactured from plastics; and, I haven't looked back since!

In over 25 years of providing technical product design services, I have been blessed with the opportunity to help create plastic items and parts for most every industry and market. Many of the products that I have helped launch are in production around the world today, several have run the course and are no longer commercially viable, and there were many that never made it through the process and were not commercialized. Over the course of these years, I did manage to pick up the "best" design award from the Baltimore-Washington Section of the Society of Plastics Engineers (SPE) for an automotive aftermarket product designed in 1990, made from Ultem (PEI) which eventually sold millions of units in the USA and Canada. Then in 2007, I was again selected for a design award, but this time it was for the International Design Award from SPE, The Glenn Beall Design Award, for an innovative pet product designed from glass filled polypropylene, which that same year won the best in show at the international pet products trade show. Finally, in 2012 at NPE in Orlando, one of my design projects, a pneumatic tube carrier for the health care industry, was honored by inclusion in the SPI International Plastics Design Competition at NPE in Orlando.

There are a great number of books filled with every sort of means to calculate nearly any load situation encountered when designing in plastics. This book is not one of them. The intent of this work is to teach the designer what to be thinking about during the design effort, how to think when designing; what to consider and when to consider it. Yes, design calculations will need to be made, but they are far less important than the correct selection of the process, or the materials, early in the design process. The material presented herein is not an academic research work, rather it is a compilation of the experiential lessons learned from day-to-day work; that is finding solutions, designing parts, and commercializing them. In addition, practical and common sense methods for managing the risks ever present during the product development cycle are presented.

After undertaking this writing effort, my hopes are that this book is useful for any new designer or engineer tasked with the design or development of a part or product from this amazing class of materials we refer to as plastics.

Mark T. MacLean-Blevins
April 2017

Acknowledgments

First and foremost, I would like to state that without the support and encouragement provided to me by my family during this writing adventure, the efforts required could not have been produced. Special thanks to Caroline for assistance with the illustrations.

Industry colleagues who deserve my gratitude for their encouragement and assistance include, Brian Wiederin, Tom Boyd, Eric Larson, Mark Wolverton, Glenn Beall, Ed Probst, Michael Paloian, Jay Waddell, Roger Kipp, and many others who took time to listen and advise—thank you.

Section I

Background Knowledge

1. Introduction 3
2. Material Selection – Which Plastic to Use? 19
3. Process Selection – Which Plastics Process to Use? 51
4. What Tooling Will be Required? 79

Introduction

The process of designing a part from plastics, or from any material for that matter, follows a fairly routine path no matter the end product. However, each different product or part will have unique requirements that will cause the path to be different, some longer, some steeper, some harder, and some just difficult, but each will go through similar steps or phases as the process unfolds. This book follows the process cycle for designing a product or part from the space-age materials we call plastics. It is arranged in three main sections, beginning with *Background Knowledge* (Section I), then moving to *Designing with Plastics* (Section II), then following the process into *Implementation of the Design* (Section III) into production. A fourth and final section is included to review the *Lessons* (Section IV) learned through the three main sections (Fig. I.1).

Working knowledge of the materials, the processes, and the tooling required are presented in this first *Background Knowledge* section and will provide the reader with, yes, you guessed it, the fundamental basics of each of the materials, the processes, and the tooling which need to be considered during the design of a plastic product or part.

The manner in which the designer thinks during the design process, what they think about, how they think about it, and why they need to consider certain things or attributes is presented in the *Designing with Plastics* section. This section provides the learner with the thinking tools necessary to allow them to simultaneously consider the part they are designing, along with the material it is to be made from, the process that will be used to form the material into the final part geometry, and the tooling required by the process selected for manufacturing the product or part. This concurrent thinking method will, with some experience, allow the user to create plastic part designs which are functional and which can be manufactured efficiently; parts which can be considered optimum.

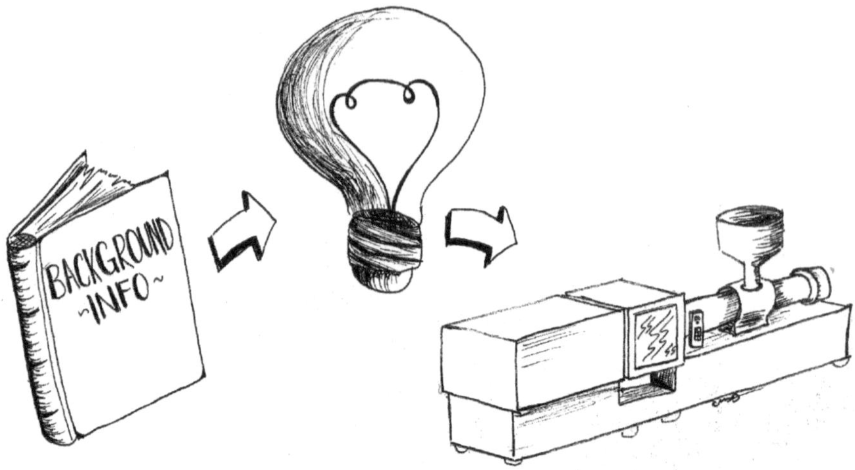

Figure I.1 Any design effort is composed of three main efforts, gaining background knowledge, active design efforts, and implementation of the design into manufacturing. This book is broken into similar sections.
Source: Illustration Courtesy of Caroline MacLean-Blevins

Once the design is completed, the development project is ready for the next section of the book, *Implementation of the Design* ready for production. This third section *Implementation of the Design* of the book covers the estimating, project verification, and design qualification steps required to balance and limit risk as the project moves to production. Make no mistake, development of any new part or product carries risk and our task is to create the next best thing on the market but to do so with zero risk.

First things first, this initial section comprising of Chapters 1–4 will cover the basic information and background needed to support the design thinking lessons presented in the section two *Designing with Plastics* of this book. This background knowledge is fundamental and is not provided as an exhaustive study in any one of the topics included, rather as a high level overview of each, meant to support the novice or young professional wanting to learn. At the end of each chapter, resources for further study are listed, providing additional opportunities for the reader to gain even more knowledge in each subject area.

Introduction

1

The ability to design a plastic part is an easy task. The ability to design an "optimum" plastic part is also an easy task, if the designer is equipped with the information and thinking skills required. As a practicing designer of plastic parts and products, I am familiar with all of the "good design practices" and "rules of thumb" that are commonly used within the plastics design community. However, as I travel from place to place, ever curious, I pick up products, various things and parts, and then look at them with a designer's eye and insight. As I look at these artifacts, many questions arise: Why did they make this feature in this manner or configuration? What caused them to choose this particular material? I wonder why they split the mold here, instead of there? Why in the world would you locate the gate there? Look at the flow marks, why didn't they add some internal features to help the material flow better? and so on.

In fact, when my children were small and with us in retail stores, as I investigated these various molded parts, they would ask what I was looking for; so, I showed them. Well, you know how this ended; every time our children went with their grandparents anywhere I would get these questions from the grandparents afterwards, "…what is a witness line? ...what is a gate? …what is an ejector mark?" and so on. The kids found it like a game, to see who could find the gate or the ejector marks, or better yet a defect.

There are so many bad examples of plastic part design out there that no one must be trying, even a little bit, to develop a part which is optimum. Hence, my desire to compose this reference text to assist, even a little bit, any designer that has the drive to design good, no not just good parts, but *optimum* parts.

What do I mean by an optimum part? Good question. Think about the last plastic part you designed; did it function as desired? Was it moldable in the most efficient process? Was the most-effective material selected and used? Was the capital required for tooling amortized quickly? Has production been flawless? Have field returns been minimal? Optimum parts perform well, can be made well, can be made repeatedly, part after part, within the working range of the tooling and equipment used in manufacture, and can be manufactured efficiently within cost targets. Easy, right?

What we often find in practice is that the product designer goes about their work and defines the product or parts in a manner which satisfies the functional and aesthetic needs of their end-user. Then they call the design final and pass the design off to the manufacturer or factory. The first thing that happens at the factory is the plastic part design specifications are sent for tooling estimates, whereupon the tool designers go about telling the factory that the part designs are indeed beautiful but they cannot be manufactured or molded as designed. The tool design team then has the part redesigned so that a tool could be built to mold the part. Now that the part has been designed twice by two different designers, how does the part perform? Like a part that can be molded well, or, like the original design intent? Perhaps both? Probably not.

Designing Successful Products with Plastics. http://dx.doi.org/10.1016/B978-0-323-44501-6.00001-5

The part designer working for the tooling supplier has a vested interest in producing a part design that can be molded so that the tool shop can build the mold with confidence (and not have their reputation tarnished by a mold that has issues) and with minimum risk to the tool shop. Hence, the part can indeed be molded well, however, the added draft, the increased corner radii, the pass-thru shut offs for forming undercuts, and so forth leave the part, well … not exactly in synch with the original design intent. Now, to be fair, most tool designers are keen to do as little to the part design as possible to make it moldable, and most will attempt to learn about the design intent and the functional requirements. So this scenario does not take place with every project every day, however if the design team learns to think first, think about how the tool will function, comprehend the process, with the tool and understand their limits and capabilities, then this scenario never arises.

Now to be sure this text is about plastics, but no … not really, it is about design. Perhaps more specifically, it is about how to use the resources available effectively and how to "think" as the design is being created. Hence, there will be some background information provided to support the design process thinking and learning that will be presented, but the extremely detailed design formulas and analysis tools will not be presented herein as these are well documented already in numerous texts. There is no effort within this text to recreate these strength of materials formulas and math; however, in each chapter the references recorded and recommended resources for further study will lead the way to finding that information.

The world of plastics and the design therefore is extremely large and broad. This work is focused on products and parts that are structural in nature. There is no coverage of films or flexible sheet manufacturing or design. Specifically we are dealing with three dimensional (3D) products or parts which can be held in the hand or observed by eye.

The focus of this text then is the method, or process, by which the engineer or designer thinks, creates, sculpts, and designs with this space-age modern material we call plastics.

1.1 Plastic part design versus product design

The design of a plastic part entails many rules that restrict the designer during the design process. However, the thought process that is happening as the designer is thinking through the design of the plastic part is indeed similar to the thinking that occurs during any product design. Yes, the rules cause the designer to think within the materials capabilities and inside the tool and process limitations, but once those rules become second nature in the thought process, the designer is then free to create. On the other hand, product design takes everything into an account, especially the needs of the end-user who will experience the product firsthand, without these materials or process limitations. Product designers are often more concerned, with the end users feelings and intuition in using whatever product they are designing, while plastic part designers tend to be looking at how to get the part made with the lowest first cost or the shortest lead times. I submit, however, that when you invest the time and resources to create an optimum part design, both the expectations of the end-user and the efficiencies of the manufacturer are satisfied.

1.1.1 They are different (sort of)

Product design, often referred to as industrial design, requires the full integration of all faculties—every sense must be accounted for—every interaction with the product must be analyzed and the product must accommodate every intended end-use and user. With the design of a plastic part, we only get one (sometimes two) materials to work with, so the choices are indeed limited, but given that said material, we can design the part to be smooth, to be gripable, to be aerodynamic, to be a sail, to be tough, to be flexible, to be fracturable, to be clear, to be opaque, and so on. Go ahead name the attribute, the designer can either hit it dead-on or approach it with the choice of material and process, and most importantly, the proper design of the part. No other material class gives the designer so much flexibility and freedom to create. So yes, the industrial designers working on product design have more overall creative freedom and can address more applications with many more materials, but designing in plastics provides many of the same opportunities.

1.1.2 They are the same

Given a functional "need," the plastic part designer sets out on the same path and in the same manner as an industrial designer might. Indeed, thinking in 3D space is a requirement for both the industrial designer and the plastic part designer. As I work alongside some hugely talented industrial designers from major international firms, we sit and sketch our thoughts as they come from our minds with pencil and paper—very much alike. Our thinking becomes graphical and that output is sketching; some extremely rough and high level, some extremely detailed and specific, but graphic output that we both can understand and appreciate.

Case Example

Once in a meeting in a South China factory, with a dozen people in the room, there was general disagreement about a certain molded feature on a part. The purchasing team was raising objections because of input from the quality control inspectors and the translator was busy in telling the Chinese toolmaker about the issues and then relaying the toolmakers questions back to the table for further comments. This went on for some time, everyone was frustrated and we were not progressing toward a solution. I made a quick pencil sketch of the problem and of the proposed solution, and shaded in the mold components and highlighted the part features within the mold—I used no words or numbers just pencil lines, shading, and a colored highlighter. The page was slid across the table to the interpreter with a verbal explanation of the color and shading meanings, she then passed it onto the toolmaker that looked at it, immediately smiled and understood, and relayed back to the interpreter that the changes could be made that evening and we could run mold trials again in the morning. One two minute sketch replaced almost an hour of verbal back and forth. The graphical communication during this meeting was not focused on design, but it shows the ability to transcend languages with a common graphic language.

The end result desired from the plastic part designer is indeed the same as the end result desired from the industrial designer, except the focus is much broader for the industrial designer. The design process steps, thinking, creating, verifying, detailing, qualifying, testing, validating, and so on, are all the same for both professions. Hence, they are indeed much more similar than most would guess.

1.2 Why choose plastics?

Why would anyone choose to use plastics for developing a new part or product? Every day there are more and more news clips about how bad plastic is for humans and for our environment. If you listen to all of the hype you will soon understand that if we had never invented and developed plastic into the array of materials we use today, mankind would certainly be far better off. Just think, we would not need all of the landfills, the oceans would be crystal clear and clean, wildlife would be thriving, everything around us would be wonderful—just like in the movies and fairytales.

Well, reality tells a better story. Food is protected and stored for longer periods of time, medicines and supplies for medical treatment are kept sanitary and sterile, and all kinds of liquids are stored and transported without risk of glass breakage and spoilage, not to mention the economic side of the equation. Just think if we could not use plastics, what materials we would be using instead, designing in steel, aluminum, glass, ceramic, and so on. Now, remember back to high school physics class, we learned all about these silly laws of nature, remember the one about conservation of energy? Energy, or the use of energy, is an expensive endeavor and conserving energy in any form is a worthwhile goal. Reduction in use of energy, for any product results in not only lower costs but also lower overall environmental impact. If we want to really be environmentally conscious we would always seek to design products and parts with the lowest total energy needs or usage over the total lifespan of the product. Plastics are certainly made from fossil fuel feed stocks, among others, but total energy input to process and manufacture a plastic part is much lower than that of most any other suitable replacement material for that same part for the same application. In fact, a 2014 study by the American Chemistry Council and The Canadian Plastics Industry Association found that within the packaging sector use of plastics materials, the energy savings from the use of plastics (for packaging only) in the USA annually is on the order of 80% and the reduction in possible CO_2 emissions was estimated to be 130% over the use of any other material as a substitute [1]. Reclaiming and recycling plastics from either the primary or secondary waste streams will once again use less energy than most any suitable replacement materials, providing further downstream energy savings. In fact, in 2012 alone the American Chemistry Council reports that over 5 billion pounds of plastics were recycled in the United States [2].

So, what if? What if, we as designers, did our very best to design parts and products that were seen as true solutions, not as problems? True sustainability can only be achieved with good design practices. Perhaps we can help to turn this bad press around and show the world that plastics are not the problem; in fact, they are, or could be, a large part of the solution. The problem lies in how we, as humans, treat our planet. Better concern by the end-users for their actions resolves the problem; banning plastics does not.

1.2.1 Materials benefits

With over 85,000 different variations of polymers available for use in creating a product or part, plastics provide a wide pallet of possibilities for the product or part designer [3]. Given valuable properties like high elongation, toughness, resilience, formability, high dielectric, chemical resistance, and so on, a polymer can be produced to provide just about any combination of properties needed for any application. There are drawbacks to using plastics of course, but the wide selection of available materials gives us an arsenal to use in solving any design challenge (Fig. 1.1).

In addition to the wide selection of properties available, plastics materials can provide extremely effective economic efficiency when compared to alternative materials and processes for any given part or product. Weight reduction, processing speed, minimized in-process waste, ability for low-energy input recycling, and the list goes on; plastics materials provide economic efficiency.

Figure 1.1 Polymers, or plastics, are available in many forms, colors, strengths, transparencies, conductivities, densities, costs, and so on.
Source: Photo from iStock.com/XXLPhoto.

1.2.2 Processing benefits

With dozens of potential processes utilizing plastics, any potential product or part in any volume can be contemplated. From single piece computer numerical controlled (CNC) machined parts to high volume multicavity injection-molded parts, there are processes to produce any kind or any type of part from a plastic material. Proper selection, based on usage requirements, or capacity, and part functional application is primary to an optimum part. Each process has benefits and drawbacks, and each will produce a part or product to different dimensional tolerances, hence the designer must be versed and consider these facts throughout the design process. Process equipment controls and closed-loop feedback in-use with those controls allows a repeatable cycle, given good tooling and properly prepared materials.

Other process benefits include the ability to get a near net shape part in a color and finish which is ready to use—no painting, finishing steps, or operations required; and, creating complex hollow part geometries and shapes. Providing color throughout the part thickness (not just at the surface) so that when in use, a worn or scratched part does not show instantly is a side benefit, but in many appliances and hand held consumer devices it is an extremely valuable side benefit. One processing benefit that can be easily overlooked but is used each and every day helping to allow developed nations to compete with developing nations with far lower-labor costs is automation of the process. These processes are extremely easy to automate, including quality assurance, testing, packaging, and so on.

1.2.3 Tooling benefits

Aside from material removal processes, as in computer numerical controlled machining, laser cutting, water jet cutting, and the like, most plastic parts are formed though some thermal process, changing the state of the material from a solid to a flowable or formable state and then back to a solid after the forming process. Tooling to support these processes ranges from porous foam and wooden thermoforming molds to high precision stacked cavitation injection molds; from low cost in the thousands of dollars per tool to high cost in the millions of dollars per tool. Regardless of cost, a good tool can manufacture parts for years with proper maintenance and care. Precision parts can be manufactured in a volume process, that is, part-to-part (cycle-to-cycle) consistency can be achieved, with proper tool design, controlled processing and proper material preparation. Couple this with a part design that supports a good tool design and you are approaching an optimum part.

In addition, since most plastic processes involve forming of a near net shape final part, there is often little or no waste; and in many processes the waste that is produced can be recycled instantly and put back into the feed stream to be reused within hours or sometimes minutes. Depending upon the particular part, the tooling can support extremely rapid processing cycles, resulting in extreme production capacities and efficiencies. I was once in a factory producing PET water bottle preforms, not an insignificant weight part (like a closure might be considered), in a 96-cavity mold with a second shuttling ejector mold half (each cycle the ejector half of the mold would move out of the press and the second ejector half would move in). The parts cooled and ejected from the ejector half outside the press during the fill cycle for the half in the press, then they would switch back again; the bottom line was a 4 or 5 s cycle to produce 96 parts, and there were several machines running side-by-side simultaneously. As I said earlier, extreme production capacities can be achieved.

1.2.4 Design opportunities

Number one on my list of design opportunities is the freedom to sculpt a part in 3D—right out of the starting gate. This means when I conceive a shape or geometry in my mind, I can recreate it without being encumbered with any preexisting conditions, such as availability of stock material sizes or percent of material removed as waste. If you couple the freedom to think and create in 3D with the ability to combine functional features of two or more parts into one, the efficiencies available when designing in plastic begin to prove their intrinsic value. Take it a step further, create the part with features for locating during secondary assembly or inspection processes to incorporate "Design for Assembly" techniques and really drive the value up. Add to that the fact that the net shape can be designed for minimal part weight, to reduce material usage and to save on transportation-related costs both for incoming resin and for outgoing finished product. Incorporating features to combine with or to enhance packaging, for strength or protection value, can reduce further energy input by decreasing the packaging requirements for the product; and the list goes on. No other modern material gives the designer so much creative freedom.

Other design opportunities include the ability to create a finished part in one or two process steps in a final finish and color, the ability to produce a part that has surface conductivity, or to produce a part that has inherent EMI/RFI shielding, or that can withstand extreme temperatures, or that can contain hazardous chemistries, or that can protect food from oxidizing, the list is truly endless. Combining materials in a two-step process, we can create part that might have a tough strong structural core with a tactile and sticky soft-touch surface for enhanced gripping or just for a nice feel when in use. Using the same two-step approach we can provide flexible seals on rigid structural machine or appliance components, and so much more. Other processes take the two material (or more) equation in different directions, coextruding multiple materials to provide structure, with one material and oxygen barrier properties with another, then utilizing the coextruded sheet to form containers for food or pharmaceutical packaging. Exploiting the various combinations of materials, processes, and tooling allow the designer to really think through the puzzle and create solutions that can be made by no other competitive means.

Costing and competitive production of any product or part is almost certainly a main driver for designing with plastics materials. The designer has, with effective use of the selected material and process, the ability to create some of the most cost-effective solutions available from any material available for use in commerce today.

1.2.5 When not to choose plastics

There are of course functional areas where plastics will not be the answer. Strength of materials will always play a part in determining the suitability of a plastic material over some other alternative. One certainly would not be thinking about plastics for a bridge truss for use on an interstate highway. Long-term static loads are another instance that is most often not a good fit for plastic materials either. Sure within reason they do fine but I always hedge my designs when long-term loads are to be encountered.

Chemical compatibility and the combination of stress with chemistries which might be present [what we call environmental stress cracking (ESC)] are certainly an application where plastics may not be the best fit. Exposure applications where long-term UV and ozone exposure (think outdoors) are present is one of the application areas where caution is also advisable—not many materials do well in these conditions. To be certain, qualification testing is required for any long-term outdoor application, or application with chemical or radiation exposure, especially when used where harm to humans, pets or other wildlife could result from product or part failure. As shown in Fig. 1.2, electrical or radiation extremes, thermal extremes, time of use or age, and chemical attack are, as Eric Larson points out, the Four Horsemen of the Plastics Apocalypse—the four areas where evaluation of the use of plastics is truly required.

Figure 1.2 The four horsemen of the plastics apocalypse by Eric Larson from his book, *Thermoplastic Material Selection: A Practical Guide* as published by Elsevier in 2015. *Source*: Illustration Credit by Eric Larson.

1.3 The *Four Pillars* of plastic part design

As we have discussed thus far through the introduction, all design considerations must be mindful of materials, processes, tooling, and the specific functional requirements of the intended application. We call these the *Four Pillars* of Plastic Part Design: Materials, Process, Tooling, and Design. No design decision can be made without impacting all four and, most importantly, every design decision is influenced by all four. The product or part designer who carefully considers all *Four Pillars* in every decision throughout the design process will provide a detailed part design that should, by all accounts, result in an optimum finished production part; one which should meet, or exceed, all specifications (Fig. 1.3).

1.3.1 Application

The "need" or application is the driver of the design, somewhere, somebody, has this "need" and we are tasked to find a solution to satisfy the "need." Creating a finished product that fulfils the "need" is a challenge which motivates us as designers. The specific functional requirements of the proposed solution for this "need" become our

Figure 1.3 A graphic illustration showing the "Optimum" part design as supported by the *Four Pillars* of plastic part design. With the application being the ground level foundation upon which the *Four Pillars* are based.
Source: Illustration courtesy of Caroline MacLean-Blevins.

guideposts, they act to constrain the design to exactly what is required, nothing more. From the designer's seat the application steers and directs the design but there is always careful consideration for each of the *Four Pillars* throughout the design process. The application becomes the ground-level foundation, which the pillars build upon to create the "Optimum" part or product.

1.3.2 Materials

The choices are many but the material selected must be suitable for the intended application and the selected material will have discrete processing considerations and tooling requirements to produce a quality part. The material also has a huge role to play in supporting the functional requirements of the application for the intended user. Materials from different base polymers or structures act differently when processed and behave differently and when processed by different forming methods or processes. It is important to understand the particulars about a material that is under consideration for a design up front as you begin to create the part. Part volume/weight of the finished part design combined with the selected material (and process) provides the framework for determination of the final part cost.

1.3.3 Processes

Just as important as the material selected, is the process considered for forming the part using the selected material. Processes vary and are often chosen based on anticipated capacity needs for the part and on the first costs involved in tooling and equipment for the process under consideration. Once a process and a material are chosen, it is important to consider the alternatives that might be used for back up, in case the selected materials were somehow not available. I always recommend having a backup plan, or parachute, just in case. In addition, look for incremental process development stages; for example, perhaps a part can be thermoformed in ABS to begin production, as the capacity needs will be low to moderate, then if and when capacity requirements grow, move to an injection-molded ABS design. These incremental steps allow lower cost of entry and a planned process for increasing capacity and lowering total overall costs.

1.3.4 Tooling

As with process and material, no design can be manufactured without some form of tooling and the particular part design will often determine the complexity, cost, and long-term consistency of the tool. Understanding the tooling requirements for each of the available processes will assist the designer in correctly selecting the process for a given solution. In addition, this understanding will allow the competent designer to create a part design, which provides for a robust tooling design to manufacture the given part from the selected process. Tooling is an investment in part quality and can be an important deciding factor in determining final product quality.

1.3.5 Design

Design makes use of the foundation and works cooperatively with the remaining three pillars to support the "Optimum" part. A good design considers the "Application" foundation and the remaining three pillars in every decision, using experience and reason to set the balance when there are competing attributes. Certainly application requirements drive the design, but material choice, process choice, tooling design to support the process, and part design all weigh heavily when creating an optimum plastic part or product. With the application as the foundation and with each of the *Four Pillars* supported and considered carefully during the design process, the optimum part or product can be realized and commercialized.

1.4 Integration of the *Four Pillars* in every decision during the design process

Given a "need" the designer will often jump to a solution almost immediately. The process of vetting the solution and detailing the design to be production ready involves many decisions, most built upon earlier decisions. When the designer thinks through each decision from a vantage point which includes all *Four Pillars*, the decision is most often a sound decision which can be built upon as the design moves forward. For example, when a new part layout is being generated, where do you start? I will often start at a plane where there is some mating part or reference that is meaningful to the function of the part; however, I always "see" the finished part in my mind first, then contemplate where the parting line might be best located to accommodate the gate, the venting requirements, and how certain features may need to be built within the tool. If I know the material shrinks differentially with flow versus against flow, then that information will play a part in my starting point as well. Each of the *Four Pillars* must be satisfied to create an optimum part and all four must be included as we are thinking through the design from the very beginning of the process. Each and every decision during the design process has an impact on each of the *Four Pillars* and every decision is impacted by each of the *Four Pillars*, period.

1.4.1 Early and cooperative involvement of stakeholders from each of the pillars

Experience has taught me that early involvement of the material supplier, the processor or manufacturer (injection molder, blow molder, thermoformer, extruder, etc.), and the tooling source(s) for the part(s) is a wise practice. While it is not always possible to know the chosen vendors in a competitive bidding situation, we typically have a current vendor resource for a given process that we can go to for early-stage discussions. By sharing the conceptual part geometry, material selection choices, and part functional performance specifics early on, the suppliers can share their respective experiences with similar parts, similar features to be tooled, and similar materials usage from other projects. This combined experience from the stakeholders from all

Figure 1.4 Early and active involvement by stakeholders responsible for each of the *Four Pillars* is recommended for enhanced experiential input and reduction of overall project risk.
Source: Photo by alotofpeople—stock.adobe.com.

the pillars will often provide far more background for further thought and design iterations. In many cases these early discussions have shifted our materials choices, causing a shift in the structural design to accommodate the requirements of the material, causing us to reevaluate the features relative to the gate location and material flow, causing us to think through how certain features might need to be designed to accommodate the tooling. This accumulated experience and early-on opinion gathering has resulted in not only better final parts, but also more efficient processing cycles, and lower-cost tooling with shorter lead times. However, early involvement also puts the design at risk of becoming designed by committee. As the designer, you must gather the information and opinions, but not cave in and sacrifice the design for the sake of one particular pillar. Careful balance and deliberation is required while making design decisions which support or satisfy each of the *Four Pillars*. It is similar to building the ultimate jigsaw puzzle, have the experience from each of the *Four Pillars* to create and place the correct pieces to finish the "*optimum*" plastic part puzzle (Fig. 1.4).

1.4.2 Concurrent design engineering—a parallel path to minimize timelines

If the project can be organized to have early involvement of stakeholders from each of the *Four Pillars*, then another interesting opportunity can be realized. The design team can actively share design data with the processing source and with the materials supply source, allowing them to be working on their respective portion of the project simultaneously. This may be a gated activity, milestone by milestone, or design review by design review, but the processor could certainly be lining up tooling resources, the tooling resources could be researching mold base suppliers or components (hot runner systems, etc.), and the materials supplier could be getting colorants and additives moving for sample preparation, etc. This assumes that the stakeholders will remain stakeholders throughout the development process and can be engaged early in the process. Although not necessarily a subject for this particular text, overall project management, and reduction of project timeline is always critical. Hence, anything our design methodology can provide to assist will be beneficial for the project in the long run.

1.4.3 Quality considerations—early and active involvement

Although quality assurance has not been identified as one of the *Four Pillars* discretely, it is implied that the quality function is present in each of the pillars. For example, within the design pillar, quality assurance will be reviewing each design and providing input on the controls and testing which might be used for validation or in-process control. Within the materials pillar the quality group may be instrumental in locking down the specifications and controls for insuring the materials are supplied with adequate certifications to specification. Within the process pillar the quality team will certainly be assisting in defining and specifying the in-process controls and checks to ensure the process stays within tolerance and that finished parts meet all criteria. Within the tooling pillar the quality team will be reviewing tool specifications and design to ensure the finished tooling will indeed be capable of providing parts to specification repeatedly cycle after cycle. Early and active involvement will, as with the stakeholders from each of the pillars, assist in reduction of the overall project timeline with improved final product results.

1.5 Conclusions—forward looking lessons

Design consumes a fraction of the total time to market and a fraction of the total development and engineering investments preproduction, yet determines, not only final product costs, but also final product quality, reliability, suitability for the application; in short, the success of the product or part is almost entirely determined by design; and this before the first cut is made in producing the first tool. However, design alone cannot bring all of the elements necessary for successful product development. The design team must request input and capture the experience of those who will be tasked with implementation of the design into a production product.

Active early involvement of representatives or stakeholders from each of the *Four Pillars* should always be considered as a means of gathering team experience for use in the design, as a means of cooperatively reducing overall project timeline, as

Figure 1.5 Just like the pawn who desires to be the King—internal drive, proper knowledge, and discipline will allow the design team to create optimum (King-like) parts and avoid producing mediocre (King wannabe) parts.
Source: Photo from iStock.com/Altayb.

a means of increasing design confidence and reducing overall project risks. Good design efforts require a knowledgeable background in each of the *Four Pillars*, not an expert in each, but enough of a background to understand the trade-offs and benefits attainable with each material, process, or tool option. There remains no better way of acquiring this background information than firsthand, face-to-face interaction with the processor, material supplier, or toolmaker.

Ultimately, the discipline and motivation of the design team will dictate the success or failure of the product or part (Fig. 1.5). Hence, let's do our best as engineers and designers to create not just nice parts from plastics, but let's strive to create "Optimum" parts.

1.6 Scope of this text

It is important to note that this text is written to introduce the reader to the methods and processes used when designing products and parts from plastic materials, specifically thermoplastic materials. However, it is written based upon the direct experience of the author with discussion of many, but not all, of the processes used to process thermoplastics. Although thermosetting materials are mentioned, they are not discussed or covered in any detailed manner within this edition of this text. Additionally, flexible films from plastics are not a subject of the book, only those applications with structure are within the scope of this text.

As this book is provided as a direct experiential text and not as a research-based work, most of the information contained in the text originated within the author's brain; hence, the cited references will be few; however there will be many resources listed for further study or reference at the end of each of the chapters of the text. The authors experience, spanning 30 plus years of designing plastic parts, some successful and some not so much, provides the reader/student with many case examples and ample elderly (grandfatherly) advice.

The intention of this text is to assist the graduate engineer or novice designer with comprehending the complexities faced when creating a product or part from plastics. Further, the focus is aimed at management of the multitude of risks constantly present when developing products and parts from plastics. If this is to be considered a "*How To*" text, then the correct title would be "*How to Think When Designing With Plastics*" and would be directed at the practicing designer using plastics to create real parts and products for successful commercialization.

References

[1] "Impact of Plastics Packaging on Life Cycle Energy Consumption & Greenhouse Gas Emissions in the United States and Canada—Substitution Analysis," American Chemistry Council and The Canadian Plastics Industry Association, Franklin Associates, 2014.

[2] "Is Plastics the Most Sustainable Material for Packaging?" Plastics Today, Heather Caliendo, March 3, 2014.

[3] Michael Sepe in UL/IDES Prospector database—as copied from the 3rd edition of Pocket Performance Specs for Thermoplastics published in 2006 by Injection Molding Magazine.

Resources for further study

S. Levy, J.H. DuBois, Plastics Product Design Engineering Handbook, second ed., Chapman & Hall, New York, 1984.

S.E.M. Selke, J.D. Culter, R.J. Hernandez, Plastics Packaging—Properties, Processing, Applications, and Regulations, second ed., Carl Hanser Verlag, Munich, 2004.

J. Rotheiser, Joining of Plastics—Handbook for Designers and Engineers, third ed., Carl Hanser Verlag, Munich, 2009.

G.L. Beall, Rotational Molding—Design, Materials, Tooling, and Processing, first ed., Carl Hanser Verlag, Munich, 1998.

E.R. Larson, Thermoplastic Material Selection—A Practical Guide, first ed., Elsevier, Oxford, 2015.

R.A. Malloy, Plastic Part Design for Injection Molding: An Introduction, second ed., Carl Hanser Verlag, Munich, 2010.

P.A. Tres, Designing Plastic Parts for Assembly, seventh ed., Carl Hanser Verlag, Munich, 2014.

Web links of note

Great Design information from DuPont can be found here:

http://www.dupont.com/content/dam/dupont/products-and-services/plastics-polymers-and-resins/thermoplastics/documents/General%20Design%20Principles/General%20Design%20Principles%20Mod%201.pdf.

Sabic provides many tools for designers here (you may need to register):

http://www.sabic-ip.com/webtools/toolsHomePage.faces.

The UL Prospector site is a valuable tool for any designer reviewing materials for any application (you will need to register):

https://plastics.ulprospector.com/.

Solvay provides a great design guide for using their Udel polysulfone materials:

http://www.solvay.com/en/binaries/Udel-PSU-Design-Guide_EN-227550.pdf.

Bayer has a nice design guide for RIM systems:

http://reactioninjectionmolding.com/wp-content/uploads/2013/09/RIM-PartMoldDesign-Guide.pdf.

Covestro offers several design related guides, including part and mold design, snap fit design, and joining techniques for plastics:

http://www.plastics.covestro.com/Library/Technology-brochures.aspx.

Material selection—which plastic to use?

2

Material selection can be a daunting task, given there are over 85,000 different commercially available plastic materials from which to choose [1]. Basically, we all tend to choose materials that we have been successful with in the past or that we have seen deployed or used successfully in similar applications. However, there are a myriad of mini decisions that goes into selecting the correct material for a particular application. Each material will have different physical properties that will affect its performance for the application, each has differing aesthetic properties that will affect how the part is perceived by the user, each has different availability which may infer supply side risks, and each has different costing which affects final product costs and overall margins for the product. Some materials will process better with less waste; some have an ample stream of recycled grades available, while some are extremely special and are literally polymerized to order (Fig. 2.1). In every application, thorough testing of the final part design in the final material is recommended prior to releasing the part or product for use in the stream of commerce. Testing and approval against the functional specifications is paramount to limiting risk once the part or product is in use by the consumer or end-user.

For the manufacture of most processed plastic parts (think structural parts—not films) we deal with two broad materials classes—thermoplastic materials and thermosetting materials. Thermoplastic materials go through a phase change from a solid to a liquid and back to a solid during processing and they can be reprocessed many times; they can be remelted and reformed. Thermosetting materials go through a chemical change upon forming and cannot be remelted, sort of like baking a cake, once it's baked there is no going backward. For a common example, of a thermoset material, think about an epoxy. You purchase the epoxy as a set of two liquid components, you dispense them, mix them, and then apply them to whatever you are potting or repairing, and within some short time they form a solid (along with a little exothermic heat) and you cannot go backward, it is done.

Thermoplastics will generally fall into two broad categories of structure, that is those which are amorphous and those which are semicrystalline (Fig. 2.2). Materials within each of these two broad categories exhibit different characteristics that can often be the first selection criteria when assessing a new end-use application. In semicrystalline materials, as the material solidifies from the melt, the molecule chains tend to fold back upon themselves and try to order themselves causing greater shrink rates but sometime offering easier processing properties in certain process methods like injection molding. In general, the big difference that the designer needs to consider is that semicrystalline materials will exhibit a distinct melting point whereas the amorphous materials do not. This particular difference means that some processes will be suitable for semicrystalline materials but not so much for amorphous materials, for

Designing Successful Products with Plastics. http://dx.doi.org/10.1016/B978-0-323-44501-6.00002-7

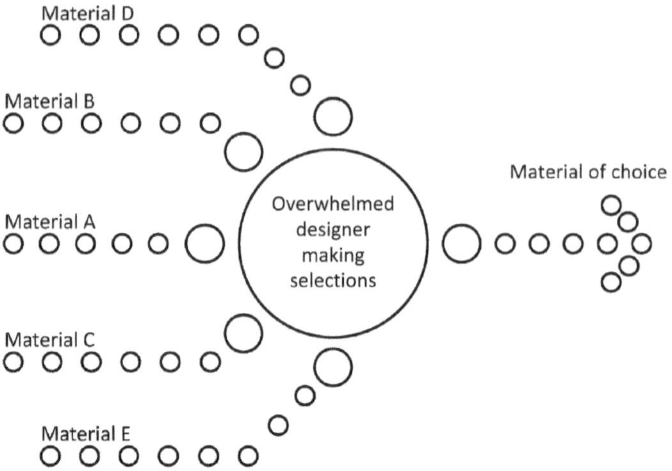

Figure 2.1 Overwhelmed by too many choices? Try and select the correct polymer from the ever expanding palate of choices for any given application.

example, thermoforming works much better with an amorphous material whereas rotational molding does not tolerate amorphous materials as well. Properties like chemical resistance are generally better in semicrystalline materials than in amorphous materials, but not always.

The oriented portions of the semicrystalline structure in the solid polymer will unfold and become long polymer chains again during thermal heat application and

Figure 2.2 Amorphous and semicrystalline molecules shown as a solid, note the random ordered (folded) portions within the semicrystalline model. The amorphous molecular array has been compared to a bowl of wet spaghetti noodles, or has been demonstrated as 10 lb. of rubber bands stuffed into a structure large enough to hold 5 lb. of rubber bands. *Source*: Illustration by Caroline MacLean-Blevins.

Properties	Amorphous	Semicrystalline
Chemical resistance	Poor	Very good
Creep capabilities	Very good	Good
Elongation at yield	Average 0.4%–0.8%	Average 0.5%–0.8%
Fatigue strength	Poor	Very good
Mechanical properties	Good	Very good
Melt temperature	Softening range	Define point
Notch sensitivity	Poor	Good
Service temperature	Good	Very good
Shrinkage	Very good	Poor

Figure 2.3 Comparison of some important properties between amorphous and semicrystalline materials.

melting to a fluid. Then as the material cools back to a solid the polymer chains will reorder themselves into the semicrystalline patterns (as illustrated in Fig. 2.2). Other properties differ between amorphous and semicrystalline materials as well. These are highlighted in Fig. 2.3 to be used for initial selection considerations.

2.1 Selection by material properties

If you go to any plastics compound or resin material supplier and begin to look at their offerings, you will first be provided with their specification sheet, full of information regarding the materials various physical properties as determined through a series of standard tests. In this manner, we can gain an instant understanding of the material characteristics, relative to other potential choices. However, no application in the field is like any of the standardized lab tests and no lab test can predict how the material will perform in real-world (outside the lab) conditions. Hence, testing of the end part or product in the specified materials is extremely important; testing in the desired geometry, with the same exposure, at the same temperatures, and same loading as the intended application. For many of my projects this testing cannot really be accomplished until we have tooling capable of producing the part in the final geometry at production process conditions. This makes full comprehension of the finished product performance an elusive quantity until late in the development process, putting most clients at some degree of anxiousness until we get real parts and verification testing can be accomplished. In Section III *Implementation of the Design*, of this book we will review and discuss verification and qualification in depth. Real part and product development require this incremental process of creative efforts, decisions and selections, then most importantly, the verification and qualification of the finished goods prior to release for production. The material selection step is but one of many to be made during development, but it may be one of the single most important (and riskiest) of the lot.

The purpose of this section is to familiarize you with the many facets of the materials selection process that you as a designer will need to consider as you are designing with these materials. This is not an exhaustive reference on materials; however, you can find an excellent reference in *Thermoplastic Material Selection*, written by my colleague and good friend Eric Larson, also published by Elsevier.

2.1.1 Typical published physical properties of common plastics

Do some research on any thermoplastic material and you will eventually turn up the specification sheet for the material. On this sheet there will be an inordinate number of values to use to compare the material to other potentials for the intended end-use. Some of these might make sense and other not so much. In Fig. 2.4 there is an example of a generic ABS specification sheet, which we will use to describe and discuss some of the properties shown [2]. Please note, since this is a generic ABS table the properties are expressed in ranges; and, some are relatively wide ranges. If one was to select a discreet ABS polymer grade from a particular manufacturer, the specification property values would be either exact numbers, or extremely narrow ranges (Fig. 2.4).

2.1.1.1 Density

The material density will most always be listed in the specification, sometime as a range. Most often this will be listed as specific gravity and will be in dimensionless units; for example, when the specific gravity for generic HDPE is listed, it is shown as 0.947–0.977 (the implied units are g/cm^3) and will be typically identified under the ASTM D792 test method. Sometimes the supplier will provide another comparative specification called apparent density, which is based on the ASTM D1895 test method and indicates the material bulk density or volume change, which may be experienced after processing the part. Hence, a HDPE resin that has a density of 0.94 in pellet form (extruded and granulated) may exhibit a different bulk density in final processed part form, after rotational molding or injection molding for instance.

2.1.1.2 Melt mass flow rate

The melt-flow rate (MFR) or melt-flow index (MFI) of a material is one of those properties that are ignored by many designers; it is a measure of the materials ability to flow in the melt. Certainly not something anyone but the toolmaker and processor need to be concerned with. Well, in reality, not so—as a designer when you understand how the material will behave in a given process you can make use of this knowledge and create part designs that can be processed better. For example, creating intricate thin wall injection-molded parts and specifying a material with a low MFR is asking for trouble in production. Likewise producing parts with precise seal-off features like fluid metering orifice in valves with a high MFR material might be creating a situation where flash might cause defects (flow rate altering orifi area changes due to flash). If a particular material is selected, for whatever reason, I will always try and understand the MFR range for the material as I create the part geometry.

MFR is typically shown as some volume per 10 min as called for under the ASTM D1238 test method. For example, for our generic HDPE example the MFR is quoted at 0.020–1.8 g/10 min. It is important to note that the MFI is inversely proportional to

(A) **Generic ABS**

Acrylonitrile Butadiene Styrene

Generic

PROSPECTOR®

www.ulprospector.com

Technical Data

Product Description

This data represents typical values that have been calculated from all products classified as: Generic ABS

This information is provided for comparative purposes only.

General

Material Status	• Commercial: Active		
Availability	• Africa & Middle East • Asia Pacific	• Europe • Latin America	• North America

Physical	Nominal Value (English)	Nominal Value (SI)	Test Method
Specific Gravity			
--	1.03 to 1.11	1.03 to 1.11 g/cm³	ASTM D792
73°F (23°C)	1.04 to 1.11 g/cm³	1.04 to 1.11 g/cm³	ISO 1183
--	0.0379 lb/in³	1050 kg/m³	ISO 1183 [2]
--	1.01 to 1.10 g/cm³	1.01 to 1.10 g/cm³	ASTM D1505
Apparent Density	0.25 to 0.36 g/cm³	0.25 to 0.36 g/cm³	ASTM D1895
Melt Mass-Flow Rate (MFR)			
220°C/10.0 kg	1.0 to 36 g/10 min	1.0 to 36 g/10 min	ASTM D1238
220°C/10.0 kg	0.49 to 36 g/10 min	0.49 to 36 g/10 min	ISO 1133
Melt Volume-Flow Rate (MVR)			
220°C/10.0 kg	0.0793 to 2.52 in³/10min	1.30 to 41.2 cm³/10min	ISO 1133
--	1.36 in³/10min	22.3 cm³/10min	ISO 1133 [2]
Molding Shrinkage			
Flow : 73°F (23°C)	4.5E-3 to 7.4E-3 in/in	0.45 to 0.74 %	ASTM D955
Across Flow : 73°F (23°C)	3.7E-3 to 9.1E-3 in/in	0.37 to 0.91 %	ASTM D955
73°F (23°C)	0.48 to 0.71 %	0.48 to 0.71 %	ISO 294-4
Water Absorption			
73°F (23°C), 24 hr	0.20 to 0.31 %	0.20 to 0.31 %	ASTM D570
73°F (23°C), 24 hr	0.26 to 0.31 %	0.26 to 0.31 %	ISO 62
Saturation, 73°F (23°C)	0.30 to 1.0 %	0.30 to 1.0 %	ASTM D570
Saturation, 73°F (23°C)	0.10 to 1.6 %	0.10 to 1.6 %	ISO 62
Saturation	0.19 %	0.19 %	ISO 62 [2]
Equilibrium, 73°F (23°C)	0.30 to 0.31 %	0.30 to 0.31 %	ASTM D570
Equilibrium, 73°F (23°C), 50% RH	0.10 to 0.37 %	0.10 to 0.37 %	ISO 62

Mechanical	Nominal Value (English)	Nominal Value (SI)	Test Method
Tensile Modulus			
73°F (23°C)	246000 to 410000 psi	1700 to 2830 MPa	ASTM D638
73°F (23°C)	251000 to 400000 psi	1730 to 2760 MPa	ISO 527-2
--	311000 psi	2150 MPa	ISO 527-2 [2]
Tensile Strength			
Yield, 73°F (23°C)	4940 to 7420 psi	34.1 to 51.2 MPa	ASTM D638
Yield, 73°F (23°C)	5200 to 9210 psi	35.8 to 63.5 MPa	ISO 527-2
Yield	6030 psi	41.6 MPa	ISO 527-2 [2]
Break, 73°F (23°C)	3830 to 7260 psi	26.4 to 50.1 MPa	ASTM D638
Break, 73°F (23°C)	4190 to 6080 psi	28.9 to 41.9 MPa	ISO 527-2
73°F (23°C)	4660 to 7750 psi	32.1 to 53.5 MPa	ASTM D638
73°F (23°C)	5800 to 7030 psi	40.0 to 48.5 MPa	ISO 527-2

1 of 6

Form No. TDS-117948-en

UL Prospector | 800-788-4668 or 307-742-9227 | www.ulprospector.com

Document Created: Sunday, February 14, 2016
Added to Prospector: July, 2008
Last Updated: 5/14/2014

The information presented on this datasheet was acquired by UL Prospector from the producer of the material. UL Prospector makes substantial efforts to assure the accuracy of this data. However, UL Prospector assumes no responsibility for the data values and strongly encourages that upon final material selection, data points are validated with the material supplier.

Figure 2.4 (A–C) The first three pages of a 6 page generic ABS specification sheet as found on the UL Prospector web database. The remaining three pages show information relative to processing for several processes, global availability for the material, and where to purchase the materials [2].

Source: Courtesy by UL Prospector.

(B) **Generic ABS**
Acrylonitrile Butadiene Styrene
Generic

PROSPECTOR®
www.ulprospector.com

Mechanical	Nominal Value (English)	Nominal Value (SI)	Test Method
Tensile Elongation			
Yield, 73°F (23°C)	2.0 to 12 %	2.0 to 12 %	ASTM D638
Yield, 73°F (23°C)	1.8 to 3.6 %	1.8 to 3.6 %	ISO 527-2
Yield	5.0 %	5.0 %	ISO 527-2 [2]
Break, 73°F (23°C)	1.0 to 57 %	1.0 to 57 %	ASTM D638
Break, 73°F (23°C)	4.6 to 27 %	4.6 to 27 %	ISO 527-2
Nominal Tensile Strain at Break			
73°F (23°C)	8.8 to 36 %	8.8 to 36 %	ISO 527-2
--	16 to 50 %	16 to 50 %	ISO 527-2 [2]
Flexural Modulus			
73°F (23°C)	223000 to 417000 psi	1540 to 2880 MPa	ASTM D790
73°F (23°C)	270000 to 415000 psi	1860 to 2860 MPa	ISO 178
Flexural Strength			
73°F (23°C)	7540 to 11800 psi	52.0 to 81.6 MPa	ASTM D790
73°F (23°C)	7830 to 12500 psi	54.0 to 86.0 MPa	ISO 178
Yield, 73°F (23°C)	7950 to 11200 psi	54.8 to 77.5 MPa	ASTM D790
Break, 73°F (23°C)	7820 to 10600 psi	53.9 to 72.8 MPa	ASTM D790
Coefficient of Friction	0.080 to 0.46	0.080 to 0.46	ASTM D1894
Impact	**Nominal Value (English)**	**Nominal Value (SI)**	**Test Method**
Charpy Notched Impact Strength			
73°F (23°C)	2.7 to 14 ft·lb/in²	5.6 to 30 kJ/m²	ISO 179
-22°F (-30°C)	4.03 ft·lb/in²	8.46 kJ/m²	ISO 179/1eA [2]
73°F (23°C)	17.1 ft·lb/in²	35.9 kJ/m²	ISO 179/1eA [2]
Charpy Unnotched Impact Strength			ISO 179
73°F (23°C)	2.9 to 87 ft·lb/in²	6.0 to 180 kJ/m²	
Notched Izod Impact			
73°F (23°C)	1.7 to 7.7 ft·lb/in	92 to 410 J/m	ASTM D256
73°F (23°C)	4.3 to 17 ft·lb/in²	9.0 to 37 kJ/m²	ISO 180
Unnotched Izod Impact			
73°F (23°C)	1.1 to 31 ft·lb/in	59 to 1600 J/m	ASTM D256
73°F (23°C)	3.2 to 48 ft·lb/in²	6.8 to 100 kJ/m²	ISO 180
Instrumented Dart Impact			
73°F (23°C)	25.0 to 351 in·lb	2.82 to 39.6 J	ASTM D3763
73°F (23°C)	2.84 to 24.3 ft·lb	3.85 to 33.0 J	ISO 6603-2
Gardner Impact (73°F (23°C))	24.0 to 359 in·lb	2.71 to 40.5 J	ASTM D3029
Hardness	**Nominal Value (English)**	**Nominal Value (SI)**	**Test Method**
Rockwell Hardness			
73°F (23°C)	97 to 115	97 to 115	ASTM D785
73°F (23°C)	95 to 117	95 to 117	ISO 2039-2
Ball Indentation Hardness	10700 to 16100 psi	74.0 to 111 MPa	ISO 2039-1
Thermal	**Nominal Value (English)**	**Nominal Value (SI)**	**Test Method**
Deflection Temperature Under Load			
66 psi (0.45 MPa), Unannealed	179 to 213 °F	81.5 to 100 °C	ASTM D648
66 psi (0.45 MPa), Unannealed	161 to 222 °F	71.8 to 105 °C	ISO 75-2/B
66 psi (0.45 MPa), Annealed	210 to 217 °F	98.7 to 103 °C	ASTM D648
66 psi (0.45 MPa), Annealed	185 to 228 °F	85.0 to 109 °C	ISO 75-2/B
264 psi (1.8 MPa), Unannealed	166 to 213 °F	74.3 to 101 °C	ASTM D648
264 psi (1.8 MPa), Unannealed	161 to 212 °F	71.8 to 99.8 °C	ISO 75-2/A
264 psi (1.8 MPa), Annealed	185 to 212 °F	84.9 to 100 °C	ASTM D648
264 psi (1.8 MPa), Annealed	193 to 214 °F	89.6 to 101 °C	ISO 75-2/A
Continuous Use Temperature	140 to 167 °F	60.0 to 75.0 °C	ASTM D794
Glass Transition Temperature [3]	220 °F	110 °C	ISO 11357-2 [2]

2 of 6

Form No. TDS-117948-en

UL Prospector | 800-788-4668 or 307-742-9227 | www.ulprospector.com

Document Created: Sunday, February 14, 2016
Added to Prospector: July, 2008
Last Updated: 5/14/2014

The information presented on this datasheet was acquired by UL Prospector from the producer of the material. UL Prospector makes substantial efforts to assure the accuracy of this data. However, UL Prospector assumes no responsibility for the data values and strongly encourages that upon final material selection, data points are validated with the material supplier.

Figure 2.4 (*cont.*)

the materials molecular mass; for example, a high molecular weight material will have a low MFI and vice versa. Since we typically associate a higher-molecular weight material, with better properties, it is common to assume that materials with lower average MFR numbers (relative to pier materials in the same basic polymer family) will exhibit better properties than a corresponding part made from a material with a

(C) **Generic ABS**
Acrylonitrile Butadiene Styrene
Generic

PROSPECTOR®
www.ulprospector.com

Thermal	Nominal Value (English)	Nominal Value (SI)	Test Method
Vicat Softening Temperature			
--	182 to 240 °F	83.6 to 116 °C	ASTM D1525
--	190 to 231 °F	87.6 to 110 °C	ISO 306
50°C/h, B (50N)	202 °F	94.6 °C	ISO 306 [2]
Ball Indentation Temperature	167 to 176 °F	75.0 to 80.3 °C	IEC 60598-1
CLTE			
Flow	4.4E-5 to 5.0E-5 in/in/°F	7.9E-5 to 9.1E-5 cm/cm/°C	ASTM D696
Flow	4.2E-5 to 5.8E-5 in/in/°F	7.6E-5 to 1.0E-4 cm/cm/°C	ASTM E831
Flow	4.4E-5 to 5.1E-5 in/in/°F	7.9E-5 to 9.1E-5 cm/cm/°C	ISO 11359-2
Flow	4.6E-5 in/in/°F	8.2E-5 cm/cm/°C	ISO 11359-2 [2]
Transverse	4.4E-5 to 6.5E-5 in/in/°F	7.9E-5 to 1.2E-4 cm/cm/°C	ASTM E831
Transverse	3.2E-5 to 5.6E-5 in/in/°F	5.8E-5 to 1.0E-4 cm/cm/°C	ISO 11359-2
Transverse	4.4E-5 in/in/°F	7.9E-5 cm/cm/°C	ISO 11359-2 [2]
Specific Heat (73°F (23°C))	0.397 to 0.400 Btu/lb/°F	1660 to 1670 J/kg/°C	ASTM C351
Thermal Conductivity			
73°F (23°C)	1.0 to 1.5 Btu·in/hr/ft²/°F	0.15 to 0.22 W/m/K	ASTM C177
73°F (23°C)	1.2 to 1.3 Btu·in/hr/ft²/°F	0.17 to 0.18 W/m/K	ISO 8302
RTI Elec	138 to 142 °F	59.0 to 61.3 °C	UL 746
RTI Imp	138 to 141 °F	58.8 to 60.5 °C	UL 746
RTI Str	138 to 142 °F	59.0 to 61.3 °C	UL 746
Electrical	**Nominal Value (English)**	**Nominal Value (SI)**	**Test Method**
Surface Resistivity			
--	4.0 to 1.0E+15 ohms	4.0 to 1.0E+15 ohms	ASTM D257
--	1.0E+10 to 1.0E+16 ohms	1.0E+10 to 1.0E+16 ohms	IEC 60093
Volume Resistivity			
73°F (23°C)	0.15 to 1.3E+16 ohms·cm	0.15 to 1.3E+16 ohms·cm	ASTM D257
73°F (23°C)	1.0E+12 to 2.5E+16 ohms·cm	1.0E+12 to 2.5E+16 ohms·cm	IEC 60093
--	3.8E+14 to 3.9E+14 ohms·in	9.8E+12 to 1.0E+13 ohms·m	IEC 60093 [2]
Dielectric Strength			
73°F (23°C)	390 to 890 V/mil	15 to 35 kV/mm	ASTM D149
73°F (23°C)	610 to 950 V/mil	24 to 37 kV/mm	IEC 60243-1
Dielectric Constant			
73°F (23°C)	2.80 to 3.21	2.80 to 3.21	ASTM D150
73°F (23°C)	3.10 to 3.20	3.10 to 3.20	IEC 60250
73°F (23°C)	2.95	2.95	IEC 60250
Dissipation Factor			
73°F (23°C)	4.0E-3 to 0.015	4.0E-3 to 0.015	ASTM D150
73°F (23°C)	4.8E-3 to 0.015	4.8E-3 to 0.015	IEC 60250
Arc Resistance	5.00 to 8.88 sec	5.00 to 8.88 sec	ASTM D495
Comparative Tracking Index	581 to 600 V	581 to 600 V	IEC 60112
High Amp Arc Ignition (HAI)	195 to 200	195 to 200	UL 746
High Voltage Arc Tracking Rate (HVTR)	0.00 to 0.147 in/min	0.00 to 3.73 mm/min	UL 746
Hot-wire Ignition (HWI)	13 to 34 sec	13 to 34 sec	UL 746
Flammability	**Nominal Value (English)**	**Nominal Value (SI)**	**Test Method**
Burning Rate	1.4 to 2.7 in/min	36 to 69 mm/min	ISO 3795
Glow Wire Flammability Index	1190 to 1760 °F	642 to 960 °C	IEC 60695-2-12
Glow Wire Ignition Temperature	1020 to 1760 °F	550 to 960 °C	IEC 60695-2-13
Oxygen Index	21 to 27 %	21 to 27 %	ASTM D2863
Optical	**Nominal Value (English)**	**Nominal Value (SI)**	**Test Method**
Gardner Gloss	59 to 99	59 to 99	ASTM D523
Gloss	30 to 100	30 to 100	ASTM D2457

3 of 6

Form No. TDS-117948-en

UL Prospector | 800-788-4668 or 307-742-9227 | www.ulprospector.com

Document Created: Sunday, February 14, 2016
Added to Prospector: July, 2008
Last Updated: 5/14/2014

The information presented on this datasheet was acquired by UL Prospector from the producer of the material. UL Prospector makes substantial efforts to assure the accuracy of this data. However, UL Prospector assumes no responsibility for the data values and strongly encourages that upon final material selection, data points are validated with the material supplier.

Figure 2.4 *(cont.)*

higher MFR number. Of course this is a generalization and every application must be analyzed in its own merits. For example, in lowering overall costing, it is popular to keep the part walls as thin as possible. In this case the lower MFR materials may not be appropriate and a higher MFR material may be needed to gain the processing window necessary to fill the thin wall part.

Case Example

For one project we selected and specified a "high flow" polysulfone material for a thin wall injection-molded part subject to high temperatures and caustic fluid exposures. The material had a MFR range of 17–24 as compared to the standard grade with a MFR top end of 15. The initial pilot and production runs were from a material lot with a high MFR and the molds ran well and the parts were nice. Subsequent lots did not reach the upper end of the MFR range and were actually near the bottom of the range; the parts were difficult to process and had minor, almost unperceivable defects, which raised the potential failure risk for the parts in the intended application. Through many mold trials we learned that anything at a MFR of 22 or higher processed well and anything below that magic 22 MFR would cause trouble. The situation became worse, as the materials supplier could not predict the resulting MFR from batch to batch of raw materials as they were polymerized, and they had difficulty in attaining the upper end of the MFR range. Our remedial action was to make tooling modifications to allow the lower MFR resin to run more efficiently thereby producing parts without the undesirable defect potential. Understanding the potential processing challenges that could be encountered by the magnitude of the MFR range for a particular resin or compound should be considered early in the design process.

It is important to note that the MFR test applies a uniform load during the test, making it a constant shear stress test and not a shear rate test (since the rate is essentially zero). Since there are a number of conditions at which materials can be tested within the MFR test specification, care must be employed when making comparisons between multiple materials. Be sure the test conditions were similar to ensure an accurate comparison.

Another test method sometime quoted relative to MFR is the spiral flow test where a narrow spiral cavity in a mold is filled at a specific temperature and pressure. The measurement of the distance the flow front will travel along the spiral channel before freeze-off is used to determine the spiral flow MFI for the material. Spiral flow test methods and numbers have historically been used primarily for thermosetting materials.

2.1.1.3 Mold shrinkage

This property directly relates to the adjustment required for the tool dimensions required for a given post-molding part dimension. It is usually quoted as a percentage, for example, for a generic ABS resin (as shown in Fig. 2.4) the shrink rate might be quoted as 0.45%–0.74% and is provided from the ASTM D955 test method. The percentage number is sometime confusing so it is often quoted in linear terms, for example, the shrink rate for the generic ABS reference might be described as 0.0045–0.0074 in./in.

Shrink rates are sometimes provided in two flow directions, as with the flow and as across the flow. Some materials will exhibit a differential shrink rate based on polymer orientation. Along the line of the polymer chain they will most often shrink a little bit more than perpendicular to the polymer chain direction. Sometimes in parts with complex geometries it become difficult to know which flow direction might be located in a tight tolerance functional feature on the part. Programs, such as SolidWorks Plastics®

Shrinkage rates for various polymeric materials			
Material	Structure	Shrink rate (in./in.)	Shrink rate (%)
ABS	Amorphous	0.004–0.009	0.4–0.9
Acetal	Semicrystalline	0.018–0.025	1.8–2.5
Nylon (6-6)	Semicrystalline	0.007–0.018	0.7–1.8
Polycarbonate (PC)	Amorphous	0.005–0.007	0.5–0.7
Polyethylene (PE)	Semicrystalline	0.015–0.040	1.5–4.0
Polypropylene (PP)	Semicrystalline	0.010–0.025	1.0–2.5
Polystyrene (PS)	Amorphous	0.004–0.007	0.4–0.7
Polysulfone (PSU)	Amorphous	0.0069–0.008	0.69–0.8
Polyvinylchloride (PVC)	Amorphous	0.0035–0.0041	0.35–0.41
PPO	Amorphous	0.005–0.008	0.5–0.8

Figure 2.5 A small sample of representative shrink values for various amorphous and semicrystalline materials [3].

or MoldFlow® can provide detailed information showing flow line projections based upon gate locations and material selection. This can be extremely useful when trying to determine the best gate locations to use for a part with a tight dimensional requirement (Fig. 2.5).

While shrink rate is often considered a tooling issue, good design using uniform walls, corner radii, and so on, will assist in providing a part that will shrink uniformly. Tool design, including uniform filling and uniform part cooling will also assist in providing a process window that results in uniform part shrinkage (which minimizes warpage) [3].

2.1.2 Typical published mechanical properties of common plastics

Along with physical properties, the specifications will usually include a section or listing of mechanical properties, with numerical findings for mechanical testing to various specifications. Some of the more popular mechanical properties are included in subsequent sections.

2.1.2.1 Tensile modulus

The tensile modulus or modulus of elasticity for the polymer is provided as a result of ASTM D638 and is usually specified in units of MPa or psi. For our generic ABS example (as shown in Fig. 2.4) the tensile modulus is estimated within a range of 246,000–410,000 psi or 1700–2830 MPa. It is important to note that these values are only found at the temperature used during the ASTM test method, which is 23°C or 73°F, at any other temperatures these values will be different; hence always understand the end-use conditions for the part or product before using these specification values to calculate some critical load reaction. If the performance is critical, ask the material supplier for a graph showing the modulus with respect to temperature. From the UL Prospector database Fig. 2.6 was extracted for a specific ABS material named Lustran 248. In this figure the stress versus strain is illustrated at two different temperatures showing the relative temperature dependence for the material.

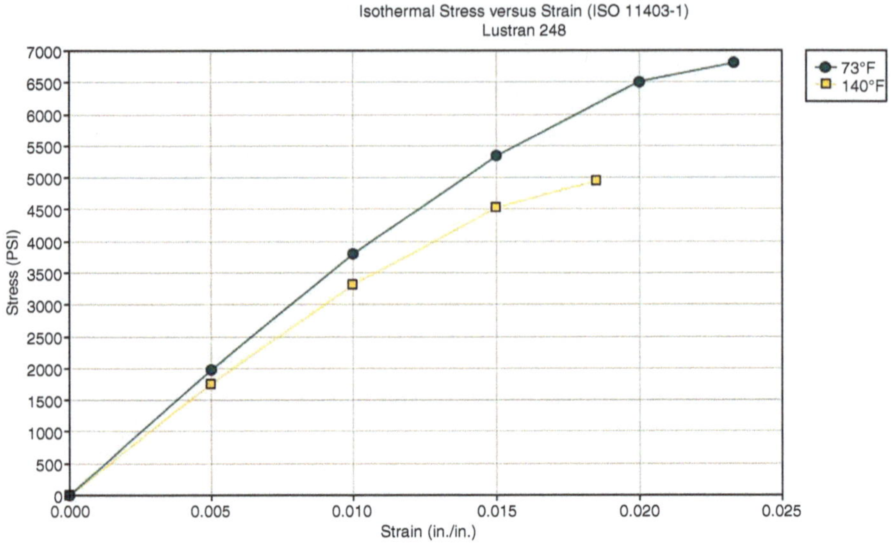

Figure 2.6 A typical stress versus strain curve plot showing the plot for Lustran 248, a general purpose grade of ABS.
Source: Courtesy by UL Prospector (https://plastics.ulprospector.com/) [2].

2.1.2.2 Tensile strength

Again, as with the numbers published for tensile modulus, the tensile strength numbers are determined at a specific temperature and will change greatly with temperature; hence, always understand the application end-use temperature and calculate with a fair degree of safety. The numbers for tensile strength are determined based on ASTM D638 and will typically be shown in two forms; as a yield strength number and as a breaking strength number. The yield point is where the increase in strain does not cause, or result in, a corresponding increase in stress. The break point is the stress (from applied strain) at which failure or rupture occurs. For our ABS generic example (as shown in Fig. 2.4), yield strength is published at 4940–7940 psi, or 34.1–54.2 MPa, and these numbers were recorded at 73°F (23°C). The breaking strength of the generic ABS example, again at 73°F (23°C), is shown at 3830–7260 psi, or 26.4–50.1 MPa. Again, most material suppliers will have plots available showing the numbers as a function of temperature which will allow more precision during design calculations.

2.1.2.3 Tensile elongation

Similar to the tensile strength numbers, tensile elongation is determined at specific temperatures and will be shown as a yield number and as a break number. Similarly, the elongation at yield number corresponds to the yield point (an increase in strain does not cause, or result in, a corresponding increase in stress) and the break point is the point of failure or rupture. This elongation number is shown as a percentage and will vary widely even within a given material class. Many additives and processing techniques can be used to tune these key properties, so be careful to select and use the

Figure 2.7 A tensile test sample (dog bone) being loaded into a tensile testing machine, which will apply a strain to the sample and record the force versus strain until permanent deformation occurs.
Source: Photo from iStock.com/JazzIRT.

properties for the exact grade of material from the exact supplier in any calculations you might be doing. For example, in our generic ABS example, the tensile elongation at yield is published at 2.0%–12% at 73°F (23°C) and tensile elongation at break is published at 1.0%–57% at the same 73°F (23°C), extremely wide ranges.

2.1.2.4 Nominal tensile strain at break

Similar to the numbers published for tensile elongation, tensile strain at break reflects the percentage change in strain at the point of failure or rupture. For our generic ABS case example (as shown in Fig. 2.4), the tensile strain at break numbers are published at 8.8%–36% at 73°F (23°C). Note this number is usually quoted as an ISO 527-2 result not as the ASTM D638 under which most other tensile properties are cited (Fig. 2.7).

2.1.2.5 Flexural modulus

Flexural properties fall under the ASTM D790 test standard and help to define the properties of the materials when bent or flexed. The test uses a simply supported beam structure and applies a centered point load and measures the stress at the outer fibers of the test sample (Fig. 2.8). The flexural modulus is similar to the tensile modulus as it is the ratio of stress to strain but for the flexural case it is specified as outer fiber stress to outer fiber strain. Remember this is a bending situation and as the sample is bent,

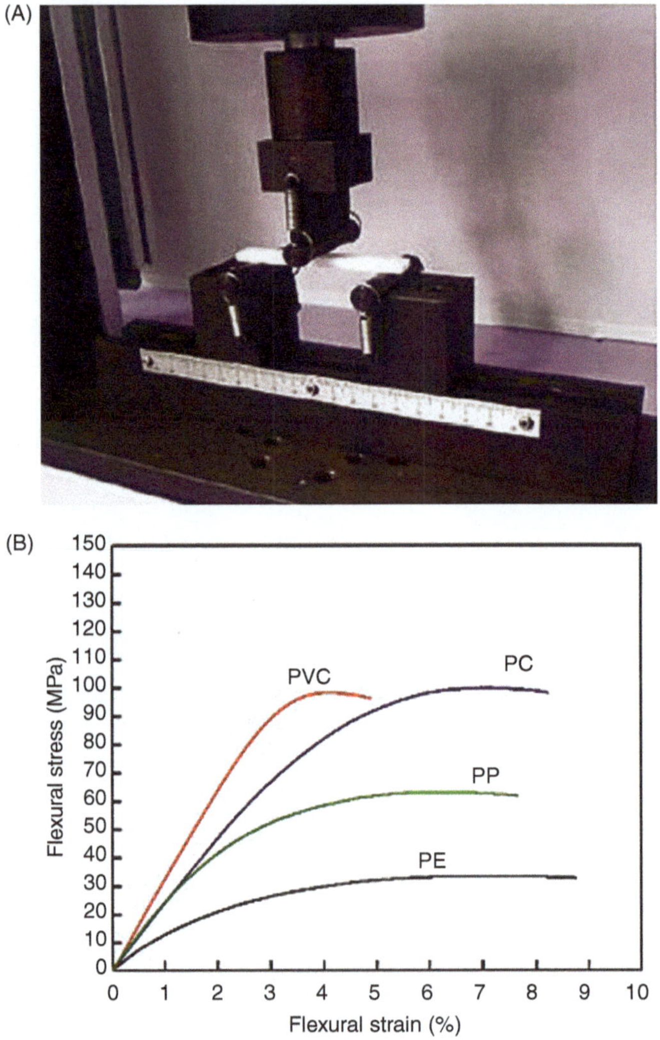

Figure 2.8 (A) A flexural test specimen mounted in the flexural text machine. (B) Flexural stress versus flexural strain charted for several plastic materials.
Source: Photos Courtesy of Shimadzu Corporation.

compression will be created at the inner fibers and tension will be created at the outer fibers. For our generic ABS case example (as shown in Fig. 2.4), the flexural modulus numbers are published at 223,000–417,000 psi, or 1,540–2,880 MPa, at 73°F (23°C).

2.1.2.6 *Flexural strength*

Like tensile strength, flexural strength is shown both as a yield strength number, and as a breaking strength number, determined at a defined temperature. For our ABS generic example (as shown in Fig. 2.4), flexural yield strength is published at 7,950–11,200 psi, or 54.8–77.5 MPa, and these numbers were recorded at 73°F (23°C). The flexural break strength of the generic ABS example, again at 73°F (23°C), is shown at 7,820–10,600 psi, or 53.9–72.8 MPa.

2.1.2.7 *Coefficient of friction*

The friction properties of plastics materials will sometimes be published and other times not published as the process, and surface finish utilized may indeed enhance or diminish the frictional characteristics of a given material. When published it is usually associated with the standard testing under ASTM D1894 and will be shown as a dimensionless number. For our generic ABS example (as shown in Fig. 2.4), a value of 0.080–0.46 is published as the coefficient of friction.

2.1.2.8 *Hardness*

Usually published as a Rockwell hardness number under the test standard ASTM D785 for plastics, the dimensionless number represents the materials ability to resist penetration by a steel ball of a specified diameter. There is a system of ball diameters and applied loads that has been standardized to accommodate materials across a wide range of hardnesses. For our generic ABS case example (as shown in Fig. 2.4), a Rockwell hardness value of 97–115 at 73°F (23°C) is published.

2.1.3 *Typical published impact properties of common plastics*

Along with standardized mechanical properties, impact properties are usually included as an indication of the failure limits at various impact conditions. Since parts and product made with plastics have failure conditions which are a function of temperature, these impact tests are adequate for comparative review; however, detailed testing and analysis of the exact part at the exact temperature and expected use conditions is required for complete assurance of the suitability for the application.

2.1.3.1 *Notched impact strength*

Often called Notched Izod impact strength, this is a relative measure of a materials ability to resist impact. It is measured under standard test ASTM D256 and is performed using a fixed specimen which is hit by a swinging pendulum device. The notched specimen is cantilevered and struck by the pendulum on the notched side

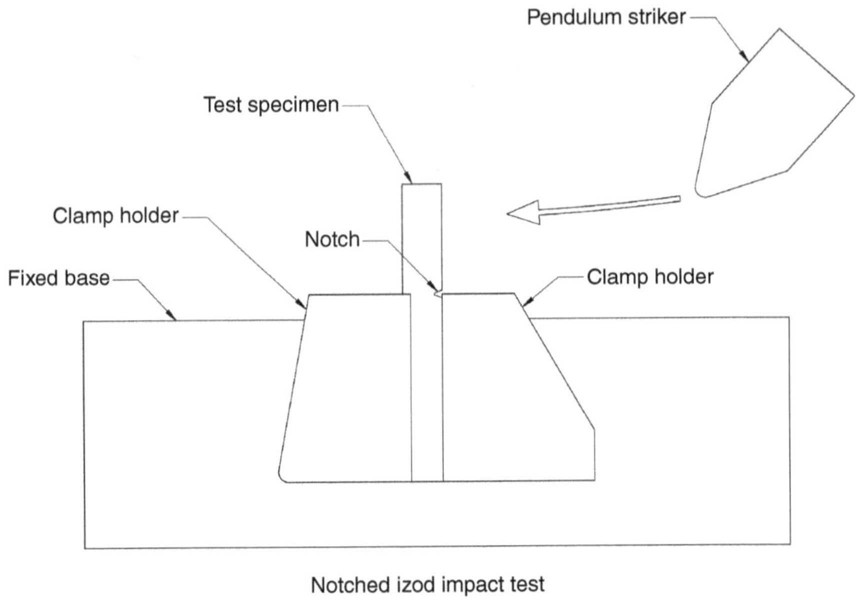

Figure 2.9 An illustrative view of the notched izod test set-up, showing the pendulum striker path and the test specimen held with the notch facing the pendulum. The distance from the notch to the striker impact line is defined within the test specification.

(Fig. 2.9). As with other mechanical properties this test is performed at a standard temperature and all published values will cite the testing temperature. Care must be exercised when applying standard test data at other end-use application temperatures or conditions. For our generic ABS case example (as shown in Fig. 2.4), the Notched Izod impact values are published as 1.7–7.7 ft. lb./in.^{2}, or 92–410 J/m^2 as 73°F (23°C). These values represent the energy value, or magnitude, required to break the specimen.

2.1.3.2 *Unnotched impact strength*

Similar to the Notched Izod discussed previously, except without the notch in the specimen. For our generic ABS case example (as shown in Fig. 2.4), the unnotched Izod impact values are published as 1.1–31 ft. lb./in.^{2}, or 59–1600 J/m^2 at 73°F (23°C). Cited at per ASTM D256 but tested under ASTM D4812.

2.1.3.3 *Dart impact strength*

Another energy measurement technique test, this Dart Impact test uses a falling dart of a specified 0.5 in. hemispherical head dart geometry and mass to reach a specified impact velocity and puncture the specimen. As such the test can be an indicator of the materials behavior in multiple directions during the impact and can infer the sensitivity of the material, with respect to the time rate of the impact event. For our generic ABS case example (as shown in Fig. 2.4), the Dart Impact values are published as 25–351 in. lb., or 2.82–39.6 J at 73°F (23°C). These values are gathered using the standard test as described within ASTM D3673.

2.1.3.4 Gardner impact strength

The Gardner impact test uses a similar falling or guided dart methodology but under ASTM 3029 calls for the values to be representative of the energy required for 50% of the test specimens to fail. For our generic ABS case example (as shown in Fig. 2.4), the Gardner Impact values are published as 24–359 in. lb., or 2.71–40.5 J at 73°F (23°C).

2.1.4 Typical published thermal properties of common plastics

Again, as with impact and mechanical properties, thermal properties are listed in most specifications for most thermoplastics. These allow the designer to access the suitability of the materials for the given application and intended in-use environment. Most of these thermal properties (and electrical properties as well) are cited with reference to ASTM specifications or standards. However, Underwriters Laboratories (UL) has published and uses an enhanced set of testing methods which are often the cited reference standards used on plastic part specification drawings [5].

2.1.4.1 Deflection temperature under load (DTUL)

As the name implies, this is a property which measures the movement of the test specimen while under stress and indicates the temperature required to allow 0.25 mm of movement when stressed in a bending or flexural mode. According to designers this number is a clue for how our parts may behave in real-world applications, knowing the anticipated loading and temperature which might be encountered. This property is often published at different outer fiber stresses of 66 psi (0.455 MPa) and 264 psi (1.82 MPa) allowing us to compare the anticipated movement at both stress conditions. Sometimes, depending upon the material, the published numbers will include results for annealed and unannealed test specimens. For our generic ABS case example (as shown in Fig. 2.4), the deflection temperature under load published as 179–213°F, or 81.5–100°C at 66 psi (0.455 MPa) and as 166–213°F, or 74.3–101°C at 264 psi (1.82 MPa) for unannealed samples.

2.1.4.2 Continuous use temperature

This property specification generally will indicate the highest in-use temperature which can be attained before loss of properties. Not specified on all materials but when specified is a good indicator of the end-use thermal range for any part or product manufactured from the given material. For our case example generic ABS (as shown in Fig. 2.4) the published values are 140–167°F, or 60–75°C. Note that this seemingly relative attribute does not indicate any particular stress level, which must be considered when evaluating performance at any temperature.

2.1.4.3 Glass transition temperature

Thermoplastics in particular will change state, that is melt, or go from a solid to a liquid. However this process does not happen instantaneously, like ice to water, rather the material might change from a brittle solid to a rubbery or gel-like mode before going totally into a viscous liquid form. The temperature at which this change from a brittle or hard state to a flexible condition is known as the glass transition temperature, and it might not be at a precise value, rather a range and the published values are generally estimated at the midpoint of the measured range for a given material.

2.1.4.4 Vicat softening temperature

A similar property to the glass transition temperature, the Vicat softening temperature is an indication of the heat-softening characteristics of different materials. The standard ASTM D1525 test uses a 1-mm^2 flat tip and measures the temperature needed to allow the tip to protrude 1 mm into a sample at a given load and rate of thermal change, for example, a load of 10 N (1 kg) at a thermal rate of 50°C/h. For our case example of a generic ABS (as shown in Fig. 2.4) the values published are 182–240°F, or 83.6–116°C.

2.1.4.5 Coefficient of linear thermal expansion (CLTE)

This property is well-understood as it is also found as a characteristic for many other solid materials including metals. The values represent the net change in length a specimen might undergo as a result of a change in temperature. For plastics we will sometime see the values represented in "flow" and "transverse" categories, corresponding to the polymer chain orientation as a result of processing. For our generic ABS example (as shown in Fig. 2.4), the values for coefficient of linear thermal expansion published are 4.4×10^{-5}–5.0×10^{-5} in./in./°F, or 7.9×10^{-5}–9.1×10^{-5} cm/cm/°C in the direction of flow and 4.4×10^{-5}–6.5×10^{-5} in./in./°F, or 7.9×10^{-5}–1.2×10^{-4} cm/cm/°C in the transverse direction (cross-flow) (Fig. 2.10).

2.1.4.6 Specific heat

As in other materials, specific heat is the energy required to raise the temperature one unit of mass of the material by 1°. Standard test ASTM C351 defines the test method and process for the determination of specific heat. For our generic ABS example (as shown in Fig. 2.4), the values published for specific heat are 0.397–0.400 BTU/lb./°F, or 1660–1670 J/kg/°C.

2.1.4.7 Thermal conductivity

The measure of the ability of heat to move through a solid is known as thermal conductivity. More precisely it is the rate at which heat flows through a unit thickness of the material per degree temperature difference across the thickness of the material. Again, this is a familiar property as it is found for many common materials used for manufacturing and fabrication of goods. For our case example of generic ABS (as shown in Fig. 2.4) the thermal conductivity is published as 1.1–1.5 BTU-in./h/ft.^{2}/°F, or 0.15–0.22 W/m/K.

Coefficient of linear thermal expansion Comparative reference values					
Common plastics			Other common materials		
Material	m/m/°C (× 10^{-6})	in./in./°F (× 10^{-6})	Material	m/m/°C (× 10^{-6})	in./in./°F (× 10^{-6})
ABS	73.8	41	Aluminum	22.2	12.3
ABS-GF	30.4	17	Brass	18.7	10.4
Acetal	106.5	59.2	Brick (masonry)	5.5	3.1
Acrylic	75	42	Bronze	18	10
Epoxy	45–65	25–36	Cast Iron (gray)	10.8	6
EVA	180	100	Concrete	14.5	8
FEP	135	75	Copper	16.6	9.3
Nylon, type 6	85	47.2	Glass (plate)	9	5
Nylon, type 6-6	80	44.4	Gold	14.2	8.2
Nylon, type 11	100	55.6	Granite	7.9	4.4
Nylon, type 12	80.5	44.7	Iron (forged)	11.3	6.3
Polycarbonate	70.2	39	Lead (Pb)	28	15.1
Polyester	123.5	69	Marble	5.5–14.1	3.1–7.9
Polyethylene	200	111	Mercury	61	33.9
PET	59.4	33	Nickel	13	7.2
Polypropylene	100–200	56–112	Silver	19.5	10.7
Polystyrene	70	38.9	Steel	12	6.7
Polysulfone	55.8	31	Tin	23.4	13
PVC	50.4	28	Zinc	29.7	16.5

Figure 2.10 Comparative values for the coefficient of linear thermal expansion between several plastics materials and other common materials [4].

2.1.5 *Typical published electrical properties of common plastics*

When the plastic part or product is used in connection with or as an insulator for anything electrical or electronic, the electrical properties in most specifications will be useful. These are for comparative evaluation use. Any final part qualification must be performed, with the finished part geometry and processed using the production process with the production-specified materials. In this manner, any part or product to be used with, or in connection with, electricity can be qualified prior to placing any consumer or end-user at any potential risk.

2.1.5.1 *Surface resistivity*

Defined as the resistance to leakage current along a surface of an insulating material, surface resistivity is determined through standard test ASTM D257. For our case example of generic ABS (as shown in Fig. 2.4) the published values for surface resistivity is 4.0E + 15–1.0E + 15 Ω.

2.1.5.2 *Volume resistivity*

Similar to surface resistivity except the volume resistivity is through the body or wall of the insulating material. For our generic ABS example (as shown in Fig. 2.4), the volume resistivity value published is 0.15E + 16–1.3E + 16 Ω.

2.1.5.3 Dielectric strength

From ASTM D149, the dielectric strength of a material is defined as the voltage per unit thickness at which the material will conduct electricity; hence, the higher the number the greater the insulating properties of the material. For our generic ABS example (as shown in Fig. 2.4), the dielectric strength is published as 390–890 V/mil, or 15–35 kV/mm.

2.1.5.4 Dielectric constant

Defined as the ratio of a capacitance of a capacitor (two metal plates with the plastic test insulator in between them) with the given (plastic) test material as the dielectric, to the capacitance of a capacitor with vacuum (in between the two metal plates) as the dielectric. For electrical insulators the dielectric constant should be low and for use in a capacitor the value should be high to reduce the size of the given capacitor. For our generic ABS example (as shown in Fig. 2.4), the dielectric constant is published as 2.80–3.21.

2.1.5.5 Dissipation factor

Defined as the ratio of power dissipated in a test material to the power applied to the sample, the dissipation factor is determined using ASTM D150. The dissipation factor for our generic ABS example (as shown in Fig. 2.4) is published as 4.0×10^{-3}–0.015.

2.1.5.6 Arc resistance

Defined by ASTM D495 and further clarified by UL, arc resistance is defined as the number of seconds an insulating material resists forming a surface-conducting path when subjected to an intermittently occurring arc of high-voltage and low-current (amperage) characteristics. For our generic ABS example (as shown in Fig. 2.4), the arc resistance is published as 5.0–8.88 seconds.

2.1.5.7 High amp arc ignition (HAI)

Stemming from UL test standard 746, high amp arc ignition is simple a count of the number of arc exposures required prior to the start or ignition of the material under evaluation. For our case example of generic ABS the values for high amp arc ignition are 195–200.

2.1.6 Typical published flammability properties of common plastics

Flammability and/or combustion properties come into play any time a part or product will be used in a regulated space or application, such as aircraft, residential living units, school use, automotive use, and so on. Again, as with most all of these standardized test specifications, these are used for comparative use. Final qualification of any particular part or product should be completed with the actual part geometry, in the actual material, from the actual production process. Again, as with thermal and electrical properties, flammability properties are often cited from the UL® test methods [6].

2.1.6.1 Burning rate

This is an ISO standardized test use primarily for materials to be used in a passenger compartment of a vehicle as defined within the ISO 3795 specification. It is a horizontal burn test and will not be truly reflective of how a given material may perform in a real-world application, especially when coupled with other materials in other orientations. However, it does provide some criteria that can be used in selection for these specific end-use applications.

2.1.6.2 Glow wire flammability index

This is an IEC test (IEC 60695-2-12) which replicates the possibility of a plastic material to combust when subjected to close proximity to a heat source. The test uses an electrical resistance element to replicate the heat which might be generated by a malfunctioning piece of electrical equipment (too much current through too small a wire) wherein the test specimen is held against the glowing hot resistance element at a specified temperature with a force of 1 N for 30 s. The wire is removed at the end of the 30 s interval and the time until cessation of flames and burning drops is recorded. In addition, there is a cotton element placed beneath the test specimen such that dripping materials will reside on the cotton. The glow wire flammability index is the highest temperature where there is either: (1) no flame and no ignition, or (2) the burning or glowing ignition time is less than 30 s after removal of the hot wire element *and* there is no ignition of the cotton element.

2.1.6.3 Glow wire ignition temperature

Similar to the glow wire index, the glow wire ignition test (IEC 60695-2-13) uses an electrical resistance element to press against a specimen for 30 s at a force of 1 N at a specified temperature. This test uses a tissue paper located beneath the test specimen to catch any material droplets. The published value for the ignition temperature is 25°C above the maximum temperature of the resistance element tip, which does not cause ignition of the material and wherein the maximum allowable flame development time is less than 5 s.

2.1.6.4 UL Flammability rating

The UL-94 flammability-rating test standard has many parts and can be confusing, but it is really a simple set of tests with results which are simple to interpret. A vertical sample is exposed to a certain specified flame for a specified duration, and then the flame is removed for a specified duration and reapplied for a second specified duration. The test also used a cotton element under the test specimen to observe any dripping of the materials especially when the flame will drip with the material. For UL 94-5V tests, a horizontal sheet is used as a means to determine the possibility of burn through occurring. The results are then compared against the published standards and the flame rating is determined. Fig. 2.11 shows a chart of the comparative results for the tests.

UL class	Pretreatment	Flame application	Characteristic	Results		
94-V				V-0	V-1	V-2
94-V	2 days/23°C/50% RH 7 days /70°C/hot air oven	20mm high Tirill burner flame 2 × 10 s	Burning time (sec) after flame application	≤10	≤30	≤30
			Total burning time (sec) 10 flame applications	≤50	≤250	≤250
			Burning and afterglow times (sec) of specimens after second flame application	≤30	≤60	≤60
			Dripping of burning specimens (ignition of cotton batting)	no	no	yes
			Specimens completely burned	no	no	no
94-HB				HB (mm/min)		
94-HB	48 h/23°C/50% RH	20 mm high Tirill burner flame 30 s	Test specimen thickness 3–13 mm	≤40		
			Test specimen thickness 3–13 mm	≤75		
			Flame is extinguished before first mark	0		
94-5V				94-5VA	94-5VB	
94-5V	2 days/23°C/50% RH 7 days/70°C/hot air oven	125 mm high Tirill burner flame 5 × 5 s at 10 s intervals	Burning and afterglow times of specimens after the fifth flame application	≤60 s	≤60 s	
			Dripping of burning specimens	no	no	
			Burn through-hole forms in sheet	no	yes	

Figure 2.11 Comparison of each UL-94 classification rating showing the sample preparation, the flame orientation and duration, and the result range for each classification [6].

2.1.7 Typical published optical properties of common plastics

Optical properties as with all of the other specified properties are used for comparative evaluation of one material against the other in the prospective selection pool. Final results will be determined by final part geometry, material selection, proper tool design and implementation, and processing conditions.

2.1.7.1 Gardner gloss

Gloss surfaces can be a difficult characteristic to achieve, however a clear (no pun intended) understanding of these optical properties can assist the designer in narrowing the range of possible materials which might be considered for the application. Gardner gloss is a measure of refractive index at a specified angle to the surface and relative to a reference surface of polished black glass. Based on ASTM D523 Gardner gloss is similar to, with some discrete differences, the ASTM D2457 gloss measurements. For our generic ABS reference material (as shown in Fig. 2.4), the Gardner gloss is published at 50–99 relative to the polished black glass with a score of 100.

2.1.7.2 Transmittance

The percentage of incident light that is able to pass through the material. The higher the transmittance value the more transparent the material. Based on standard test ASTM D1003, our reference generic ABS example, shows a published value for transmittance of 86%–93%.

2.1.7.3 Haze

Haze is a measure of the light scattering and measures anything more than 2.5 degrees from the incident beam, through the test specimen. This value is again a result of testing under ASTM D1003 and for our example generic ABS material the Haze value published is 1.5%–4%.

2.1.8 Typical published process-specific properties of common plastics

Process condition properties are also included in many specification, although these will be ignored by many design teams, they do assist the mold designer in preparing the mold design to best suit the processing conditions specified.

2.1.8.1 Melt viscosity

Not usually a designer number, melt viscosity is usually a number the processor will be concerned with; however, with the advent of fill analysis being used by designers to assist in proving the fill behavior of the part, designers are becoming more familiar with the number. It is, at a specific or given shear stress or shear rate, the ratio of shear stress to shear rate. It is important to comprehend and understand that melt viscosity is sensitive to various factors like molecular weight, melt stability, and additives and fillers (colorant, lubricants, flame retardants, etc.). For our case example generic ABS the melt viscosity is published at 155–1550 Pa-s and is tested under ASTM D3835.

2.1.8.2 Apparent viscosity

Tested under ASTM D3835, similar to melt viscosity, the apparent viscosity differs in that it describes a material exhibiting non-Newtonian behavior in the melt. It is typically corrected so that it can be correlated against or compared with melt viscosity.

2.1.9 Special formulations of materials for specific applications

Many thermoplastic materials are customized to provide specific benefits, for example, with the family of plastics known as ABS there are formulations (often referred to as grades) which can be electroplated, grades which resist organic growth, grades which have greater impact resistance, and so on. Within the family of plastics known as polypropylene, there are grades which are clarified, grades which can be sterilized by certain processes, grades which are toughened, grades which are cross-linked, grades which are high flow, and so on.

Once you have a general family under consideration, you can dive in and see exactly what custom options might be available. Be wise and ensure that the distinct customized formulation you are looking to select is indeed a commercially available material and that it is stocked in some quantity at some location. Often a special grade or formulation will provide the exact properties you are looking for, however it may be special order or it may be that there is a minimum buy required to purchase the material—both are cost drivers, so be vigilant and make sure before you specify.

In addition, the specific properties or attributes required might be available through secondary sources, such as a compounder using additives for specific applications. Rather than source a name brand special grade, a compounder can source a generic material and then design a formula of additives which when compounded in and repelletized can provide a material with the desired attributes.

2.1.9.1 Grades with fillers for stiffness and/or strength

Most materials can be modified by the addition of fillers to act as stabilizers and to add stiffness or strength. There are always tradeoffs when you add fillers; stiffness goes up but ductility goes down, and so on. Fillers are basically foreign material in the polymer matrix and, as such, the designer must be careful to understand the properties of the filled polymer are vastly different than the specifications for the virgin polymer. Most polymer suppliers will offer specification data for each of the filled grades within their offering. If you can't find the specification information, call the supplier and ask for it, many times it is available but not always published.

2.1.9.2 Grades with fillers for electrical or thermal conductivity

Plastic materials are used in many applications simply because they are great insulators. However there are applications where the part needs to have some conductivity, either bulk conductivity or surface conductivity. Fillers are available that will provide some degree of conductivity, not good conductivity but some conductivity. These are used mainly for providing some degree of EMI/RFI shielding. Surface conductivity can assist when a product or part must dissipate a static charge, for example, for an explosion proof electronics enclosure. Many of the same filler materials and techniques used for electrical conductivity will also enhance thermal conductivity. Compounders and name brand material suppliers will be able to assist with specific applications.

Another trick to gain EMI/RFI shielding is to paint the part with a conductive paint or to electroplate to gain the surface conductivity needed. Again, if the part is to be plated or painted, the base material must be capable of supporting the secondary operation.

2.1.9.3 Grades that contain additives for lubrication or reduced friction effects

Materials made for low coefficient of friction use are available. Be sure to understand what, if any, external lubricants might be used. Friction can be a potential failure mode, since friction causes heat and heat alters the properties of the plastic materials, especially under some load. In any friction application, it is wise to choose materials carefully and then to fully vet the loading and thermal situation that will be encountered. Adding a lubricant or other fillers to lower the apparent coefficient of friction can indeed help, but they will not change the overall physics of the load and the temperature changes (at least not in any meaningful way) so be careful. If possible build models in exact materials, with exact surface finishes and then test them to exact loads for exact durations to fully vet the design. Electroplating has been used to enhance and lower the coefficient of friction, and to some limited effect assist in heat dissipation.

Sometimes other polymers are introduced to assist with friction reduction, for example, PTFE or Teflon (DuPont) can sometimes be an additive. Other additives for reduction of friction can be waxes or fats, these tend to have some lifespan and tend to work well initially but loose effectiveness with time as the fat or wax leaches out from within the polymer matrix. Most suppliers of these additives can (if asked) provide some reference test examples showing this change in apparent coefficient of friction over time (or useful life). Temperature variations in-use will also tend to either accelerate, or retard this phenomena; as with all attributes, be sure to fully understand the end-use conditions and evaluate accordingly.

2.1.9.4 Grades which can withstand sterilization processes

Sterilization of parts or finished products is a requirement for most anything to come in contact with a patient (human or otherwise). Every major supplier of raw materials has grades specifically formulated to be used with each of the sterilization methods in use in commerce today. Most will supply specific test results for each of their grades for each sterilization method. Again, as with other recommendations regarding selection of materials, be careful, understand the testing methods, and ensure that your designed part or product would react or process in the same manner as whatever test you are using as a qualifier. Then, once you have first molded parts, put your parts through the test and compare results to ensure there is no degradation of performance from the sterilization method employed.

2.1.9.5 Grades modified for flexibility or enhanced elongation

Impact modifiers will typically be added to materials to give them a higher elongation and will enhance flexibility in some designs. These can be extremely simple additives which will work well in any number of applications. The caution for these fillers is to specify the filler and the desired finished properties well. Specify them in a manner that allows the finished part or product to be tested to ensure the modification has not made the part too soft/flexible or that the tolerance range for the modifier will always add enough elongation for the intended application functionality.

2.1.9.6 Grades modified to have enhanced UV stability

Products and parts to be used outdoors for extended lengths of time need to be able to withstand the sun and ozone effects. Some polymer families do this naturally, such as PVC, but others will degrade rapidly in an outdoor exposure. Several means for combating this UV degradation exist, most notably adding carbon black or titanium dioxide to help absorb and reflect the UV. Other UV stability fillers are available which act to bind to the reactants released when the UV begins to break the polymer chains, thereby slowing the rate of degradation. However, once these binders, or scavengers, are completely used then the degradation process will run at full speed. One trick that is used to enhance the lifespan of a plastic part or product in an outdoor application is to use a secondary surface covering, such as painting or electroplating. Urethane-based paints or acrylic-based paints will adhere to most molded part surfaces and provide an excellent barrier to UV entry to the polymer matrix in the part. You see examples of this everyday on most automobile body part manufactured from plastics, such as bumpers, trim, and so on.

2.2 Selection by aesthetic properties

Plastics are used as packaging and as enclosures for many consumer goods and for consumer electronics, as well as for automotive interiors, exteriors, and so on. All applications which are considered to require some aesthetic properties; that is, they need to look good, feel good, convey feeling, and have a perceived "high quality" characteristic about them. Product designers love using plastics, they can get high gloss, deep textures, engraved or embossed text or graphics, they can get soft touch, or lensing effects; you name the aesthetic characteristic and there is a material and process combination that can provide it (within reason of course). However, each material brings a little something different to the party; some are good at high gloss while others are not, some are good for lenses and clarity and others are not, so the designer needs to understand and select the ideal material based on the desired characteristics, while still considering the process and tooling requirements, as well as all the other functional requirements for the application.

2.2.1 Appearance properties

Appearance properties, including surface finish, texture, gloss, color, transparency, and so on, are some of the main reasons that industrial designers choose to use plastics for their product designs. In most instances the desired color, surface finish, and texture can be achieved in one process step and without any added per-unit labor cost. Each material has certain inherent surface capabilities. Be sure to research and find the materials which can support the appearance properties you desire for the finished part or product.

2.2.1.1 Surface finish and texture

Surface finish, whether glossy or textured is a function of the quality of the tooling and the materials used. Some materials, like ABS, can provide a deep rich gloss surface in most any color, other plastics like PP can provide a gloss surface but the color and gloss richness is just not there. Textures can be applied to most any molded part surface and most materials will show texture well. Be mindful to allow adequate draft on surfaces with texture, otherwise risk seeing drag lines and marks from the ejection of the part from the tool.

2.2.1.2 Secondary processes for appearance (painting and plating)

Painting and electroplating are common secondary processes for parts made from plastics, however some materials will allow better adherence of the paint than others and some cannot support paint adherence at all. Understanding the appearance needs as you are selecting materials is prudent, ask the material supplier or processor about their experience painting or plating a specific material, if you have plans to do so. Careful consideration of part fixturing or holding during the secondary painting or plating process is advised. Think about how the paint shop or plating vendor might best hold the part to ensure perfect application of the coating materials. Sometimes adding a feature or two to provide a mounting or holding point for this secondary operation can improve the finished quality of the parts, while reducing the scrap, and increasing overall production efficiency.

2.2.1.3 Color and special effects

Materials can be colored with any number of pigments and also special effects pigments with metal flake or pearlescent appearance. These "special effects" can be an interesting feature on a part or product, but remember, final part performance is based on the materials used and when additives are blended in the resulting part is not 100% polymer, so some properties will be different. Be sure to test samples with the intended colorant and other "special effects" materials to ensure that the final part measures up to the functional specification. Post-molding part shrinkage rates can be affected by colorants and other additives. Be sure to understand the full range of potential colorants or additives before determining the shrink rate or final tool cavity dimensions.

Case Example

On one project the client produced a device for liquid dispensing that was used for retail packaging. The parts were designed and developed in a PP using a shrink of 0.016 in./in., and the molds tested fine and the initial parts performed well. Then the marketing team got involved and decided upon an array of many colors for each of their different liquid products. Most of the colors worked well but one in particular did not shrink as expected; leaving one internal seal fit too loose resulting in a low frequency potential leak from the package. For that particular color dispensing device, we had to "grow" the internal mating part to affect the correct seal squeeze; totally unplanned and costly from a set-up and debug standpoint. The lesson herein is to determine the colorant and other additives up front, then talk with the material suppliers and determine if there will need to be any adjustment to the shrinkage over the entire spectrum of the variations expected for the part.

2.2.2 *Tactile properties*

Tactile properties of plastics can range from downright slippery (especially when wet) to a robust graspable surface. This is yet another reason why these materials are so widely used by the industrial design community. In a single, or perhaps a two-step, process the finished part can be ready for use with no additional parts or secondary operation required. Grip surfaces can be molded from a soft-touch material directly over a more rigid structural material, sometimes in a two-mold two-step process and sometimes within the same mold with a dual injection cycle. Texture can be added selectively to areas to allow a user to feel the area required for an action, gloss can be added in areas where the part needs to glide over a surface without dragging, and so on. The best example is the common everyday manual hold in your hand toothbrush; most every toothbrush on the market has these features (gloss to slide between your lips and textured soft-touch materials to provide an adequate grip when wet).

Single tool dual injection process soft-touch second material overmolding has traditionally been accomplished using two thermoplastic materials. However, with extremely balanced volumes of liquid injection-molded silicone materials, single tool solutions have been developed to allow a thin silicone (thermoset) portion to be overmolded onto a thermoplastic first shot. Since there are different processing conditions for thermosetting silicone rubber (inject cold and then heat to vulcanize) versus the typical thermoplastic cycle (inject hot and cool to freeze the part to a solid), thermosets like silicone were traditionally applied as a second tool, second operation, overmolding process.

2.2.3 *Auditory properties*

Did you ever want to hear a click when the door snapped shut on one of your products? Some materials provide the hardness and density to provide great auditory feedback and some, well just don't. Find a part that exhibits the sound qualities you are looking for and figure out what it's molded from to start the process. Alternatively, when looking through the specifications look at hardness, density, and elongation as metrics for the materials ability to propagate or dampen sound. Many times plastic materials are chosen as sound barriers where we do not want to attenuate any vibrations, we want to deaden the sound. Soft touch materials, thermoplastic elastomers, and the like, can work well for these applications, as can materials that have been processed to have foam cores or are foamed throughout. Structure, to be sure, can affect auditory properties, but the combination of well-designed structure and a correctly selected material will work in unison to make the magic happen.

2.2.4 Clarity and transparency properties

Selecting materials for parts that require transparency limits the pool of candidates, but there are still many different families of polymers that can produce clear and glass like parts. These are typically amorphous resins, as the semicrystalline resins appear to be cloudy in their natural state. Issues with transparency abound; be mindful that both tooling and processing requirement will be more stringent to get high quality transparent parts. Toughness and scratch resistance will also be an issue; transparent things get dirty and needs to be cleaned so we can see through them—the act of cleaning will, more often than not, scratch the surface and begin the process of surface degradation. Some materials fare better than others regarding scratch resistance, so do your homework if you have a visual application where the transparent surface will need to be cleaned.

2.2.5 Other additives for aesthetic properties

Perhaps not necessarily and aesthetic property, but certainly a human factors issue is static surface charge. Ever watch a child slide down a playground sliding board manufactured from plastics; you know the big long yellow slides? What do you notice? The girl's hair is sticking straight out from her head as she departs the slide (similar to placing one's hand on a Van der Graaff generator), then she surprises her friend and "snap" the charge is dissipated as her friend yelps. Any time your part or product will be in frictional contact with some other material(s) and it is held or grasped by a human, be mindful that a static surface charge can be carried and transferred from the plastic surface to the unsuspecting human. I am reminded of this every time that I remove my polyester fleece jacket during the winter months and then reach to grab the hangar to put it away—"snap" the static is discharged at my expense (Fig. 2.12).

2.3 Selection by economic properties

Plastics materials are not unlike any other materials or products in that their costs are a function of raw material costs, and supply and demand. Commodity resins like polyethylene, high impact styrene, and polypropylene can be found from many different suppliers. While specialty resins, some with patent or trade secret protections, are only available from a single supplier, hence much higher costs per unit that the commodity

Figure 2.12 Static charge can be an interesting phenomenon. Plastics in frictional contact with other materials can build a static charge quickly. Some materials will hold a static charge much longer than others.
Source: Photo by Michael Ireland—stockadobe.com.

resins. Manufacturing location for the resin or plastic relative to the manufacturing location for the part being designed also plays a role. Designing a product to be made in Asia from a material only produced in Houston, TX is a cost driver.

2.3.1 Manufacturing efficiencies

Be aware of your anticipated capacity requirements, many times the requirements are not large enough to warrant truckload pricing for the resin to be used, if this is the case, look for a supplier (molder, processor, etc.) who is already using the material for some other customer—they are already stocking and buying it in some quantity, adding more quantity to their purchases, even in small increments helps them to be more efficient and may lower your costs as well. Another area for efficient operation is using regrind material, that is taking what would be waste, sprues and runners, and regrinding them and mixing them back into the feedstock. Some materials are tolerant of large amounts of regrind without any noticeable sacrifice in properties; other materials are not tolerant at all and will degrade quickly with repeated heat cycles. Total cycle time is another area where the material selected can have an impact on final unit costs—high heat capacity materials take longer to cool and set ready for ejection, hence longer cycle times resulting in less capacity per hour and higher per part costs.

2.3.2 Total energy input discussion

Remember total energy input drives cost, hence if a high-temperature material is selected when a lower-temperature material would suffice, every part molded costs more than it should—more energy is invested to make each part. In addition, higher molding temperatures mean longer cooling times, which in turn means longer cycle times and less parts per hour—higher costs per unit. Likewise, if the process can support multiple simultaneous part production, for example, multiple head extrusion blow molding, multiple part thermoforming, high cavitation injection molding, and the like, then the energy input per cycle gets utilized over a greater number of parts or finished units, thereby increasing efficiency. Total energy input thinking can also include the transportation and original raw materials manufacturing energies required prior to receiving and processing the raw materials for your part or product.

2.3.3 Weight reduction effects

One of the largest benefits when using plastics as a material for any part is the ability to reduce the weight of the part as compared to other suitable materials, for example, metals. This reduction of weight has far reaching benefits both in primary manufacturing and in end-use application. The total energy input for the finished end-user benefit is also reduced as compared to any suitable replacement materials for any application.

2.3.3.1 Part handling benefits

Initial part manufacture with thermoplastics is usually a fairly closed loop process—that is the raw materials used are efficiently processed and most waste is captured, and reintroduced into the feedstock. Finished parts are at net shape and weight in one, or sometimes two process steps. This means no extra energy is required to lift, move, or position the part in subsequent manufacturing operations or assembly steps.

2.3.3.2 Part shipping benefits

Energy input for transportation of the raw materials into the manufacturing location will be utilized efficiently since there is little waste from most thermoplastic processes. In addition, parts being transported from the primary processing site to a second assembly or final use site will, in a similar fashion, be efficient in the use of the energy input for transportation, since only the part itself is transported. Unlike, for example, a cast aluminum part where the casting is produced at the foundry, then transported to a machining site where the excess aluminum is machined away, then the part gets transported to the finishing (plating or painting) site and the waste gets transported to the recycling site. Sometimes, as in cut sheet thermoforming operations, the waste does get transported back to the sheet extrusion site for reprocessing. However, this is still far more efficient in total overall energy input than most of the processes for most suitable replacement materials.

2.3.3.3 *Final unit in-use benefits*

Once again, the finished assembly or product will be lighter as a result of using the plastics material, as compared to most any other material that could have been used. This results in a lower weight finished packaged item that gets transported through the distribution system, through the retail system, and to the end-user, resulting in lower costs of transportation throughout. Once in the hand of the end-user the lower weight unit can be held or used for longer periods, or can be moved easier, or can be lifted repeatedly, and so on, when compared to similar products made with alternate materials.

2.4 Final "How To" select a material

As illustrated earlier in this chapter, the use of a material selection database, such as the UL Prospector database for searching is invaluable. Not only are there thousands and thousands of specification sheets available, there are comparative analysis tools, articles regarding selection of materials, tutorials on selection, and so on. One of the most useful tools in their offering is the material properties spider graph, as illustrated in Fig. 2.13, where several basic material properties are plotted simultaneously against the generic form of the material. For this illustration a grade of ABS is compared to generic ABS and we can see that this grade has a slightly higher tensile strength than the generic comparative and has a slightly lower–melt mass flow rate. Each variation

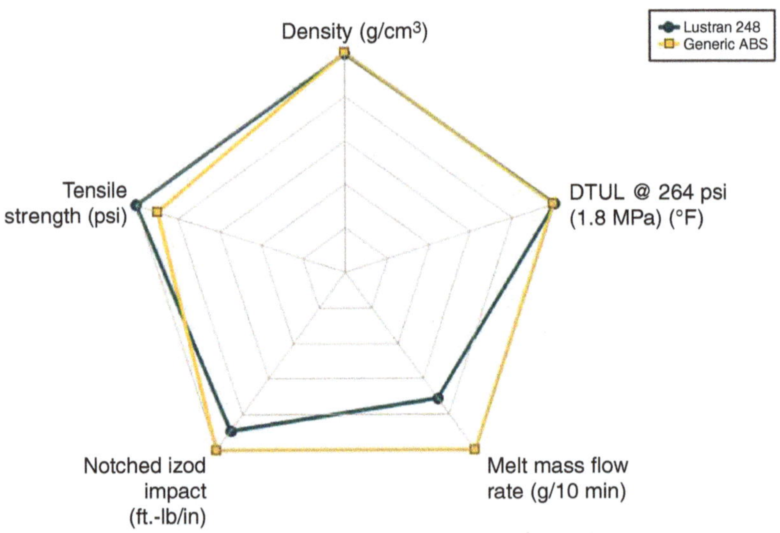

Figure 2.13 A spider graph for Lustran 248 ABS showing comparative values for five different properties simultaneously.

Source: This is from the UL Prospector database and when viewed on their website the actual numerical values for each data point on the graph are displayed when the user places their cursor over the graph data point (https://plastics.ulprospector.com/) [2].

of each material will have different values for all of these basic properties and the spider graph allows one to search quickly and digest lots of information quickly.

2.4.1 *Risk management*

Material selection is perhaps one of the trickiest choices that any designer will make on any new part or product development project. There are so many things which can seem insignificant but can become critical depending upon the intended end-use, or worse yet, the unintended end-use. Our job as designers of products and parts made with plastics is to plan for the expected and the unexpected. When a part or product fails, the manufacturer is the guilty party; and if there is harm done because of the failure, you can be sure that legal action will follow soon thereafter. As the designer, I will NEVER specify the material. I will work with the material supplier to identify the family and the specific attributes for additives which may be used and will make recommendations to the manufacture or customer. However, in the end, I make sure that the material selection is a collaborative effort, with each supplier or processor taking responsibility for the selection as a team. Most importantly, as a team the recommended material selections are presented to the customer and they need to understand their choices and direct the team as to which they wish to use. This top to bottom buy-in from every supplier and the customer ensures that: (1) a good decision has been made, and (2) that the responsibility for the selection is spread across the entire team.

Risk management, both for the finished product or part and for the development team, is a necessary task for any development project. It may not be foremost in the designers mind when working on some new cutting edge design, but it will certainly be worth the small time investment required when the product is misused once in the stream of commerce.

References

[1] Michael Sepe in UL/IDES Prospector database—as copied from the 3rd edition of Pocket Performance Specs for Thermoplastics published in 2006 by Injection Molding Magazine.

[2] https://plastics.ulprospector.com/.

[3] https://www.lyondellbasell.com/globalassets/documents/polymers-technical-literature/tech-topic-mold-shrinkage.pdf.

[4] http://www.engineeringtoolbox.com/linear-expansion-coefficients-d_95.html.

[5] http://www.ulttc.com/en/solutions/testmethods/electrical.html.

[6] http://www.ulttc.com/en/solutions/testmethods/combustion-fire.html.

Resources for further study

S. Levy, J.H. DuBois, Plastics Product Design Engineering Handbook, second ed., Chapman & Hall, New York, 1984.

S.E.M. Selke, J.D. Culter, R.J. Hernandez, Plastics Packaging—Properties, Processing, Applications, and Regulations, second ed., Carl Hanser Verlag, Munich, 2004.

J. Rotheiser, Joining of Plastics—Handbook for Designers and Engineers, third ed., Carl Hanser Verlag, Munich, 2009.

G.L. Beall, Rotational Molding—Design, Materials, Tooling, and Processing, first ed., Carl Hanser Verlag, Munich, 1998.

E.R. Larson, Thermoplastic Material Selection—A Practical Guide, first ed., Elsevier, Oxford, 2015.

R.A. Malloy, Plastic Part Design for Injection Molding: An Introduction, second ed., Carl Hanser Verlag, Munich, 2010.

P.A. Tres, Designing Plastic Parts for Assembly, seventh ed., Carl Hanser Verlag, Munich, 2014.

Web links of note

Great Design information from DuPont can be found here:

http://www.dupont.com/content/dam/dupont/products-and-services/plastics-polymers-and-resins/thermoplastics/documents/General%20Design%20Principles/General%20Design%20Principles%20Mod%201.pdf.

Sabic provides many tools for designers here (you may need to register):

http://www.sabic-ip.com/webtools/toolsHomePage.faces.

The UL Prospector site is a valuable tool for any designer reviewing materials for any application (you will need to register):

https://plastics.ulprospector.com/.

Solvay provides a great design guide for using their Udel polysulfone materials:

http://www.solvay.com/en/binaries/Udel-PSU-Design-Guide_EN-227550.pdf.

Bayer has a nice design guide for RIM systems:

http://reactioninjectionmolding.com/wp-content/uploads/2013/09/RIM-PartMoldDesign-Guide.pdf.

Covestro offers several design related guides, including part and mold design, snap fit design, and joining techniques for plastics:

http://www.plastics.covestro.com/Library/Technology-brochures.aspx.

Process selection—which plastics process to use?

3

Process selection for the experienced designer becomes sort of second nature during the initial conceptual size-up of a potential solution for a given "need" however, good design practice teaches us to always evaluate each potential process for the given design solution first, then proceed based upon carefully considered benefits for the chosen process. Sometimes a design solution might be plausible when produced from more than a single process and the process selection can linger until enough design progress has been made to show clearly which process is best for the particular design solution. Hence, keep an open mind and ensure the best process selection is made for the given design solution. There are several ways to consider which process might be best suited for a particular design solution, by functional requirements, by capacity requirements, by cost economics, by the secondary operations which might be required, and so on.

As discussed earlier in this book, the focus herein is on design using thermoplastic materials and processes. While some thermosetting materials and processes are mentioned as references or as comparisons, with the exception of reaction injection molding (RIM), they are not covered in depth in this chapter.

3.1 Selection by functional requirements

The application in which the part or parts being designed must function will possibly determine the process that must be or should be used. Some functional requirements will dictate that a part be processed in a certain manner, for example, if we were to design a part to hold a fluid, we could use mostly any process, however, if the functional requirements also called for long-term sealed storage of the fluid, then the part becomes regulated to perhaps two processing methods, or perhaps a process combined with an assembly method. For this example, the obvious process would be blow molding or rotational molding; however, many tanks are produced using injection-molded halves with a hot plate weld producing a sealed vessel. If tight tolerances are needed to fit or mate to other parts, then the process selection might come into discussion, as some processes are better suited to tighter tolerances than other processes; for example, injection molding tolerances will generally be less than tolerances in thermoforming and machined parts can be held to tolerances even tighter than injection molding. Table 3.1 shows a brief comparative review of the attributes achievable using the most popular plastics processes.

Designing Successful Products with Plastics. http://dx.doi.org/10.1016/B978-0-323-44501-6.00003-9

Table 3.1 **Comparative attributes of several popular plastics processing methods used for creating product or parts that a designer or design engineer might specify**

Plastics process	Functional	Capacity	Costs	Secondary operations	Aesthetics
Machining	Capable of producing tight tolerance solid parts that can be strong and durable.	Typically lower volumes are produced by machining from solid, extruded, or cast materials.	High—small tooling costs but high per unit costs and little if no efficiency at higher quantity production.	Painting or other surface modifications are sometimes required.	Similar to any other parts machined from stock. Can be polished to a high shine when required.
Resign casting or vacuum casting	Can produce tough and durable parts; however, part-to-part tolerances are somewhat wider than other processes.	Typically a lower-volume process used for short-run production or prototyping. Mold life is usually 25–50 pour cycles.	Moderate—tooling costs are low relative to other potential processes, however the tooling has a lifespan of 25–50 parts and must then be replaced. Casting process is long—sometimes a 24 h cycle.	Resins can be mixed with pigment reducing the need for secondary operations. However, since the molds are extremely simple, some secondary machining can be required.	Good—the process is capable of nice surface finishes.
RIM	Can produce strong parts in exacting tight tolerance geometries. Materials also allow for softer more resilient parts as well.	Moderate—can be used for high volume parts, especially larger parts with specific exposure requirements. Smaller parts are usually have lower-volume capacity as injection molding becomes more efficient at some annual volume usage.	Moderate—this low pressure process can be supported with aluminum molds costing less than the steel counterparts needed for injection molding. Part production cycle is longer to accommodate the chemical reaction time and postreaction cooling time.	Minimal—sometime these parts will need painting or decorating, similar to injection-molded parts.	Good—the tooling allows exact surface finish attributes, much like injection molding.

Plastics process	Functional	Capacity	Costs	Secondary operations	Aesthetics
Injection molding	Can produce strong and durable parts to exacting tolerances. High-pressure process can be used to fill complex geometries.	High—injection cycles are fast and cooling is rapid allowing one of the shortest cycle times of any plastics process.	Tooling first costs are some of the highest of any of the processes. Per unit production part costing can be the lowest of any of the plastics processes.	Minimal—sometimes threaded inserts are added, painting is specified, or decoration is required.	Excellent—the high-pressure process is capable of exact surface finishes and molded-in colors.
Profile extrusion	Can produce complex profile geometries in materials to provide rigidity or in softer materials used for resilience.	High—the extrusion process is a continuous in-line process ideally suited for higher volume usage.	Tooling costs can be extremely reasonable—to even being considered low cost. Profile production costs can be low, depending upon volume. The process set-up and debug time is amortized into the production run volumes.	Extruded profiles must be cut to length and then other features (holes, slots, etc.) must be added postprocessing.	Mildly frosted or textured surfaces can be achieved however deep patterned textures are much more difficult. Sometimes this is material dependent.
EBM	Can produce strong hollow shapes and other geometries and can be used with a multiple extruded layers of different polymers. Offset neck geometries are possible, as are handles and other more complex geometries.	Moderate to high—process can be augmented by multiple extrusion heads and multiple molds for higher capacity products.	Tooling costs are moderate as aluminum can be used for the mold materials because of the low pressure nature of the process. Part costs can be on the low to moderate end depending upon secondary operations required.	Flash removal is standard and is most often automated to some degree. Surface treatment and labeling can be in-line secondary operations as well.	Surfaces can carry textures and engraved text or graphic elements.

(*Continued*)

Table 3.1 Comparative attributes of several popular plastics processing methods used for creating product or parts that a designer or design engineer might specify (*cont.*)

Plastics process	Functional	Capacity	Costs	Secondary operations	Aesthetics
IBM	Can produce extremely strong hollow shapes. Typically used for containers with a centrally located neck geometry.	High—the two-step process demands higher volumes to justify the tooling investments. Preferred for smaller containers over the EMB process.	First costs of tooling for the injection-molded preform and for the secondary blow molds is high. Part costs are low since multiple cavities are processed simultaneously and since the blow operation is a reheat process with zero flash removal required.	Usually no secondary operations are required after the blow operation. Surface modification for labeling or decoration may be required depending upon the material.	Surface textures and text or graphic elements can be molded into the surfaces. Highly polished molds can produce near transparent final part conditions.
SBM	Can produce highly pressure resistant thin wall containers—typically used for carbonated beverages. The stretch portion of the cycle serves to assist in orientation of the polymer chains prior to the blow process.	High—the two-step process demands higher volumes to justify the tooling investments. Preferred for smaller containers over the EMB process.	First costs of tooling for the injection-molded preform and for the secondary blow molds is high. Part costs are low since multiple cavities are processed simultaneously and since the blow operation is a reheat process with zero flash removal required.	Usually no secondary operations are required after the blow operation. Surface modification for labeling or decoration may be required depending upon the material.	Surface textures and text or graphic elements can be molded into the surfaces. Highly polished molds can produce near transparent final part conditions.

Plastics process	Functional	Capacity	Costs	Secondary operations	Aesthetics
Thermoforming—heavy gauge cut sheet	Can produce heavy gauge strong and durable parts.	Cut sheet heavy gauge processing is a low to moderate volume capacity process.	Tooling costs can be low to moderate, with part costing being moderate to high as compared to injection molding.	Routing of the excess sheet and part features is required.	Raw materials can be purchased extruded in colors and surface texture can be processed as well. Fine detail cannot easily be molded into the surfaces.
Thermoforming—rotary die	Can produce thin gauge strong and/or flexible parts.	Rotary die thermoformed parts can be an extremely high capacity process—think coffee cup lidding, for example.	Tooling costs can be high. Part costs are low at high volumes. Parts can be processed extremely thin and the in-line rotary extrude, form, and trim process is highly efficient. Die cut waste is ground and returned to the feedstock stream.	Automatic stacking and sanitary packaging is usually found immediately downstream of the rotary die equipment. Otherwise no other secondary operations are usually required.	Thin wall parts can carry fine details, text or graphics and the extruded sheet can be extruded in colors or transparent polymers.
Rotational molding	Can produce extremely strong and durable hollow vessels or structural geometries. Very large parts can be produced using this low-pressure process.	Considered to be a low to moderate capacity process. Long cycle with lots of handling of the mold and the part.	Tooling costs are low to moderate—tools are typically cast from aluminum alloy. Part costs are moderate to high due to the labor input and longer cycle times.	Flash removal and routing of part functional features are typically encountered on Roto-molded parts.	External surface of the part can be textured and can carry some text or graphics. The resins can be compounded to allow molded-in colors.

EBM, Extrusion blow molding; IBM, injection blow molding; RIM, reaction injection molding; SBM, stretch blow molding.

3.1.1 Functional end-use

Every part or product to be designed must perform some functional end-use. Often this end-use will determine, or at least partially determine, the process which needs to be used. As in our opening example, the design of a container to hold a fluid will point us to two or possibly three ideal processes to select from. Likewise, other end-use criteria will limit or determine the process as well; for example, when designing a precision optical lens element, one would not even remotely be thinking about thermoforming or rotational molding, rather injection molding would be at the top of the list. Another example might be a tubular geometry used to contain electrical or communication cabling; one would certainly jump right to selecting profile extrusion and not consider any other process.

3.1.2 Strength considerations

Every application has unique requirements and the strength requirement is often the first and perhaps easiest to quantify and use as a criteria for selection. In addition, the in-use temperature range needs to be detailed, as many polymeric materials have very different yield points at different temperatures. Extremes of either hot or cold can limit the choices in materials available for the application. Likewise, the process selection may be driven by strength and/or in-use temperature range specifications. Certain processes, like heavy gauge thermoforming, will produce extremely high-strength and rugged parts, where other processes, like blow molding can produce hollow parts that can withstand large pressures but might not be so structural in an axis perpendicular to the neck. Each application will call for some feature on the part to carry some load; hence strength considerations are easy first criteria to begin the selection analysis process. Table 3.2 shows some limited examples of discreet parts produced by each of the various processes.

3.1.3 Surface properties

The desired surface finish or hardness will also be a decision driver in selecting the process. Some processes, like injection molding and blow molding, will be able to offer an almost unlimited array of surface finish texture choices, while others, like thermoforming, must rely on available sheet goods in available finishes and hardness's. Surface hardness, by the way, is a function of the material and for a given material you cannot specify a value or magnitude for hardness. You get the surface hardness that the material is capable of, nothing more and nothing less. This is unlike metals where the crystalline structures can be altered at the surface to be tougher and harder.

Table 3.2 Production capacity comparisons of several popular processing methods used for manufacturing products or parts from plastics

Plastics process	Capacity	Tooling costs	Part costs	Secondary operations	Examples
Machining	Low to moderate	Low	High	Painting or finishing	Functional parts, mostly nonvisual parts.
Resign casting or vacuum casting	Low to moderate	Low	High	Painting or finishing	Prototypes of injcetion molded parts.
RIM	Moderate	Moderate	Moderate	Painting or finishing	Large outdoor use parts for trucks and boats.
Injection molding	High	High	Low	Painting or finishing	Functional parts, electronics enclosures, appliances, closures for containers, and so on.
Profile extrusion	High	Low	Low	Adding *z* axis features—holes, slots, and so on, cutting to length.	Pipe and tubing, window casings, bumpers and trim, and so on.
EBM	Moderate	Moderate to high	Low to moderate	Flash removal and surface treatment.	Containers for fluids or beverages, larger containers, offset neck containers.
IBM	High	High	Low	Usually none	Containers for fluids, usually smaller containers.
SBM	High	High	Low	Usually none	Containers for beverages including carbonated beverages.
Thermoforming—heavy gauge cut sheet	Low to moderate	Low to moderate	Moderate	Routing of the excess sheet and part features is required.	Large structural or functional parts, enclosures for large appliances, and so on.
Thermoforming—rotary die	High	High	Low	Usually none	Lidding for consumer disposable cups, packaging, and so on.
Rotational molding	Low to moderate	Low to moderate	Moderate to high	Flash removal and routing of part functional features.	Large tanks, garbage containers, tanks for agricultural equipment, sprayer tanks, and so on.

EBM, Extrusion blow molding; IBM, injection blow molding; RIM, reaction injection molding; SBM, stretch blow molding.

3.2 Selection by capacity (parts per year)

Annual capacity requirements probably determine more process selection decisions than any other criteria, since capacity also is a fundamental component in cost recovery. More units per year will typically result in recovery of capital first costs in a shorter time span. However, each part and project is different and each must be analyzed independently. Some processes are inherently low-volume processes and others are polar opposites, demanding high volume to justify not only the capital for tooling but also to justify the use of the equipment resources required as well. Table 3.2 shows some relative comparisons for the capacities attainable from the various processes discussed.

3.2.1 Low-volume processes

Low volume is a relative qualifying term; what is low volume? Well, producing one of any item is indeed a low-volume process, but what is the upper limit? Well that depends on the application, the part size, and many other factors, but generally we think of low volume as less than 100,000 or so parts annually.

3.2.1.1 Machining

Most common plastics families are manufactured in shapes and forms for fabrication by standard machine shop practices. For prototyping or for short-run needs, simple fabrication from sheet, rod, or tube may well be the most expeditious. No tooling is required and the parts can have features, including thicknesses that cannot be attained in high volume processes. Of course, the parts will have no cost advantage over other materials fabricated in a similar manner, or at best just a slight advantage because the feeds and speeds can be somewhat faster than in some metals.

Sometimes a combination of two processes is required to gain efficiency in manufacture, for example, extruding a shape to gain the detail in the x–y direction, then machining the shape to finish the detail in the z direction. Fabrication by welding, solvent bonding, and so on, can also be combined with the machining process to produce complex parts or prototypes.

Waste from machining processes can be collected and recycled but since it is mostly a prototype and short-run process, the great majority of chips and waste are not reprocessed (Fig. 3.1).

3.2.1.2 Thermoforming (vacuum forming, pressure forming, and twin sheet forming)

In general, cut sheet vacuum forming is thought of as a low-volume process. This variety of thermoforming is defined by the use of vacuum to allow atmospheric air pressure to force the softened plastic sheet over or into the mold. This process starts with a sheet of a selected material, in a selected thickness and color, and sometimes with a selected surface finish. The sheet is cut to size to fit in a holding frame which is a part

Figure 3.1 A black Delrin (Acetal) part being machined in an end milling machine.

of the processing equipment. The tooling for this process is a die block form which defines one side of the finished part; typically it would define the inner or nonaesthetic side of the part. The die block form will either be porous or it will have small diameter venting holes bored through in specific and selected locations and it will be mounted firmly in the process equipment on top of a sealed chamber. The process equipment will move the sheet of material into an oven portion of the equipment, heat it to a specific temperature, then move the hot sheet to a position over the die block form tool. The process equipment will then draw the material down onto the die block form and simultaneously a vacuum or suction is pulled on the sealed chamber beneath the tooling. This simultaneous draping of the hot sheet and suction from beneath and through the tool will pull the hot sheet onto and against the die block form taking the exact shape of the form, then the formed sheet cools while on the form.

The die block materials vary according to the anticipated capacity of production required. Prototype and short-run projects can sometimes use a rigid porous foam material which is inexpensive material to purchase and easy to machine on standard multiaxis CNC equipment. Another prototype and short-run tooling method for small-to-medium sized parts is the up and coming fused deposition 3D printing technology. For production tooling, aluminum alloys are most often specified because of their long-term stability and their ability to assist in cooling the part rapidly for reduced cycle times.

Once formed and cooled, thermoformed parts will require secondary operations to finish the part. This is typically accomplished with multiaxis high-speed routing equipment. Other secondary operations might be the addition of mounting blocks and mating features to the part. This is typically done using solvent or adhesive welding of premachined or formed blocks to the part in fixtures to precisely align the blocks and mating features correctly relative to features on the part.

Pressure forming is a variation of thermoforming that is a bit more equipment and tooling intensive, but can produce even nicer finished parts from an aesthetic point of view. In pressure forming, the process is essentially the same as vacuum forming, except that when the hot softened sheet is drawn over the die block form, another separate portion of the process equipment closes off the space on the opposite side of the hot softened sheet. When the vacuum is initiated to pull the sheet over the die block geometry, the opposing side of the process equipment will then pressurize the hot softened plastic sheet to greater than one atmosphere. The hot softened plastic material is then being pulled onto the die block geometry from one side and being pushed onto the die block geometry from the other side, resulting in the ability to create very exacting details in the part from the die block not possible with vacuum alone.

Another variation capable with this process is twin sheet thermoforming. For this process the arrangement is similar to those discussed, but now we will use two die block forms and two heated sheets of material, arranged such that the two hot sheets of material end up between the two die block forms. Vacuum is applied to both dies as the two dies are brought together to pinch the two hot sheets together at the perimeter of the part. The resulting part is a hollow part, with detail formed in the upper and lower surfaces by the two die block forms. Sometimes pressure can be introduced in between the two sheets during the forming process as well to add inner pressure similar to blow molding.

Still another twin sheet variation is used to laminate two materials in a first step then use the twin sheet laminate in a second reheat and draw step to form the resulting part. Sometimes this is accomplished via a coextrusion method during the primary manufacturing of the sheet goods used for the cut sheet thermoforming process.

Waste from the thermoforming process is collected and stored in a clean manner, sent back to the sheet extruder, ground and reprocessed with virgin resin into new sheet. The exception to this would be for the coextruded or laminated materials which could be either different materials or at least different colors of the same material (think red solo cup) (Fig. 3.2).

3.2.1.3 Rotational molding

Like thermoforming, rotational molding is considered a low-volume process, although there are instances where it is used in some higher volume applications. Whereas thermoforming is usually used to make a planar or open box sort of part, rotational molding is almost exclusively used to make hollow parts. Many of these hollow formed

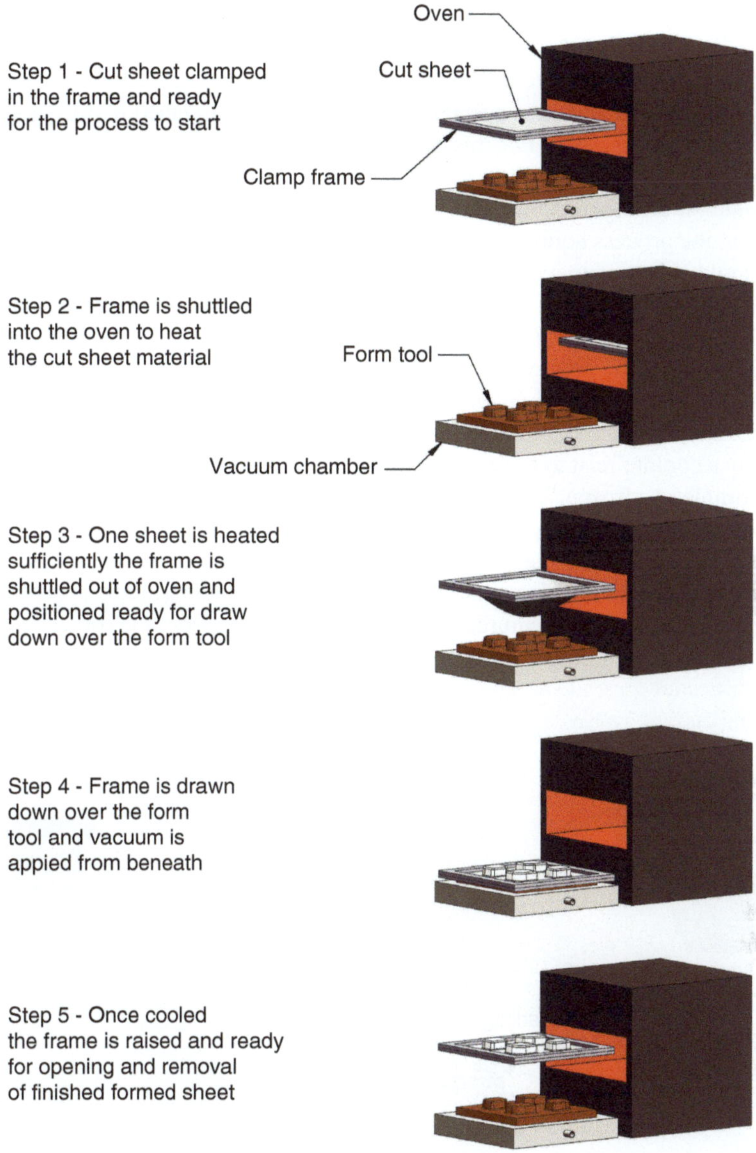

Figure 3.2 A graphic illustration showing the process steps required for the cut sheet thermoforming process.

parts will have secondary operations performed to turn them into deep containers like garbage receptacles, hoppers for storage, and the like. However, the process is mostly known for its ability to make large tanks for use with all sorts of chemistries. These tanks are seen on agricultural equipment, floor polishing machines, landscaping and irrigation equipment, and in all sorts of industrial applications.

The process is however somewhat different from thermoforming and starts with resin in granular or powder form, not sheet. The tooling is usually in two pieces and is most often made from sand cast aluminum alloys. The two mold halves are arranged to split at the parting line in a manner that will allow the finished part to be removed without damage to the part. For this process, the tooling is opened, a metered portion of the resin is loaded into the tooling and then the tooling is clamped shut, usually with some bolt arrangement around the perimeter of the parting line. The tool is then loaded into the process equipment which shuttles the mold into a heating oven portion and the mold is heated to a predetermined temperature. During this heating cycle, the dynamic rotation portion of the process is initiated. The rotation molding equipment provides the structure and energy to rotate the mold axially in multiple directions to distribute and coat the inner walls of the tool with the powdered or granular polymeric material, whereupon the polymers begin to become molten. Once the heating cycle is complete, the process equipment will remove the mold from the heating chamber and then apply a coolant mist to the exterior of the tooling as it continues to rotate until the part is completely frozen. The tool is then delivered by the process equipment to an unloading station and the operators will manually unclamp the tooling and go about removal of the completed plastics part.

Postrotational molding secondary operations will include multiaxis routing to add holes and other functional features to the final part. Waste from the router operations, if kept clean, can be reground and reintroduced into the feedstock resin (Fig. 3.3).

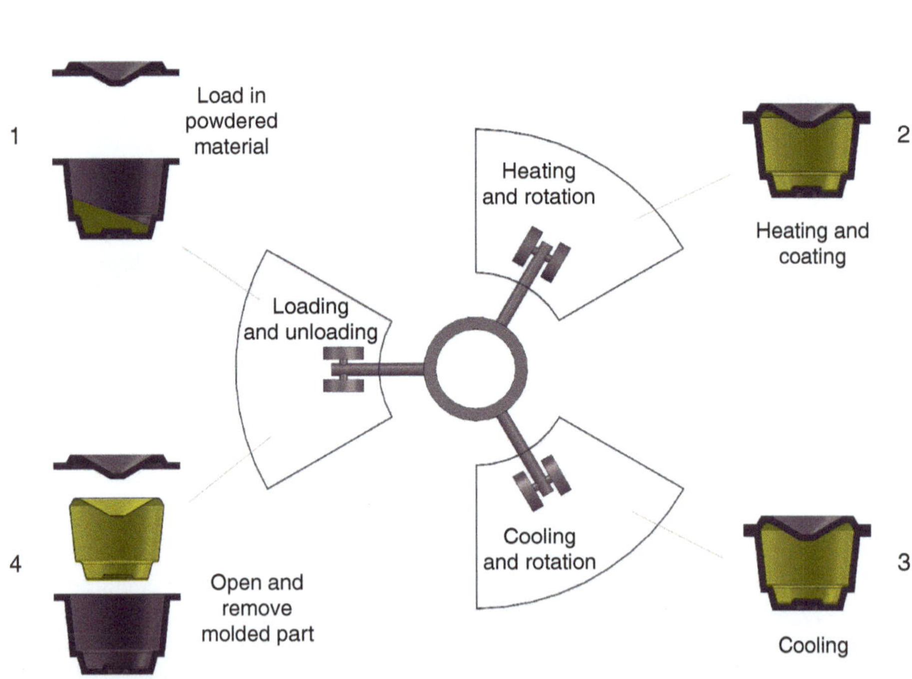

Figure 3.3 A simplified schematic view to illustrate the rotational molding process.

3.2.1.4 Reaction injection molding (RIM)

RIM is similar in many ways to injection molding with one huge exception—the materials are introduced into the mold as low temperature liquids and then, the REACTION part of RIM happens. The reaction is an exothermic chemical reaction of the two liquids injected simultaneously at low pressure into the mold cavity. Since the plastic part is formed from the reaction of two chemistries, this process is a thermosetting process; unlike thermoplastics, parts made from this process cannot be reprocessed and will not melt back into some other form. In addition, they cannot be heat processed for secondary operations like welding or heat staking, and so on.

This low-pressure process allows construction of the mold from aluminum, allowing lower-mold fabrication costs and faster heat transfer when cooling the part for ejection. Likewise the low-pressure nature of the process leads to lower tonnage presses with lower-power consumption resulting in an overall lower-cost operation when compared to injection molding. So, the benefits are lower first costs of tooling and lower-operational overhead costs. However, the process cycle is determined by the thickest portion of the part and is a longer overall cycle than a comparable injection molded part. In addition, the materials most often used are polyurethanes and will often require secondary operations (or in mold precoatings) to achieve desired surface aesthetics. Since the process is a low-pressure process, many designers look to this process when larger parts are being considered. Examples include automotive, heavy truck and bus parts, marine vessel parts, and the like. In addition since these are polyurethane and when painted with a polyurethane paint system (which has excellent adhesion) the finished parts perform well in outdoor exposure and high UV environments (Fig. 3.4).

3.2.1.5 Vacuum casting

Vacuum casting is exactly what the name implies, a casting process which uses a vacuum assist to completely evacuate all air from the mold cavity. The process requires a mold which is typically cast in a room temperature vulcanizing (RTV) silicone casting compound using a master model. The master model is sometimes fabricated by CNC machining, sometimes carved or prepared by the hand of a craftsman, or sometimes prepared using stereolithography or one of the other rapid prototyping (RP) methods in use today. Some now refer to these RP methods as additive manufacturing or 3D printing.

Once the master model is prepared, the RTV mold is cast about the model, with special allowance for a mold separation plane (parting line) so the finished part can be removed. Other allowances were planned in the preparation of the master model, including the shrinkage factor for the materials to be used, filling ports, and considerations for inserts or removable cores. Once ready for production the RTV mold is filled with the liquid resin, and placed into a vacuum chamber and the evacuation process is initiated. Once fully evacuated the mold is cured at an elevated temperature. A given RTV mold is typically useable for 25–50 pour cycles before needing to be retired. The parts can be cast with fairly accurate details and the resins can be mixed, with pigment to provide extremely nice parts with surface feature properties and aesthetics similar to their injection-molded counterparts. Since the process is a small batch process the

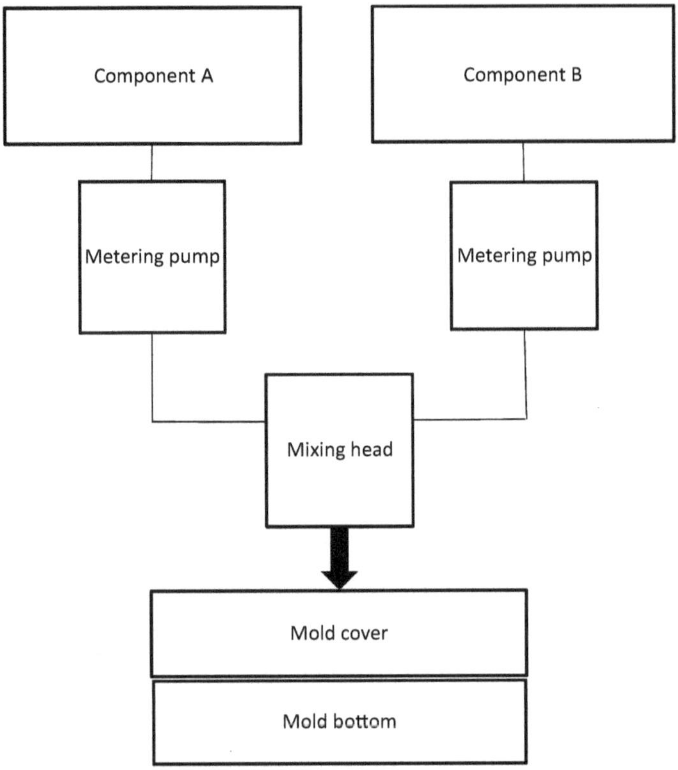

Figure 3.4 The RIM process starts with two different liquid components which are delivered with a metering pump into the mixing head and control valve which directs the flow of the mixed materials into the mold. The metering pumps allow the mix ratio to be controlled. The controlling valves within the mixing head prevent mixing of the two component liquids until the process is ready. The chemical reaction between the two liquid materials occurs once the mixed liquid has been injected into the mold, creating exothermic heat, and then undergoing a phase change to a solid.

part costs are higher than a corresponding injection-molded part, however the tooling investment and lead times are far less. In many programs this process is used to produce initial assemblies to be used for marketing studies, packaging creation, photo-models, and so on, while the production injection-mold tooling is being fabricated (Fig. 3.5).

3.2.1.6 Additive manufacturing—fused deposition modeling (FDM)

Additive manufacturing refers to a bit-by-bit or microdot-by-microdot application of a molten polymer to a substrate of already deposited bits or dots, whereupon the applied bit or dot then freezes to the substrate structure; this is done with a computer-controlled application dispenser which is driven from part file data generated from the part solid file. This method, also known as fused deposition modeling (FDM) has been used for a decade or more as a rapid prototyping (RP) method for creating first samples of plastic parts for look, touch, and feel. Ongoing developments of both the materials used and

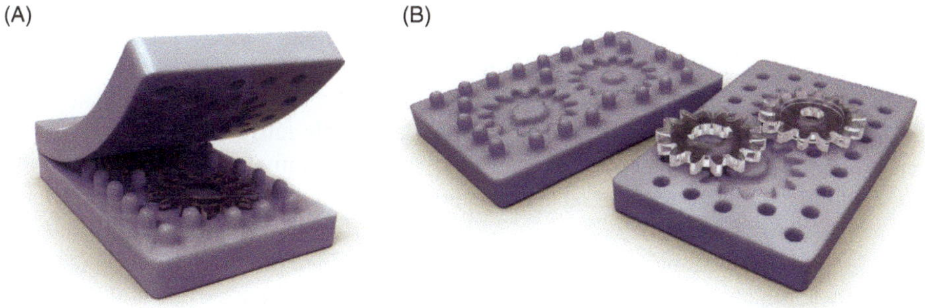

Figures 3.5 The light blue mold is fabricated from a silicone RTV casting compound (A) then the soft mold is used to vacuum cast the liquid resin materials to produce the gear parts shown (B). *Source*: iStock/coddy.

the process delivery and controls have allowed the method to achieve much greater final part densities and reduced total time input per part. Given the reduced time per part, the increased density of the structure, the ability to design parts which could not be molded from any other process, and the ever widening array of available materials, the process is now finding its way into the world of production. Not to mention the ability for discrete production specialization on a part-by-part basis, not even remotely available from other production methods; for example, an ear plug or hearing aid housing can be produced to the exact shape of the intended users ear, via the use of a laser 3D scan of the ear surfaces, converted into a part file then used to create the housing or plug device on the FDM equipment (Fig. 3.6).

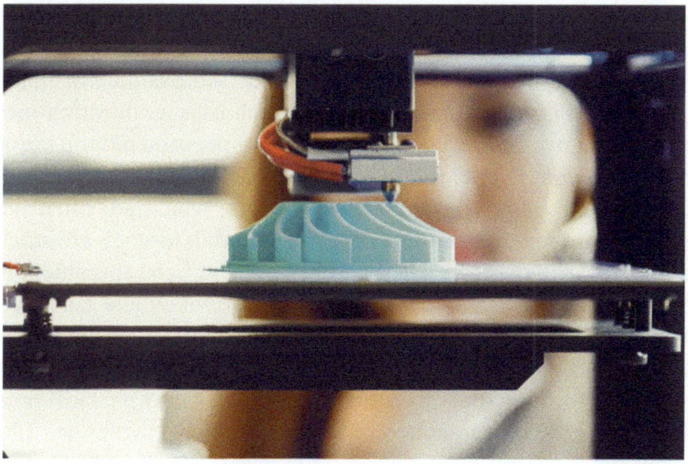

Figure 3.6 A fused deposition modeler (FDM) shown in-use depositing a thermoplastic material bit-by-bit in layers to form the solid object. Note that the thermal process is localized to the bit being placed and those immediately surrounding the most recent placed bit. Hence, molded in stresses are localized and generally do not cause warpage of the finished part. *Source*: Photo from iStock/izusek.

3.2.2 High-volume processes

As with the discussion and examples given for low-volume processes, the exact definition of high volume cannot be defined, however, depending upon each part and application, the high-volume demarcation tends to be identifiable fairly easily. Usually when the volume of parts to be made can easily pay back the capital tooling in a relatively short period of time, we tend to categorize the part as a high volume part. Sometimes, again depending upon application, this can be as few as 10,000 pieces annually and can be as high as billions of part annually; like injection-molded preforms for use in stretch blow conversion to one half liter water bottles, or insulin syringe barrels and plungers.

3.2.2.1 Injection molding

The granddaddy of high-volume processes and perhaps the most widely used of all the plastics processes, injection molding can produce parts from the widest variety of materials and can produce parts from microscopic size up to extremely large. Injection-molded parts can be extremely precise and can have a variety of molded-in features that can be used for joining and fastening with other parts, both plastic and nonplastic. In addition, there is a wide variety of additives and filling materials that can be used with the base materials in injection molding to accomplish such things as, adding color, adding flame retardant, increasing stiffness, providing electrical conductivity, providing surface charge dissipation, clarifying the base resin, adding lubricity, and the list goes on. The process also lends itself to other variations including the ability to inject two or more materials within the same tooling within the same cycle for single part production comprising multiple colors or multiple densities/harnesses; for example, adding the soft touch TPE surface material over the rigid PP structure for a toothbrush or disposable razor. In some instances, injection-molded parts are used for further processing; sometimes for over-molding in a second injection molding operation, or for use in stretch blow molding (SBM), or to be machined to add some feature that was not able to be molded in the primary shot. An example of this might be a common spatula used in commercial kitchens which uses an injection-molded nylon handle, which is then overmolded in a secondary compression molding process to add the flexible HCR silicone rubber (thermoset) spatula element to the handle.

The process begins with pelletized raw materials, which are often preprocessed prior to the injection process to dehumidify the materials so they are suitable for the injection molding process. The raw material is loaded into the equipment from a hopper or feed system and is heated and plasticized in a screw and barrel arrangement similar to that of an extruder. The material, once molten, is injected into the mold at high pressure (providing high-velocity flow rates), whereupon the material is cooled back to a solid state forming the finished part. There are many physical process events which happen during the injection molding cycle, the mold cavities are filled, the process equipment holds the fill pressure for some predetermined time to compress the polymer during the initial freeze change of state, then the pressure is dropped but the mold is held closed for the part to freeze to a condition where it can be ejected without damage or distortion, then the mold opens, the part is ejected, the mold shuts and the process cycle begins again. The part will continue cooling once ejected from the mold and can continue to change size, known as postmold shrinkage (Fig. 3.7).

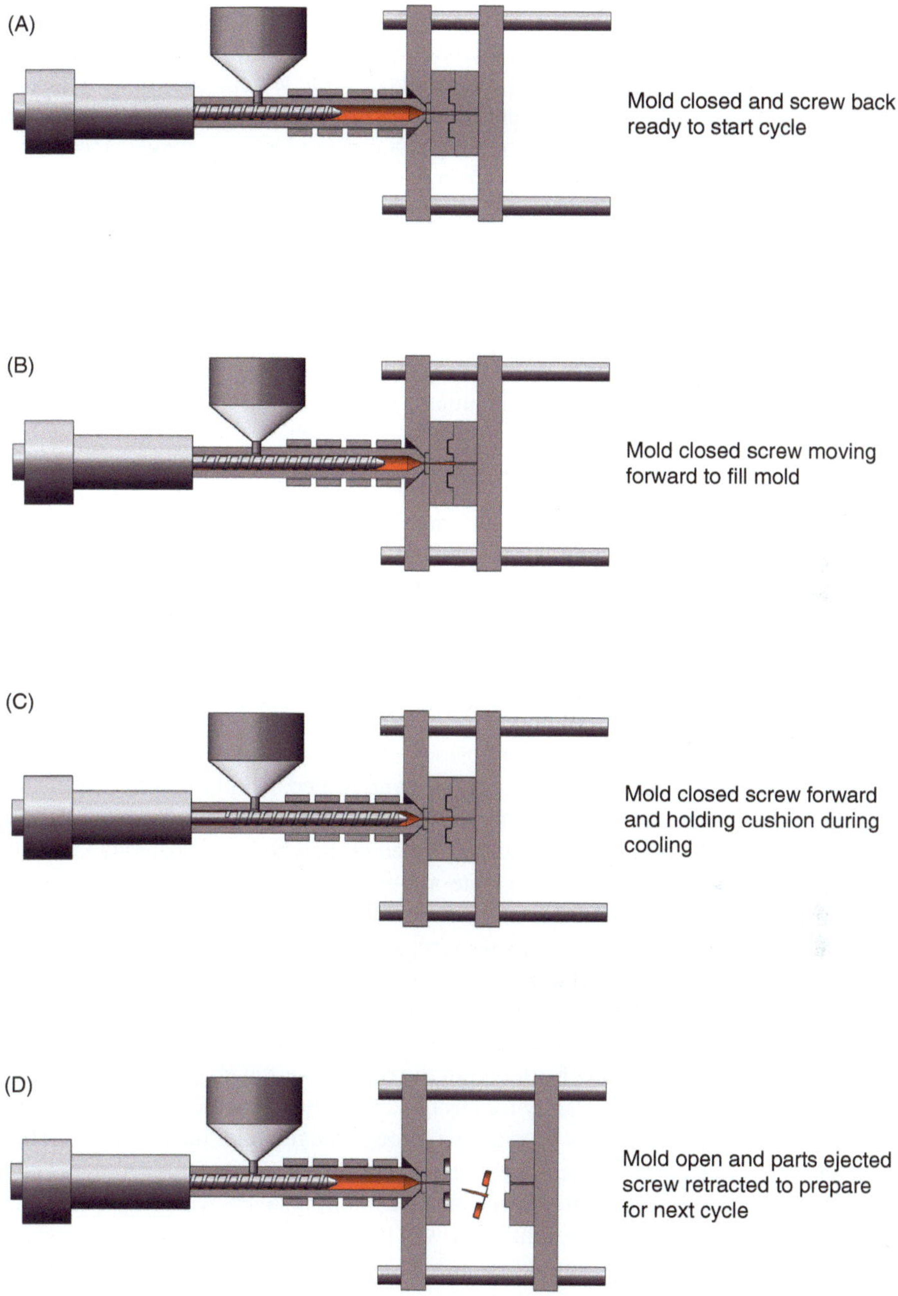

Figure 3.7 The general injection molding cycle follows the following basic steps. (A) Filling the barrel with material and preparing (plasticizing) the material for the injection cycle, (B) the mold filling portion of the cycle, (C) holding cushion (pressure) while the material solidifies within the mold, and (D) mold opening and parts ejection, then the mold closes, the screw retracts and the cycle begins again.

One advantage to this high-pressure process is the ability to simultaneously fill and create multiple parts, commonly referred to as cavitation, or multicavity molding. Indeed the multiples can be as low as two parts per cycle and the upper end is not yet defined, but we commonly see 32- and 64-cavity molds in production, sometimes as high as 128 cavities. This multicavity approach can produce enormous capacity per processing equipment unit, especially when coupled with the appropriate runner system and cooling capacity.

Runner waste from this process is typically ground and reprocessed back into the feed stream; different materials can accommodate various percentages of regrind material, most suppliers will have a recommended maximum regrind mixture ratio to avoid degradation of final molded part material properties. Purge waste, created at the start of a new tool or at the end of a tool run should not be reprocessed as it will most likely have a mixture of materials which could contaminate virgin materials when mixed back in.

3.2.2.2 Blow molding

There are several variations on the blow molding process; extrusion blow molding (EBM), injection blow molding (IBM), and SBM constitute the main three process categories. All use air pressure to form a part from within against a mold which defined the outer surfaces of the part. These processes are most often used for hollow parts and containers with threaded closures.

3.2.2.2.1 Extrusion blow molding

The EBM process begins much like extrusion or injection molding with pelletized raw materials, fed into a screw and barrel arrangement, heated and plasticized. Like extrusion, the molten material is then pressure fed through a die, except this die is arranged in a vertical position such that the resulting extruded molten parison is hanging vertically from the die. The die will typically create a tubular extruded cross-section, round or oval depending upon the final shape of the finished blow molded part. There is some relationship between the extruded parison shape, the parison wall thickness, and the final part shape and wall thickness that is an experiential learned relationship. This hanging vertical parison is extruded down and in between a mold which clamps up over the parison, pinching it off at the bottom and at the top. A blow pin is inserted through the parison to allow pressurization from within the part. Once the blow operation is complete and the part has frozen suitably for ejection, the mold opens and the part is ejected, allowing the extruded parison to be extruded through the mold for the next cycle. There are many variations of this process including equipment made to extrude multiple parisons simultaneously, equipment that can extrude multiple layers within the same parison, equipment with rotary capability that will hold several molds and can provide a continuous nonstop process, equipment which shuttles tools into and away from the parison, and the list goes on. It is important to note that the EBM process is suitable for parts from small to large size; however single material smaller parts will most often be more economically produced using the injection blow process (Fig. 3.8A–B).

Waste in the form of pinched off parison material and trimmed flash can be granulated and delivered back into the raw material stream feeding the extruder for repro-

Figure 3.8 (A) A photograph of an EBM opened up in the press; the blown part has been shuttled out for ejection (left-hand side of the photograph) and the parison can be seen extruding down between the opened mold, ready for mold close and pinch off. The part will be ejected and the blow pin (still within the part in this photograph) will shuttle back into position after the parison is cut and before the mold finishes the close stroke. (B) A photograph shows the finished part from the blow mold process illustrated in part (A).
Source: Courtesy of Blow Molded Specialties, Pawtucket, RI.

cessing for single material parts. Waste from multilayer coextruded parts cannot be fed back into the primary stream in this manner.

3.2.2.2.2 Injection blow molding

The IBM process is different from EBM in that the raw material is first injection molded into a shape which is then shuttled hot into a second process station where the part is heated and then the blow operation is initiated to form the part. This typically takes place in two tools within one piece of process equipment; hence the process is efficient and can produce thicker wall higher quality parts than the extrusion blow process variation. One benefit is the dimensional stability and thickness of the neck portion can be tightly controlled because it is formed in the injection mold cycle. Another benefit is the absence of pinched off material and flash; once blown and cooled the parts are ejected and there are usually no secondary operations required. One downside is that since the preform is injection molded, it cannot be a multilayered material; hence barrier properties cannot be easily added into the process as they can in multilayer coextruded parisons for the EBM process. Another downside for this process is the design latitude of the container shape relative to the neck portion. The final blown shape of the container needs to be fairly homogeneous in all directions from the neck feature to ensure proper material flow during the process. This means no excessive offset neck designs, and so on. First costs tend to be higher for IBM than for EBM since a high-pressure injection molding process is used along with the lower-pressure blow process.

3.2.2.2.3 Stretch blow molding

The SBM process takes IBM to the next level and typically is performed in two distinct steps, first injection molding a preform and then the preforms are loaded into a second piece of process equipment which reheats them and then performs the blow operation. This process is used to produce nearly every plastic soda bottle used and disposable water bottles also. I have seen 128 cavity preform injection molds with twin ejector halves which shuttle in and out of the press, with every other cycle running with extremely short cycle times. These preforms are then packaged and sent off to the stretch blow processor where they are automatically loaded into the process equipment, heated, stretched, and then blown to form the final bottle part. The stretch operation was an innovation developed to aid in orienting the polymer chains prior to the blowing operation. In addition, the stress of the stretch operation on the polymer chains can induce crystallization prior to the blowing sequence providing extreme strength when loaded internally with a pressurized fluid (Fig. 3.9A–B).

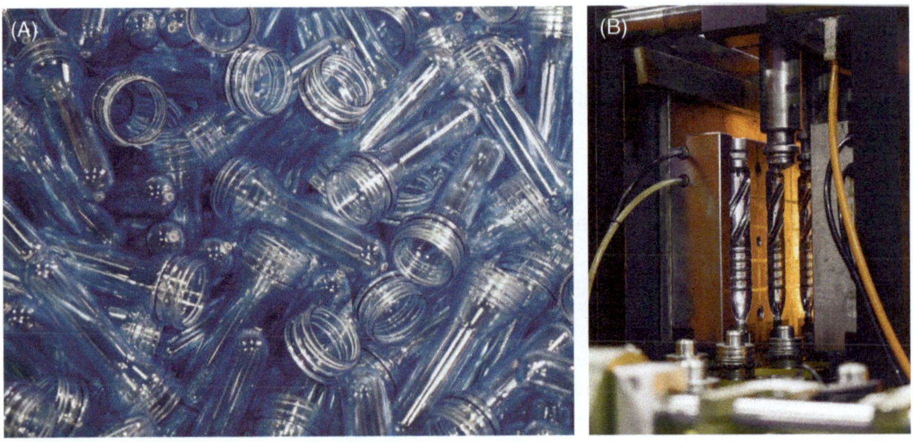

Figure 3.9 (A) Injection-molded preform parts which get reheated, then stretched, then blown into a thin walled bottle part. (B) Blow mold mounted in the press, shown open illustrating the cavity which the preform would be blown into during the process. Note the preform is moved into the mold via the conveying system which runs through the heating portion of the process then into this molding portion. Note the neck area of the preform is captured and held in a constrained state when the mold is closed for blow operation, thereby protecting the neck area and preserving the injection-molded tolerances for the threaded feature.
Source: Photos from sorapolujjin—stockeadobe.com.

3.2.2.3 Thermoforming

Although thermoforming and vacuum forming are ideal for low-volume applications, they can also transition and become valuable high-volume processes as well. For example, there are numerous packaging examples where extremely high volume parts and containers are formed using a thermoforming process; just think of disposable styrene coffee cup lidding, or form, fill and seal condiment packaging. The coffee and hot drink lidding example is formed on machinery which uses thin film from a sheet extruder coiled for shipment. The coils of thin film are fed into a heated tunnel and then into a thermoforming station in-line, and the transported (again in-line) directly to a trimming station. The lids once trimmed are ejected and fed automatically into a packaging station where they are stacked and inserted into a polybag to ensure cleanliness and sterility. The waste sheet is automatically rerolled and sent back to the sheet extruder, where it is fed into granulators and reintroduced into the feed stock for new extruded sheet.

Another high-volume thermoforming (actually pressure forming) process uses a twin extruded sheet to form and then seal the sheets together to produce hollow parts. This single process combines the features of hot plate welded injection-molded part assembly, with the simplicity of a simple process operation; however the part detail is limited to the attainable detail on one side of the tool (thermoforming process) unlike the additional detail attainable with injection-molded parts.

Case Example

Cubitainers manufactured by the Hedwin Division of Zacros America, Inc. are made by using a process described as a continuous, molten sheet extrusion (twin sheet) which is coupled with a vacuum/thermoforming process. The result is a flexible hollow part which can be used within a corrugated carton to provide a liquid storage and shipping container that competes with injection-molded pails and blow-molded containers. Except since these are thin wall, almost bags, they will use up to 83% less polymer per unit than a comparable pail container [1]. This is but one example of thermoforming process which is used in the twin sheet variation in a high-volume manufacturing application.

3.2.2.4 Profile extrusion

Other high-volume processes include extrusion of tube, pipe, and profile shapes. The extrusion process can support foamed polymers as are used for plumbing pipe and the like. Other high volume extruded products include PVC exterior architectural siding, PVC/acrylic window frame casings, and HDPE drinking straws to name just a few.

3.3 Selection by cost economics

Often projects are budget limited and cost economics will be the primary driver for a process decision, for example, a start-up company may not have the capital required to fund injection molds, or an established company may not want to invest capital for tooling until they can achieve some market research with actual product.

3.3.1 Initial costs—nonrecurring engineering (NRE) and tooling

As mentioned earlier, funding the tooling requirements will often be a challenge and selecting a process that does not have a high entry cost seems a natural choice. There are tradeoffs however, because the part costs will be higher and margins may not provide the profit required to sustain the product line. Sometimes the plan will be to start with a lower first cost process, until the market penetration is achieved and production requirements can support the higher cost tooling, then switch to the higher tooling cost process and reintroduce the product made in the new process, often with a lower cost and higher perceived quality.

3.3.2 Production per-unit costing

For established products with known production requirements, first costs can be less of an issue, while the costing per unit produced is the main focus. In these instances the process selection is typically easy. It is the variations within the process which can affect the per-unit costing that become the decision points. Things like, tool cavitation, runner systems, cooling capacity, materials and additives, ejection systems, and so on, can have large effects not only on the per-unit cost but also on overall tool capacity and annual process tooling maintenance costing.

3.4 Selection considering secondary processes

Sometimes a particular application will require a certain assembly method to accommodate some functional requirement or the mating parts of the assembly will require certain properties or precision for the required performance. In these instances, the secondary process might be the driving requirement in specification of the fabrication/manufacturing process. This secondary process may also be the driving requirement in material specification.

3.4.1 Assembly and joining

Parts designed for secondary assembly to other plastic parts or to parts from other materials require special consideration during the design and development process. Good part design always considers these secondary processes, and incorporates features and design practices to ensure maximum efficiencies and minimum risks when the part is processed correctly during these secondary operations. Process variability inevitably plays a huge part in process selection for assembly; if the process cannot provide parts within the tolerance range needed to mate with the other parts of the assembly, then it cannot be specified. For example, automotive parts required to assemble to metal parts which will undergo thermal excursions, with each engine use cycle may require a level of dimensional stability and precision not attainable in a thermoforming process, but attainable from the injection molding process.

3.4.1.1 Welding

Thermoplastic parts can be welded together by a number of different processes; these include, solvent welding, hot plate welding, hot air welding, vibration welding, spin welding, ultrasonic welding, just to name a few. The precision of the joint or weld surfaces prior to the welding operation will often determine the plastic process required. For example, a thermoformed part can certainly be solvent welded to a machined part or to an extruded part, even to a suitably designed injection-molded part; however, the same thermoformed part might prove extremely difficult to ultrasonically weld to an injection-molded part. Likewise a thin wall blow-molded part might be difficult to spin weld to an extrusion. Alternatively, the blow-molded part may hot plate weld just fine to a thermoformed part, given similar materials with similar melting temperatures and wall thicknesses.

Each plastics part welding process has different requirements for the parts being assembled, in terms of part rigidity, wall thickness, melting temperatures, and so on, that must be fully understood when determining the primary process to be used for a given part which will be subject to the welding operation during use.

3.4.2 Painting

Painting is a common secondary operation for appearance parts and a well-designed part can provide the form and features required to make this process easy and repeatable. A poor-designed part can cause tremendous problems during the painting process. The designer needs to understand the painting process and visualize the paint shop operations, including masking and multiple part racking for optimum process efficiency. Small details like a raised edge to seal off against the masking fixture that will separate without damage to the painted surface can be crucial to hitting capacity requirements and finished part cost targets. Processes which complete the part, with good surfaces requiring little or no trimming are often better candidates for paint operations because the surfaces are uniform and will not have potential chips or burrs from the trimming operations.

3.4.3 Machining

Parts from many processes and from many materials can indeed be machined after primary processing, with certain guidelines observed and with certain risks understood. Parts from the sheet based thermoforming processes and from rotational molding must be machined as part of the finishing process. Often these are machined using multiaxis high-speed router equipment. Blow molded parts are often deflashed using die cutting operations or cryogenic freeze and break methods, and are sometimes knife cut or machined as well. Injection-molded parts are perhaps the parts most likely to be a risk when machining after primary processing and certain materials will be more risk tolerant than others. Machining can potentially leave microcracks, or sometimes will produce sharp stress concentration features on the part which can be primary failure modes if not carefully considered during design. Reheating of the primary processed part from tool friction can also present issues if the part/tool interface is not controlled (cooled) adequately during the machining operations. In addition, the cooling fluids used during the machining operations must be compatible with the plastics materials in use and must not contain materials which could cause degradation of the polymer.

Hence, if a part must be machined before, during or after assembly, the process selected for the primary process, the material, and part design must all be compatible to allow the final (as machined) part to function as intended without risk of failure. In addition, the design of the primary processed part should consider the functional loading relative to the machined surfaces and arrange the design to minimize potential failure mode risks potentially induced by the secondary machining process. Further, the design of the part from the primary processing operation should provide pick-up locations for accurate positioning and adequate fixturing for the machining operation. Good design practice would lead the designer to ensure that the minimum amount of material to be removed by machining was provided to ensure an efficient secondary operation process. Again, design thinking with every operation and end-use in mind during the formation of the design solution and especially during part structure creation.

3.4.4 Electroplating

Certain grades of certain thermoplastic materials can be plated and the process requirements for these materials may influence the process selection for the part. Some materials which can be plated may not be available in cut sheet for thermoforming, and so on. The plating is used most often for appearance, but can also be used as a conductive layer for EMI/RFI shielding and the like. Once again, as in the case with painting, a good understanding of the plating process and the part handling during the process is needed to allow the designer to choose the primary process, and provide a part design that can be processed efficiently and without undue quality concerns. During the plating operations the plastic parts need to be held, racked, and masked, similar to parts as designed for painting. Providing a part design that is easy to mask and that has some features for hanging or racking where the plating appearance can be poor will assist the secondary operation processing team in providing high-quality parts in an efficient manner.

3.4.5 Annealing

Annealing as a postprocess is performed to further remove molded in residual stresses or machined in stresses; and, sometimes the annealing process can be used to promote additional postmold shrinkage. Sometimes parts are annealed after some particular assembly operation, for example, after a press-fit operation where the stresses at the press interface could be a source of failure. Any time a finished part will be subject to challenging environments while under some load or mechanical stress, the risk of part failure is always present. Sometimes the stresses are externally applied other times they are internal from the primary molding process or from assembly, like press fits or welded joints. Thinking through the design, considering the total picture of stresses and environment should bring to mind the opportunity for annealing the part to assist in relieving some of the stresses and perhaps providing a better final product solution.

3.4.6 Plasma treatment

Plasma treatment for surface tension modification can be advantageous for some adhesion operations or for some postprocessing surface wetting applications or coatings. Understanding up-front, during solution conceptual thinking, which surfaces need to be posttreated for whatever secondary operation, the designer should allow and plan for some holding or fixturing features to allow full treatment of the required surfaces. If the design requires some downstream process (printing, labeling, adhesion, etc.) that will require plasma treatment, the design team must carefully understand how the part might react to plasma and each process may produce a part which reacts just a little different. Discussion with plasma experts, including materials selections and process selections, is recommended before getting too deep into the design with any particular process selection.

3.5 Selection by other criteria

In some instances, other criteria will creep into the selection process. These could be combinations of functional requirements and downstream assembly requirements, or they could be environmental concerns, or they could be economic use of existing infrastructure, and the like. These other criteria will be known up front in most every project and they will couple, with other required attributes of the design to focus the selection of the process. For example, the product to be designed must hold some sensitive electronics that will need to be installed within the plastic housing and then potted with an epoxy compound for use submerged under water at some given depth; as such, the enclosure needs to be closed and heat welded shut to seal the electronics securely. For this example the process selected must be a thermoplastic process. Selecting a thermosetting process would be considered to match the CLTE of the potting epoxy, however the thermoset materials cannot be heat welded.

3.5.1 Part size considerations

Part size does influence the available choices in plastics processes. For example, it is common to rotationally mold containers which can hold hundreds of gallons of

Part size by process comparison		
Process	Size range	Approximate max. volume (L)
Injection molding	Extremely small to moderately large	500
Thermoforming	Small to large	1,000
Blow molding	Small to large	1,000
Rotational molding	Can produce very large parts	10,000

Figure 3.10 Each process has the capability to produce small to moderately large parts. Rotational molding can produce very large parts and injection molding can produce microscopic parts. There exists some custom built injection molding equipment for extremely large moldings (upward of a 235 lb. shot) but these are not the norm for the process.

a fluid, no blow mold equipment in the world could accomplish such a feat. Likewise, injection-molded parts can have features so small they are hard to see with the naked eye that could never be thermoformed into a part. Fig. 3.10 illustrates the general part sizes attributable to the various processes use to manufacture plastics parts.

Reference

[1] http://www.hedwin.com/products/cubitainer/compare-to-plastic-pails.

Resources for further study

S. Levy, J.H. DuBois, Plastics Product Design Engineering Handbook, second ed., Chapman & Hall, New York, 1984.

S.E.M. Selke, J.D. Culter, R.J. Hernandez, Plastics Packaging—Properties, Processing, Applications, and Regulations, second ed., Carl Hanser Verlag, Munich, 2004.

J. Rotheiser, Joining of Plastics—Handbook for Designers and Engineers, third ed., Carl Hanser Verlag, Munich, 2009.

G.L. Beall, Rotational Molding—Design, Materials, Tooling, and Processing, first ed., Carl Hanser Verlag, Munich, 1998.

E.R. Larson, Thermoplastic Material Selection—A Practical Guide, first ed., Elsevier, Oxford, 2015.

R.A. Malloy, Plastic Part Design for Injection Molding: An Introduction, second ed., Carl Hanser Verlag, Munich, 2010.

P.A. Tres, Designing Plastic Parts for Assembly, seventh ed., Carl Hanser Verlag, Munich, 2014.

Web links of note

Great Design information from DuPont can be found here:

http://www.dupont.com/content/dam/dupont/products-and-services/plastics-polymers-and-resins/thermoplastics/documents/General%20Design%20Principles/General%20Design%20Principles%20Mod%201.pdf.

Sabic provides many tools for designers here (you may need to register):

http://www.sabic-ip.com/webtools/toolsHomePage.faces.

The UL Prospector site is a valuable tool for any designer reviewing materials for any application (you will need to register):

https://plastics.ulprospector.com/.

Solvay provides a great design guide for using their Udel polysulfone materials:

http://www.solvay.com/en/binaries/Udel-PSU-Design-Guide_EN-227550.pdf.

Bayer has a nice design guide for RIM systems:

http://reactioninjectionmolding.com/wp-content/uploads/2013/09/RIM-PartMoldDesign-Guide.pdf.

Covestro offers several design related guides, including part and mold design, snap fit design, and joining techniques for plastics:

http://www.plastics.covestro.com/Library/Technology-brochures.aspx.

What tooling will be required?

4

Producing parts and products made from plastics materials requires enormous up-front knowledge and planning; none as critical as the basic design decisions which determine the material and the process. Tooling requirements follow these two basic decisions and are dependent upon both the process selected and the anticipated production capacity needed from that process. Tooling designers and part designers are two different breeds of technical thinkers and most would never want to trade vocations. However, good part and product design demands that the part designer have a thorough understanding of the tooling means and methods used, so that the part design can allow the tool designer to prepare a tool design which is adequate, robust even, for production of the part(s) in question. While capacity requirements and process selection define what tooling is necessary, part quality requirements and product life cycle also enter into the tooling decision matrix (Fig. 4.1).

4.1 Process dependent

The process used to fabricate, form, or manufacture a plastics part is a major factor in the tooling required; injection molding demands different tooling than resin casting or thermoforming, blow molding requires different tooling than extrusion or rotational molding, and so on. However, most plastics processes fall into two categories, those requiring high pressures and those requiring low pressures to form or mold the finished part. The cost and complexity of the tooling is reflected by the design of the tooling to accommodate the pressures used in the process. In all of the processes, the ability of the tooling to not only form the part required but to aid the process by the rapid removal of heat from the material cannot be understated. The faster the material can freeze or set, the faster it can be removed from the tool, and the faster the next cycle can be started.

4.1.1 High-pressure processes

Injection molding is the single high-pressure process most widely used to manufacture parts from plastics materials today. Injection pressures can run well into the 20,000 PSI range and sometimes higher, which is why injection mold tooling is built with heavy walls—built to withstand these high-internal cavity pressures. Extrusion is also a high-pressure process, whether extrusion for pipe and profiles, for sheet goods, or for extrusion of the blow molding parison. The extrusion die tooling, while much

Designing Successful Products with Plastics. http://dx.doi.org/10.1016/B978-0-323-44501-6.00004-0

Figure 4.1 Injection mold shown opened up on the bench with the parts it produces shown in the foreground. This mold produces two pieces each of two different parts (two sets) of the enclosure shown in an ABS material.

simpler than injection molds, still requires the strength to handle the high pressures and temperatures used during the process.

4.1.1.1 Injection molding

Injection molding is truly the leader of the high-pressure processes. Injection molding equipment can produce enormous internal cavity pressures and provide mold clamp forces to prevent the high internal cavity pressures from separating the mold during the process. This is sometimes referred to as blowing the parting line and occurs when the internal cavity pressures, acting on the inner surfaces normal to the parting line create forces greater than the available clamp or mold closing forces the equipment is capable of supporting. Usually, the equipment is matched, or sized, properly but sometimes a hydraulic failure or other mechanical failure on the equipment will allow this to occur. Extreme flash at the parting line and out of specification dimensions across the parting line will result. Cavity pressures are a function of the cavity geometry, the wall thickness of the part, the materials selected for the part and the process parameters used for the selected material. In general, the "old guy" rule of thumb is to use approximate 5 tons of clamp force for every square inch of projected cavity area normal to the parting line; hence if you had a part with 9.8 $in.^2$ (round up to 10) of projected cavity area, the clamp force required would be on the order of 50 tons.

Injection molds can take many forms and can be produced from many materials, depending upon the desired outcome. For most materials and for most part applications the mold will be made from a steel material, often hardened tool steel is used for longevity. Corrosive materials, such as polyvinylchloride (PVC) will require stainless steel materials to avoid premature tool surface degradation from the corrosive nature of the polymer during high-temperature processing. Geography and processing shop environment can sometimes be a cause for selecting stainless steel for molds as well; for example, a processing shop without dehumidification in a geographic location with a high-ambient humidity—think South China—would result in molds that would be ideal condensation surfaces, resulting in a rusty mold if it were not made in a stainless alloy. For prototyping and extremely short run production, composite materials, sand cast from Kirksite or zinc–aluminum composite mixtures have been used with success. For some polymers, tempered aluminum alloys can be utilized for the tool, when cavity pressures will allow. Alternatively, new FDM/additive manufacturing methods combined with insertable mold cavities, outfitted to receive these FDM produced cavity inserts, are finding their way into the mainstream of prototyping for injection molding part development.

For multicavity requirements, a new level of tooling expertise is required, since now we need to provide precise and balanced filling and cooling to each and every cavity simultaneously. If we have one cavity filling far earlier or far later; or if we have one cavity running hotter or colder than another then the part-to-part dimensional consistency will suffer, as may part properties. The standard, time tested, procedure for limiting risk with multicavity molds, is to build a prototype mold (single cavity) first and test the part from the single cavity prior to investing in 16, 32, or 64 cavities. However, be smart when you go down this road, design the multicavity mold first, and then when the single cavity mold is produced, produce it EXACTLY like one of the cavities from the multicavity design. That means, the exact steel, the exact steel dimensions (inside and out), the exact cooling line positions and directions, the exact cooling flow rates and temperatures, the exact venting features and vent to atmosphere path distances, EXACT in every way. When the prototype tool is built correctly it becomes a great learning and risk mitigation tool, when it is not built EXACTLY as the multicavity mold will be built, then nothing valuable is learned and false assumptions lead to production frustration and poor quality parts [1]. Balancing multicavity injection molds is critical for uniform cavity to cavity part performance; running a short shot analysis during initial mold trials has been shown to assist in not only balancing the cavities, but also for verification of gate location and size, venting, runner system geometry, and the like (Fig. 4.2).

The high-cavity pressures required for the injection molding process provide extremely efficient processing cycles and competitive per unit costing, but demand heavy wall tooling to withstand the pressures. Hence, injection molds are expensive to build, expensive to repair, and expensive to alter; therefore make sure the part is complete, verified, and ready for tooling before embarking on tool design and building. Do your homework!

Figure 4.2 A multicavity mold mounted in an injection press. The runner system and the venting channels are visible on the cavity insert face.
Source: Courtesy of Quality Molding, Inc., Somerset, WI.

4.1.1.2 Blow molding

As discussed in Chapter 3 there are several variations to the blow molding process, extrusion blow molding (EBM), injection blow molding (IBM), and injection stretch blow molding (ISB). Each variation requires somewhat different tooling with a wide variation in possible costs.

For EBM there will be two tools required, the extrusion die which produces the extruded parison and the actual blow mold which forms and cools the finished hollow part after the blow process is completed. Tooling for EBM is relatively inexpensive since the blow mold is typically made from aluminum, with some higher-thermal conductivity materials and harder materials used in pinch-off areas. Since EBM begins by pinching off an extruded parison, there is always some flash to be trimmed at the end of the process. Sometimes trimming die tooling is required to produce a clean finished part and to provide part-to-part consistency. One design benefit of EBM tooling is that the mold can clamp up on the parison in more than one place, allowing for handle designs and features that would be impossible from any other processing means and tooling. Tack-offs, where the opposite walls of the extruded parison are forced to touch when the mold closes on the parison and the blow operation begins, can add substantial structure to the finished part (Fig. 4.3).

Tooling for IBM and ISB are somewhat more expensive, since both use a much higher-pressure process and hence need to be stronger and harder. Both processes utilize an injection-molded part for the initial step of the process, hence the high-pressure process. The IBM process retains the injection molded preform on the core from the injection-molding step and quickly rotates the core and preform into the blow mold tooling and begins the blow process forming the hollow part. The ISB process is similar, except the preform is often removed from the injection molding core, stored, then reprocessed into a blown part at some later time. In this process, the preform is loaded in the equipment and reheated in a reheat station, then moved and loaded into the blow molding tooling, a stretch rod then stretches the preform for some predetermined

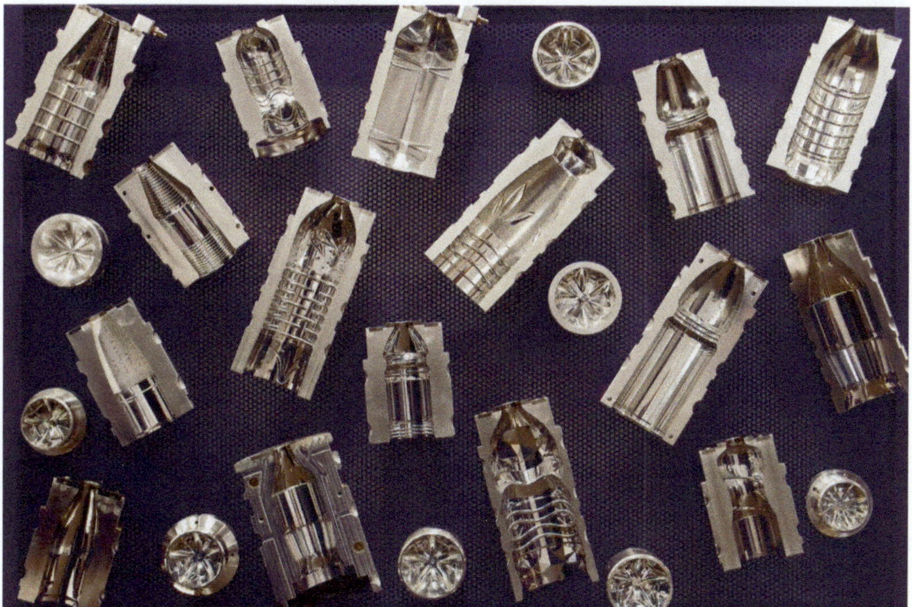

Figure 4.3 An array of various blow molding mold cavities showing different designs and different bottom configurations. Note the neck ring insert is each of these is inserted and is probably hardened acting as a pinch off surface.
Source: Photo from Alterfalter—stock.adobe.com.

Figure 4.4 A stretch blow mold tool mounted in a press, opened with the stretch rods extended.
Source: Photo from nordroden—stock.adobe.com.

distance and then the blow operation begins. This stretch then blow process results in biaxial polymer orientation leading to extreme strength and superior permeability resistance for the resulting thin walled vessel. Tooling for these two processes are somewhat more expensive than the corresponding EBM process, however, the cycle times and cavitation can be far more efficient producing parts at a fraction of the cost of EBM processing. Unlike EBM, since both of these processes expand upon an injection molded preform slug of material, no handle forms or clever pinch-off designs are possible. In addition, no residual pinch-off or flash will need to be removed as in the EBM process (Fig. 4.4).

4.1.1.3 Profile extrusion

Tooling for extrusion is somewhat less expensive than tooling for most any other plastics process, but that does not mean that the tooling is easy. Closed profiles present a larger tool design challenge than open profiles and each version of each material has a somewhat different die opening requirement relative to finished part size. Uniform wall thicknesses and corner radii on the extruded profile will aid in both tool strength and part strength. The continuous dimensional stability of the profile is also enhanced by good design practice, including the use of uniform walls, corner radius, and avoiding closed or nested features within the profile (Fig. 4.5).

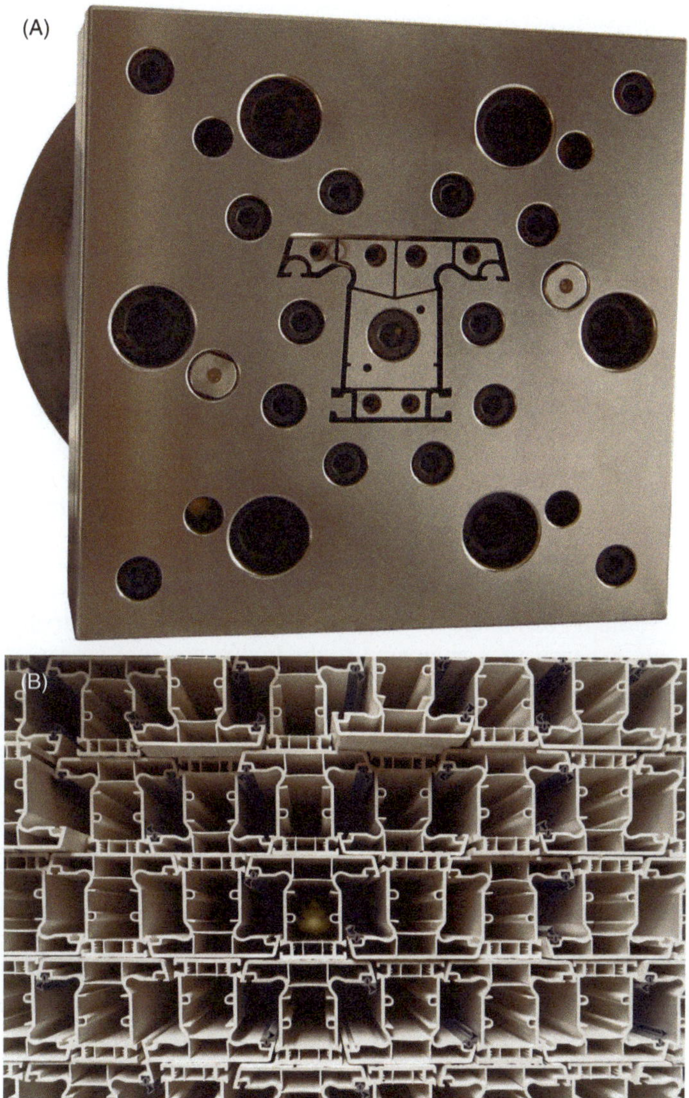

Figure 4.5 (A) A profile extrusion die. (B) Extrusions stacked and ready for shipment to a customer.
Source: Part A, Photo from iStock.com/jujuenter; part B, Photo from iStock.com/ozgurdonmaz.

4.1.2 Low-pressure processes

Rotational molding, thermoforming, resin casting, and so on, are examples of low-pressure processes used in the manufacture of parts and products from plastics. Each has different tooling requirements but all are inexpensive relative to the high-pressure processes discussed previously.

4.1.2.1 Rotational molding

For a fairly low-pressure process, rotational molding tooling can get expensive, but only in relation to part size, since roto-molded parts can get large. The tooling is typically a sand cast aluminum shell that gets bolted together for each molding cycle. Hence a pattern must be first fabricated, then sand castings made for each mold half; and, many times multiple molds are cast and fabricated so the molder can be filling one, while another is being heated, and yet a third is rotating and cooling all simultaneously. Rotational molded parts are similar to EBM parts in that they are filled from inside out, producing a well-defined exterior but having a nondefined interior surface. In addition, rotationally molded parts are similar to thermoformed parts in that they require substantial secondary trimming and finishing operations. This secondary trimming is most often accomplished by multiaxis high-speed routing equipment (Fig. 4.6).

4.1.2.2 Thermoforming (vacuum forming, pressure forming, and twin sheet forming)

Tooling requirements for the thermoforming processes, including vacuum forming and pressure forming, can range from simple and inexpensive to extremely complex and moderately expensive (as compared to the higher-pressure processes). Forming dies can be made from many materials, from wood and foam for prototype and short-run forms to cast and machined aluminum for production forms. Deep-formed parts will sometimes use a male tool, often referred to as a "plug" to help form the softened hot sheet as the tooling closes just immediately prior to the vacuum/pressure application in the process. Multipart forms can help to increase efficiencies and lower per unit part costs, similar to multicavity tooling for injection molding.

In addition to forming dies, which give the thermoformed part its shape, trimming dies or holding fixtures for multiaxis high-speed routing equipment will also be required to allow the formed part to be trimmed to final size. Sometimes secondary operation fixtures are needed for adding bosses and assembly features to the thermoformed parts. Be vigilant doing estimating for thermoformed parts and be sure that you have collected all of the tooling costs, not just the form die costs (Fig. 4.7).

4.1.2.3 Resin casting or vacuum casting

Resin casting, sometimes called vacuum assisted liquid casting, uses soft molds cast from a master model. These soft molds are often cast from room temperature vulcanizing silicone compounds (thermoset) and are typically useful for 15–25 castings

Figure 4.6 These photos show an example of (A) a rotational molding tool (mold), opened up (note the inner surfaces which form the exterior surfaces of the resulting parts) and (B) the parts which are formed from the tool.
Source: Photo courtesy of Plasticraft Corporation, Darien, WI.

before they are no longer able to produce adequate quality parts. These can be ideal for replicating prototype parts in polyurethane materials to simulate parts from injection-molded tooling. The parts can be quite complex and the finished parts can have surface finishes and hardness similar to the injection molded parts they are portraying; however, the mechanical properties are not replicated and the parts cannot be used for functional performance testing.

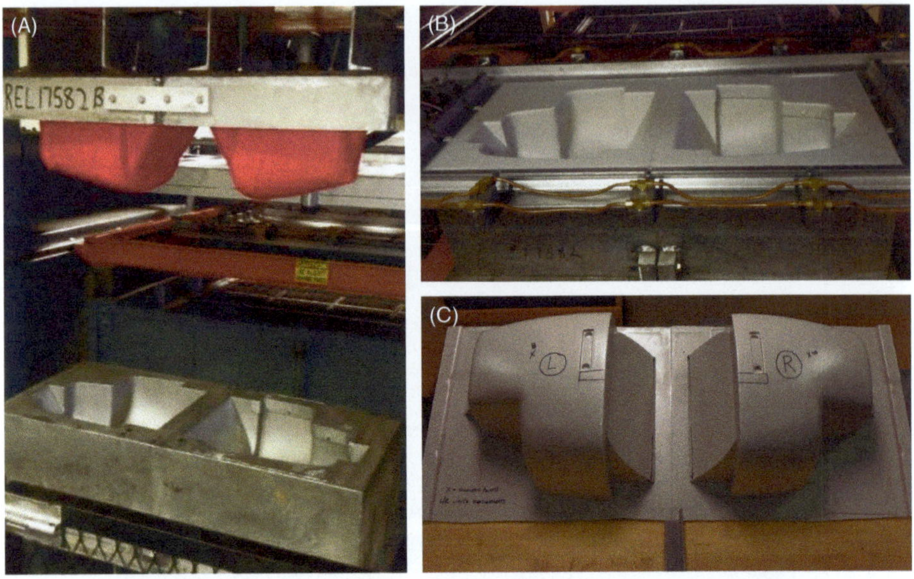

Figure 4.7 (A) This thermoforming tooling uses a male plug *(top mounted)* to help shape the molten sheet as the tooling is closing and as part of the forming process. The lower cavity shown is made from an aluminum alloy. (B) After forming but before part removal, showing the cut sheet clamp frame still in position on the form tooling. (C) Finished formed sheet as removed from the form tooling. This formed sheet will now be machined using multiaxis high-speed routing equipment to finish the part to final mechanical dimensions and specifications. *Source*: Photos from Courtesy of Roger C. Kipp and Associates, and the Plastics Innovation & Resource Center at Penn College.

4.2 Capacity dependent

Part or product capacity requirements will typically dictate the process, but can also dictate the tooling. For example, let's take a look at disposable styrene coffee cup lidding. Capacity requirements are off the charts, millions and millions are used daily, hence we think of a high-capacity process like injection molding. However, they are not injection molded, they are actually thermoformed. Thermoforming is usually associated, with a low-volume heavy wall form and cut style of manufacture, but there is another method used in thermoforming, rotary die form and cut, or in-line transfer form and cut. In these processes the sheet is extruded and fed in-line, into a thermoforming process station, where it is formed and then handed off to a die cutting station. The example lids are collected and packaged, the web is fed into granulators, and the regrind material is conveyed back into the feedstock for the sheet extrusion. Although a low-pressure process, it is none the less a high-speed process and the capacity from this kind of equipment is huge; however, so are the tooling and equipment costs. Similar processes are used for form, fill, and seal packaging equipment for foods, such as yogurt, for some single serve condiments, for dose packaging of pharmaceuticals, and the like (Fig. 4.8).

Figure 4.8 A photograph showing an in-line thermoforming operation producing portion package trays for automated pharmaceutical capsule packaging. Further downstream the pharmaceutical product will be inserted and a foil protective layer will be heat sealed to the thermoformed sheet, and then they will be die-cut into the correct capsule count packages and inserted into point of sale packaging.
Source: Photo from iStock.com/zorazhuang.

On the other end of the spectrum are the extremely low-capacity requirements. For example parts needed for medical instrumentation where the annual usage might be 100 units or in some cases as few as 15–20 pieces annually. These require a process that is not tooling cost prohibitive, hence for larger parts, thermoforming or rotational molding come to mind, for smaller parts resin casting might be a viable option. Several new processes are being developed to turn rapid prototyping means and methods into production viable processes. Fused deposition modeling (FDM), in particular, has become the leader in these efforts to become an additive manufacturing process, not just a prototyping process. Production on demand is the promise; however, several part quality and process clean-up issues are still hindering the transition. Work is progressing to alleviate these problems and new materials for the FDM process are being introduced regularly. As of the inherent porous nature, the FDM process with the correct materials can be useful for manufacturing the forming dies used in the thermoforming process.

Other capacity-related tooling decisions come into play in injection molding, for instance, when the capacity demands are extremely high as in closures for beverage packaging. Not only are high-cavitation molds utilized, but other methods, such as stack molds and shuttle molds are utilized to improve the capacity available from a single press installation. Injection molds used for high-capacity injection molding of SBM preforms will sometimes have multiple ejection halves, such that once the injection cycle is complete, one ejector mold half will be moved out of the press while a second is brought into the press. This allows the second ejection halve to close and the second shot to occur while the parts molded on the first ejection mold halve are cooling and being ejected (outside of the press) thereby increasing the number of parts per hour by a significant margin.

4.3 Material dependent

While tooling is mainly process and capacity dependent, there are some materials which require certain kinds of tooling. PVC for example is, when molten, corrosive, and will attack raw steel used in the tooling; hence any mold built to process PVC will most likely be made from a grade of stainless steel. Some materials tend to adhere to the tooling surfaces unless certain surface finishes are utilized; hence we always recommend involvement of the material supplier during tooling material selection and part design efforts, so that these nuances can be understood early in the design process.

Other material dependent tooling issues arise due to processing temperatures. Sometimes the tooling will be used to assist in the chemical reaction process. This is seen clearly in thermosetting materials, but also is a factor in some high-temperature thermoplastics, for example, PPO, PEEK, and the like. Other materials that use tooling heat for final chemical reaction are most of the elastomers (thermosets) where the materials are compression molded; that is loaded into the tool at room temperature, then heated and pressurized until vulcanization occurs. Liquid silicone molding operates in a similar process, whereby the silicone is kept cooled, mixed cold, pumped into the mold cold, then heated by the tooling until vulcanization occurs.

4.4 Manufacturing location dependent

Tooling quality can be specified and will vary according to the end-use application for the part in question and sometimes based on the location of manufacture. In developed locations with high-labor rates, it is common to build extremely precise tooling and to automate the tooling and process as much as possible; done as a means of reducing the total labor input required per unit. In some locations with lower-labor costs, the quality of the tool or the processing control precision may be less precise and final part production may require additional labor input. For example, a top quality tool for injection molding may be built such that when the part is ejected from the press it is complete and ready for use; that is there is zero flash, the gates are trimmed, and the part is finished and ready for the next assembly step. For the same part made in an area with lower-labor costs, the tool cost might be 1/4 or less of the investment for a top quality tool, but the part as molded might have flash that would need to be trimmed, would need the gate(s) removed, might have hand loaded cores, might need holes drilled from the sides, and so on. The resulting part from either set-up might be fine for the intended application and the final per unit costing might be competitive, when the tooling costs are amortized and accounted for over the life of the part.

4.5 Part design dependent

Understanding the tool requirements for each feature of a particular design allows the part designer to anticipate the complexity and costing for the tooling required. Depending upon the particular application, and yes, upon the capacity required, the part designer can opt to produce either extremely simple or extremely complex part designs, resulting in tooling on either end of the cost spectrum.

For example, for a part that has a low-annual capacity requirement, say less than 250,000 parts annually, which may be built in a single, double, or perhaps even a four cavity mold, the designer can be a bit more creative and can have some exterior slides, side actions, and so on. During design of an assembly, the goal is to reduce the number of parts and to exploit each individual part, attempting to get as much functionality as possible from each part. For parts in this low-to-moderate annual capacity range, adding features and complexity will indeed result in added tooling complexity and cost. However, the benefits and cost savings on each finished production unit will typically pay for the added tooling complexity and costs very quickly, usually within year one or two.

However, if the part has capacity requirements in the tens of millions and will be built in high-cavitation molds, 16 cavities, 32 cavities, or more, the designer must keep the part just as simple as possible and minimize the mechanical movement required within the actual cavity. On high-capacity part designs, each gram of material and each tenth of a second of process time is meaningful and the design must be focused to minimize both. The ability of the part design to provide a tooling design that can remove the heat rapidly and allow for rapid and complete ejection is a necessity. A close working relationship between part design and tooling design is highly desired for maximum project efficiency and effectiveness on these large capacity projects (Fig. 4.9).

4.6 Tool quality relationship to part quality

For any given plastics process, for any selected thermoplastic material, and for any given part design, tool quality will do more to ensure part-to-part consistency than any other factor that the part designer has influence over, except of course, the actual part design. That said, the proper design of the part which allows for good tool design, will allow the processor to manufacture good parts each and every cycle of the tool. The tendency when starting a new product project is to be as conservative as possible with first costs, keep the tooling investment just as low as possible. There is wisdom

Figure 4.9 A single cavity and core set from a multicavity mold. Note the O-ring seals isolating the cooling channel on the cavity side (left) and the ejector pin holes on the core side (right). The round brown part (center) is molded from the cavity (left) and core (right) shown.

in this approach, however, lower quality tooling can present issues over the tool lifetime; hence, be smart about tooling decisions, not cheap. Sometime the investment in a higher-quality tool is the best decision, other times a lower quality tool can provide enough good parts until the next generation part or product is ready for introduction; each project will have its own specific requirements, just be aware and be ready to demand higher-tool quality when it is the smart decision for the part and project in question.

4.7 Specifying the tool within the part design specification

Part design specifications, as contained within 3D part solid data files and within 2D part dimensioned drawings will often note everything needed to make the part such that the part will perform as required. However, the detailed design specification can also be used to convey the designer's intentions with respect to tooling design as well. Some frown upon this as it might cause tooling costs to increase, or to remove responsibility from the tool designer and fabricator.

Case in point is mold shrinkage values; whose responsibility is to determine the shrink rates used for the mold prior to mold construction? From the designer's position, it is clearly the toolmakers responsibility to deliver a tool that can be processed to provide a part which meets the specification drawing. From the toolmakers position they are not materials experts and carry tremendous risk if they misjudge the shrink rate when determining tooling dimensions. My approach has always been to recommend a shrink rate range and to attempt to get the materials supplier to buy-in and recommend to the toolmaker the correct value to use. In this manner, the "team" approach will often find the "team" agreeing on a shrink value for the particular part in the particular material. Another mold shrink note that I will often insist upon is marking the exterior of the tooling with a simple note saying something like *SHRINK* = 0.016 to provide the toolmaker making the next revision to the mold (perhaps years later, or in another factory, on another continent) to understand what shrink rate was used when the tool was originally built.

A case can be made that correctly adding notes to the detailed drawing specification that will guide the tool designer, not tell them exactly what to do and how to do it, can be extremely beneficial. Then once the tooling is built, validated and ready for production, those same notes should be updated to cite exactly what and how the tool design was implemented to accomplish the design intent, including what shrinkage values were used. This not only records the tool design attributes as a general best practice, but it will also allow the learning's from the initial development to be used later in the product life cycle when tooling is built for production at other factories expanding capacity globally.

If a company is going to build a product in several different countries, shouldn't the tooling be similar from factory to factory? This makes intuitive sense, but as a product grows into other markets, you will find that the tool build for each new market will probably be just a bit different from the previous; some better, some not. When the tool attributes required for a quality part are noted correctly on the 2D detailed part specification drawing, the best practices learned during development can be propagated as the product grows globally into each country or region (Fig. 4.10).

4.8 Sourcing the tooling

Defining and specifying the proposed part is the designer's domain; sourcing the tooling will, most likely, not be within the designer's domain. However, during the sourcing activities for the tooling, the designers and their work product will be utilized frequently. There are any number of possible arrangements for sourcing tooling for a project and several are worth noting for reference.

4.8.1 Internal sourcing—captive tool shop within the organization

Working with internal resources can be extremely beneficial for the design team and the closed loop communications can protect the organization by limiting the exposure of the project outside the immediate inner-company personnel. However, internal

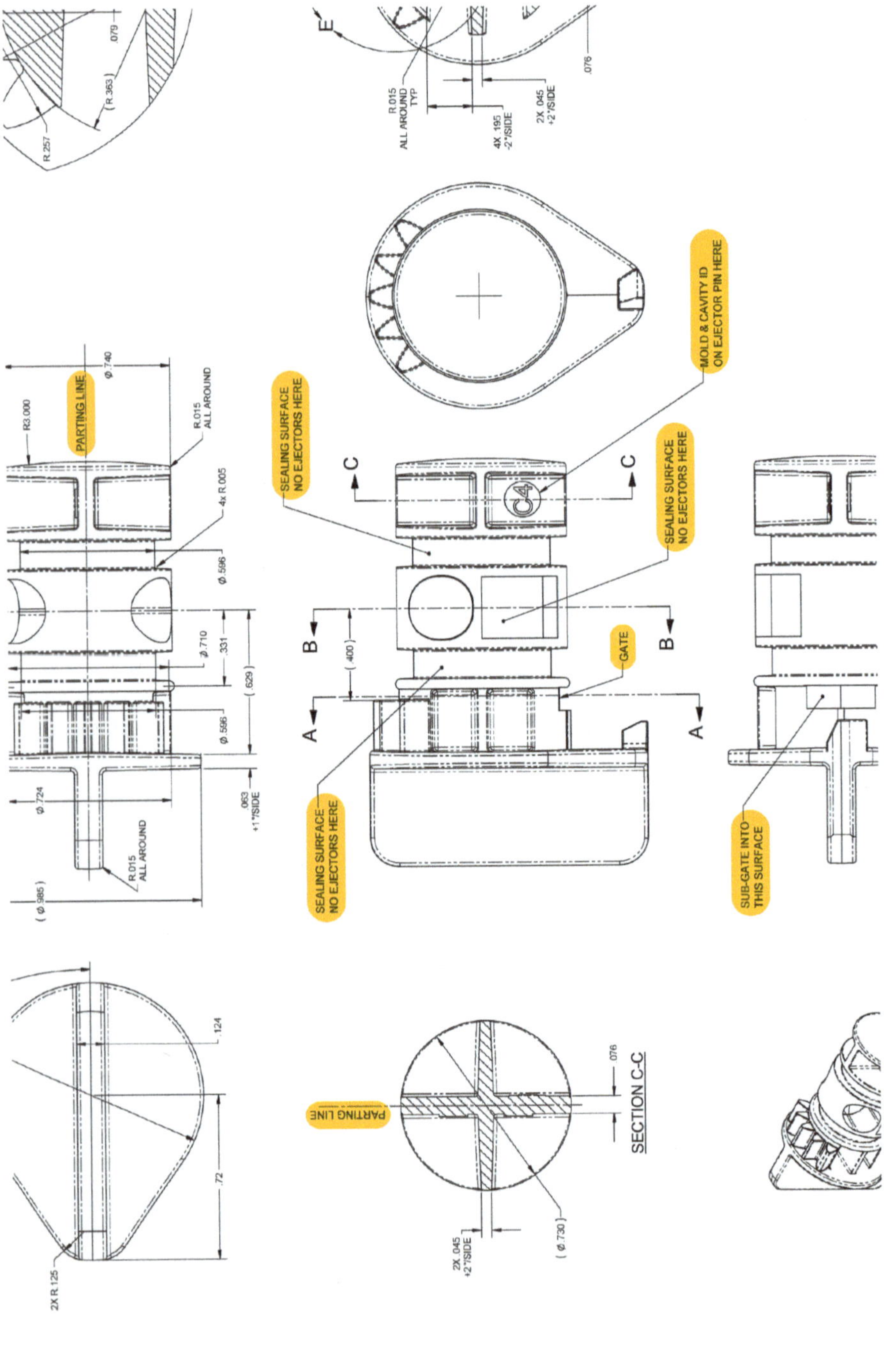

Figure 4.10 Illustrations showing an example of tooling notes inserted on the piece part specification drawing, used to advise the toolmaker of specific surfaces for sealing, cavity ID locations, the anticipated parting line location, and the proposed gate location.

resources will often have a preferred methods and means for development of tooling, which can be limiting within some organizations. Project "buy-in" and participation at design sessions and design review meetings should all be easier to attain with an internal resource. In addition, the internal tooling resources will certainly have an immense knowledge base regarding the capability and capacity of the internal processing or of contracted external processing facilities for which they have provided tooling on previous projects. This intimate knowledge is invaluable and can steer the design team toward a solution which fits the capabilities that will be utilized. However, if the capabilities do not fit well, don't be afraid to go outside and look for contract sources that have better capabilities for the design under development.

4.8.2 External sourcing—captive processing within the organization

Captive tooling shops are indeed rare within the customer base that I have been privileged to work with over the last few decades, however, many clients have had captive processing, on-site to allow better vertical integration of their manufacturing and assembly processes. In these cases, the captive process department will often have two or three outside tooling resources that they call upon for new tooling and repair of production tooling, as well as routine maintenance of tooling. The captive production team will be able to recommend a potential tooling resource based upon their knowledge of the capabilities of the various tooling resources they use; as some may be better with larger tools, some may be multicavity specialists, some may excel at short run tooling, while others may be optical part specialists, or gear specialists. Inclusion of the captive processing team and their recommended tooling resources early in the design process is always desired; the knowledge base they bring can prove extremely valuable early in the design decisions process. Make sure that all proper nondisclosures, and so on, are in place before bringing any outside resource into the project-team discussions.

4.8.3 Contract based sourcing—external contract processor sources tooling as a single source supplier

Perhaps the arrangement that I find most utilized for new product development among my clientele is the contracted processor (injection molder, thermoformer, roto-molder, extruder, etc.) who accepts responsibility for both the developing tooling for the part and providing finished parts or assemblies to meet the part specifications. This arrangement is ideally suited for small to midsized manufacturing companies that do not yet have the business magnitude for vertical integration but still need quality processed parts supplied. There are many benefits from this arrangement, including the fact that the manufacturer now has a single source for the completed part to specification. When molding trials for the new molds are completed, there is no competition between the tooling source and the processing source for which needs to pay for any remedial work required. Given the processing source has reviewed and approved the tooling designs, and given that the tooling was built to the design specification, then the lines of responsibility for any remedial work required should be crystal clear. On the other hand if a design error causes the need for remedial work, then neither external resource will bear responsibility for the remedial costs. Bringing the external processor and their chosen tooling resources in for participation in design review discussions is highly recommended as a means of reducing risks during the development process. Of course, collaborative efforts with outside resources does bring risks, hence it is wise to ensure that nondisclosures are valid and in place prior to disseminating any project information to any external or outside vendor or resource.

Reference

[1] From the school of hard knocks—experience, be it good or bad, can be teaching and learning events.

Resources for further study

S. Levy, J.H. DuBois, Plastics Product Design Engineering Handbook, second ed., Chapman & Hall, New York, 1984.

S.E.M. Selke, J.D. Culter, R.J. Hernandez, Plastics Packaging—Properties, Processing, Applications, and Regulations, second ed., Carl Hanser Verlag, Munich, 2004.

J. Rotheiser, Joining of Plastics—Handbook for Designers and Engineers, third ed., Carl Hanser Verlag, Munich, 2009.

G.L. Beall, Rotational Molding—Design, Materials, Tooling, and Processing, first ed., Carl Hanser Verlag, Munich, 1998.

E.R. Larson, Thermoplastic Material Selection—A Practical Guide, first ed., Elsevier, Oxford, 2015.

R.A. Malloy, Plastic Part Design for Injection Molding: An Introduction, second ed., Carl Hanser Verlag, Munich, 2010.

P.A. Tres, Designing Plastic Parts for Assembly, seventh ed., Carl Hanser Verlag, Munich, 2014.

Web links of note

Great Design information from DuPont can be found here:

http://www.dupont.com/content/dam/dupont/products-and-services/plastics-polymers-and-resins/thermoplastics/documents/General%20Design%20Principles/General%20Design%20Principles%20Mod%201.pdf.

Sabic provides many tools for designers here (you may need to register):

http://www.sabic-ip.com/webtools/toolsHomePage.faces.

The UL Prospector site is a valuable tool for any designer reviewing materials for any application (you will need to register):

https://plastics.ulprospector.com/.

Solvay provides a great design guide for using their Udel polysulfone materials:

http://www.solvay.com/en/binaries/Udel-PSU-Design-Guide_EN-227550.pdf.

Bayer has a nice design guide for RIM systems:

http://reactioninjectionmolding.com/wp-content/uploads/2013/09/RIM-PartMoldDesign-Guide.pdf.

Covestro offers several design related guides, including part and mold design, snap fit design, and joining techniques for plastics:

http://www.plastics.covestro.com/Library/Technology-brochures.aspx.

Section II

Designing With Plastics

5. Conceptual Design—How do we Begin the Design Process? 101
6. Preliminary Detailed Design 127
7. Final Detailed Design 197

Introduction

As introduced and discussed in Chapter 1, designing products and parts with plastics requires adequate consideration of the *Four Pillars*: Materials, Processes, Tooling, and Design. Materials, Processes, and Tooling considerations have been covered in *Background Knowledge* (Section I); now we delve into Design; the most important of the *Four Pillars* and the main focus of this section. The design process is an iterative process of defining requirements, creating solutions, validating the solution concepts, then rigorous detailing of the individual features and parts which make up the final product solution. In an effort to simplify and organize the design process functions, the design process has been divided into three distinct steps covered in Chapters 5–7, respectively. Each of these steps is discussed in detail within their respective chapter, where Chapter 5 covers *Conceptual Design*; getting the seeds for the solution by defining the requirements and applying creative thinking techniques. Chapter 6 covers the *Preliminary Detailed Design* efforts, where accurate design models are produced to allow prototyping and validation of the concepts before investing time and resources in precision *Final Detail Design*, which is covered in Chapter 7 (Fig. II.1).

As discussed earlier, no two plastics design projects will ever be the same. In fact every single one is exclusive and completely different; which is why we love the challenge of a good design project! Knowing that all design projects are different, it follows that not all of the design process steps or considerations taught in the following three chapters will be applicable to each and every design project. However, all design efforts go through the same basic evolution. We identify a need, we create some conceptual solutions, we parse through those solutions to find the best combination of features or attributes, and we use those best features and attributes to create the final solution. This same methodology is used for many other life activities and events. For example, as I am preparing

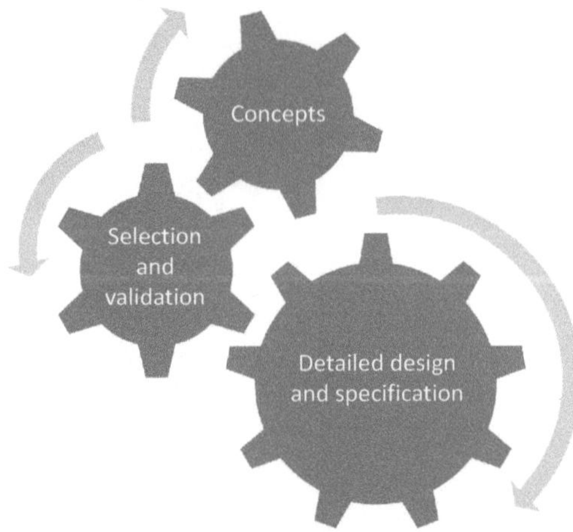

Figure II.1 The three key elements of the design effort—concepts, selection and validation, and detailed design. The focus for Chapters 5–7 covering design.

this section I realize “I am hungry”; alright, I could eat lunch with my friends, or eat alone at my desk, or go home for lunch with my family; but what would I eat? I could have a sandwich, I could have soup, I could get a piece of cake; now thinking about it, I need to stay close to my desk because I have a meeting soon, and since I have been cold all morning soup would make me feel warmer; can I get warm soup? Yes, the café downstairs always has fresh hot soup available; for today my ideal solution is to eat at my desk and just have a simple bowl of soup. The simplistic lunch example exhibits the same process methodology that we use for designing with plastics; define the problem, create solutions, chose the best from the solution pool, validate, and then implement.

Conceptual design—how do we begin the design process?

5

First, before anything conceptual can even begin, there must be a "need"; a problem or situation that necessitates the development of a new part or product. In addition, every "need" must have a "user" (also known as the customer) that has this "need" and would benefit from a solution. Once this "need" is identified and it is determined that some new solution is in order and would be beneficial to some "user" we then set out on our creative journey. To be true to the project we must first fully understand both the "need" and the "user," and then qualify and specify the full scope of the "need." Once the "need" and the "user" are fully understood and specified, the process goes into ideation, or expansive thinking, creating every possibility of a solution, no matter how remote or crazy it might appear. Often the ideation step is implemented as a defined time group event, for example, a one half-day meeting with attendees invited from marketing, engineering, design, manufacturing, potential users, and as many various groups that might provide creative insight for the market segment or application where the "need" might exist. Once finished with the expansive concept creation, each of the possibilities created in ideation is reviewed and the weed-out process begins. Weed-out removes those concepts or ideas, which have little or no apparent value for a solution, and focuses the remaining possibilities down into functional application, process, or material dependent categories. Once this weed-out is complete, the design professional(s) will take the list of possibilities and begin to make some design choices; given the scope of the "need" they will think through each of the solution possibilities from ideation and conceive just how they might arrange a solution, either though selection of the best ideas from ideation, or by cherry-picking features, attributes, or functions from several ideation concepts, and combining them into a new arrangement that might be the best single solution for the given scope of the "need." The goal at this point is not to find the exact single best solution, but rather to find the top five or the top three that might prove valuable in satisfying the "need" so some comparative and performance based analysis of each can begin (Fig. 5.1).

5.1 Understand the "Need" and the "User" for the product, part, or assembly

Starting a new design project is like the first day of a long awaited vacation, filled with eager anticipation and lofty expectations. It is an exciting time and an easy time to be so eager that small details might seem insignificant and get overlooked. One of the best pieces of advice handed down from my good friend and mentor Glenn Beall is that "The single most important thing that a design engineer does in a new

Designing Successful Products with Plastics. http://dx.doi.org/10.1016/B978-0-323-44501-6.00005-2

Figure 5.1 A graphic illustration of the wide array of potential "users" with "needs" for any potential design project.
Source: Photo from stock.adobe.com—iQoncept.

product development project is define what is required" [1]. Be wise, be vigilant, have discipline, and get ALL the facts and record them in some organized fashion PRIOR to starting to solve the "need." If you are working with a team, have everyone on the team review and agree that the complete scope of the "need" and the "user" are known and recorded. Under estimating the full and correct scope of the "need" is bad (and it will bite you). Not having a full understanding and appreciation for the intended "user" is equally bad, perhaps worse in some situations and applications. Every design decision from this point forward during the development project will be based upon this knowledge of the scope of the "need" and "user." Misdirection, wasted design resources, and loss of valuable schedule time result when a full understanding of the "need" and the "user" is not secured early and completely. This single simple first step in any project will limit downstream risk more than any other step of the development process. Don't let the project start on poor knowledge; take the time to do the simple work required to get it right!

Case Example

On one project, we were tasked with designing an electronic-based sensor for a water-heating device. The sensor needed to deal with mildly heated chlorinated pool water on one end and had a cable exit that was in the heater enclosure at the other end, which was open to the atmosphere but completely covered; no sunlight could reach the location where the sensor resided in the heater enclosure during use. The end of the sensor in contact with the pool water had to be isolated and sealed to prevent pool water from leaking past the sensor into the heater enclosure. The design was rather interesting, we overmolded a portion of a printed circuit board with polypropylene (the end in contact with the chlorinated pool water), and then that first overmolded assembly was simply inserted into a rigid injection-molded part which provided an O-ring gland used to seal the sensor assembly into the pressurized pool water system. This subassembly, consisting of the rigid injection-molded part along with the printed circuit first overmolded assembly, was then loaded into a second mold and overmolded to permanently affix the rigid injection-molded part to the overmolded circuit and cable assembly using a flexible material typically used for overmolding cable ends. A great combination of technologies used to create a finished unit protected at both ends and providing the rigid platform for the O-ring seal required in between.

Creating the second mold to properly seal off on the first over molded assembly and the rigid injection-molded part was a fine detail that the tool makers needed to work out. Features for precise locating and to allow mold seal-off were figured out and incorporated into the part designs. The injection mold for the rigid O-ring gland support part was built and the project seemed to be moving ahead well. That is, right up until .... the client found a new potential application for the sensor in an agricultural application; an application with full exterior exposure. Our original material selection for the rigid injection-molded O-ring gland support part did not need to deal with full exterior exposure—it was buried in the heater enclosure and protected from sunlight and UV exposure. This "new" application would potentially place the sensor assembly in full sunlight exposure. The exposed portion of the sensor comprised some portions of the rigid injection-molded O-ring support part and some portions of the flexible compound used for the second overmolding shot. The flexible second shot material was a black polyvinylchloride (PVC) based compound and would be acceptable for use in the "new" full exposure application, however, the rigid injection-molded O-ring gland support part was specified to be molded from a polysulfone (PSU) material; chosen for its excellent properties (hydrolytic stability) in hot chlorinated water. There was not much of the PSU material exposed, but the PSU was definitely not a good match for this full sunlight exposure (high UV) environment. Fortunately for us, we were able to select and run a rigid black PVC resin in the mold (similar shrink rates) and get rigid parts to satisfy both applications—see Fig. 5.2 for a side-by-side comparison of the part molded in the two materials. For the PSU material, the mold was run with hot oil to get the mold temperatures up to the level required for the PSU. To run the PVC materials, the mold had to have the oil lines scrubbed out, the fittings replaced, and the original oil heating lines were changed into coolant lines for running the PVC. Not an ideal situation, however, we were able to get functional parts from a rigid UV stabilized black PVC which were suitable for both potential end-use applications.

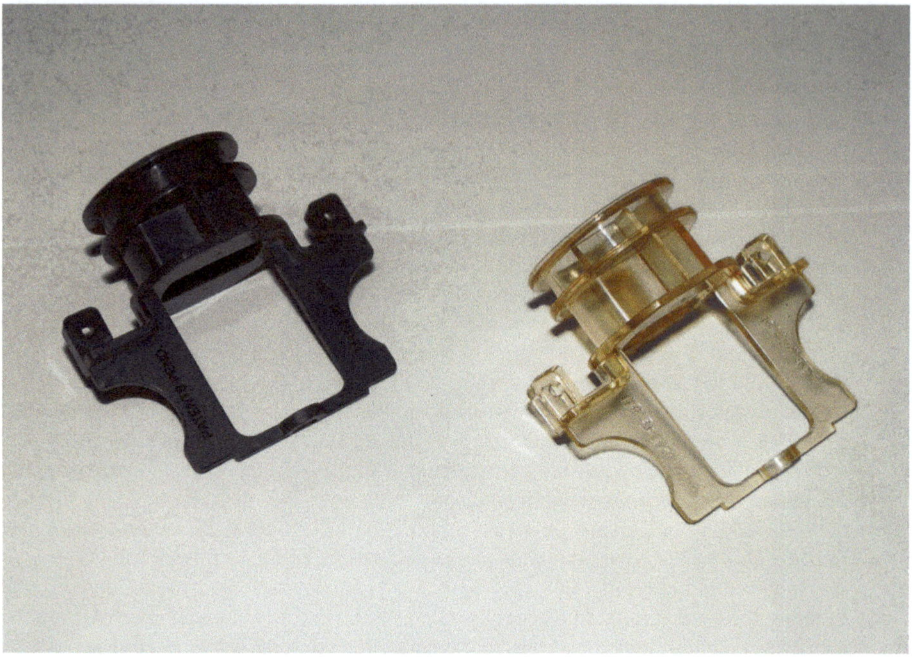

Figure 5.2 Side-by-side examples of the O-ring support part moldings, in black PVC and in translucent amber PSU.

In the sensor development case example the concept was fully understood and qualified for the intended application; however, the client, a swimming pool heater manufacturer, did not think that a secondary application (user) might exist. Only after the project was well underway did the secondary application (user) become known. Had the initial concept scope included the fact that other applications might be forthcoming (a complete understanding of the intended users), perhaps the PSU material selection might have been questioned earlier in the process allowing the tool to be built for the correct material from the start.

5.1.1 Qualify the "Need"

Due diligence in determining the extent or magnitude of the "need" often falls on the marketing or market research professionals, however, it is never wrong to do some research to see if anyone else has attempted to satisfy the "need"—often you will find that someone has. Patent research is a good place to start, not that a product or plastic part designer needs to understand patent law, but a ton of information can be learned from reading published patents relative to the field in which the "need" exists. The real substance of this research is the ability to understand the good from the not-so-good. In other words, one must learn from the good information without being side-tracked or mislead by the bad information. Sometimes, especially when reading patents, this is a difficult task.

Project #3001 - Easy-Clean Bird Feeder Assembly				
Item:	Need or Feature	Reference	Description	Filing Date:
1	Multiple Feeding Reservoirs	4327669	Teaches multiple individual tube, each having the capacity to hold a unique substance or feed material.	February 28, 1980
2		5195460	This system approach uses individual feeders hung in a gang to facilitate providing multiple feed substances at a single feeding location.	October 21, 1991
3		5406908	Teaches a plurality of stacked modular containers, each capable of housing a unique substance or feed material, allowing the user to provide many different feeds at one location.	March 15, 1993
4		5150665	Teaches and interior baffle system for separation of the interior compartment into discreet feed containers capable of providing multiple feed materials at a single location.	August 22, 1991
5		D267355	Shows the style and design elements for a multiple tube feeder from 4327669	
6		D359592	Shows the style and design elements for a feeder with a bifurcated storage container allowing two different substances to be stored and delivered at the same feeding location.	June 30, 1994
7	Feed Capacity Status	4327669	Teaches clear transparent storage tubes.	February 28, 1980
8		4355597	Teaches clear transparent storage tubes.	August 7, 1981
9		4356793	Teaches clear transparent storage tubes.	November 16, 1977
10		4607597	Teaches transparent shaped molded reservoir.	April 24, 1985
11		5215039	Teaches clear transparent storage tubes.	August 6, 1992
12		5309865	Teaches clear transparent oval window located at the half-full point on the feed storage container portion of the device.	June 9, 1993
13		5642689	Teaches clear transparent storage tubes.	August 7, 1995
14	Protection of the Food Substance	4327669	Teaches a top rain cover which overhangs the feeder by a substantial amount - however, individual feeding ports are open and not protected, allowing rain water to enter the food storage area, possible leading to mold or bacterial growth. No drainage on the feed area is provided.	February 28, 1980
15		4607597	Teaches an inverted stepped shaped reservoir which has the largest diameters/perimeters at the top and then cascading down in size at each subsequent step - thereby allowing the shape to act as a rain protectant for the next layer down feeding ports. No drain is presumed for the feed storage compartment.	April 24, 1985

Figure 5.3 Example chart showing a listing of the particular "need" or feature as found in reference patents for a proposed new "Easy-Clean" bird feeding apparatus.

I personally make a spreadsheet and have columns set up for each identified "need" and then next to it I record the idea or solution along with the reference patent—see Fig. 5.3 for an example of this method. Often I will find several patent references all attempting to satisfy a similar category of "need" and can learn from their examples. Sometimes we learn what is protected and where we cannot go with our design efforts; but most often we learn what has been attempted and failed (patents are full of failure documentation proving why the instant patent is indeed novel and deserves to be issued).

Be sure to fully question the marketing/sales professionals about their qualification of the "need"—fully understanding their process and results can often aid in the full understanding of their desired solution direction. In other words, the survey questions used by the marketing team to qualify a "need" can sometimes lead the respondent to a desired conclusion. Be sure the conclusion is valid and not driven by misleading questions. There is nothing more frustrating than investing resources and personal energy in a project only to find the qualification was tainted by some ego driven person(s) and not by good data. It is our job to provide the very best solution to the "need" and we cannot do that without adequate information, hence question everything; the final product solution will be better for it.

Likewise, examine the potential "user" identified for the "need" and be sure that all potential "users" are identified. Sometimes there will be different and distinct "user classes" or categories, each with very different requirements. For example, if the "need" was a closure for a container of a hazardous fluid, say an insecticide for gardening use, the "user" might be identified as an adult homeowner. However, in reality, the user might be a teenager (or younger) hired to help the homeowner, or an elderly adult homeowner with some physical degradation from advanced age effects, or perhaps a professional landscape service provider, and one cannot rule out that some child may have access to the container when stored within the homeowners property or when displayed for retail sale at the local garden supply center. The design concept solution must consider all these user groups.

5.1.2 Specify the "Need"

Once the "need" and the "user" are fully comprehended and understood, and once the "need" has been qualified thoroughly, it is prudent to document the findings in the form of a concept specification document. Product development specifications range from simple descriptive statements like, "We want the device to do this and such …" to complete layouts including every minutia of detail that the team can derive. No one format is preferred as each development project is different. For me, I tend to try and document as much as possible, including definition about the intended "user" and any unknowns that are floating around in my mind and causing me worry—see Fig. 5.4 for an example of a simple concept specification document. I tend to want collaboration on the concept specification and, like the team buy-in mentioned earlier for the project scope. I tend to ask for team and client (end-user, marketing, etc.) buy-in to the written concept specification document.

Some see the written specification document as extra work that does not bring value to the end product. I disagree. Written specifications deter scope creep. What is scope creep you ask? It is similar to long-term creep in parts molded or formed from this space age material we call plastics, except scope creep is more of a short-term (before the project is over) phenomena. The marketing/sales team (also known as the client) wants to invest a fixed sum of resources for the given design solution to satisfy the "need" and then, once the project is underway (with a defined timeline and budget), decides what they really wanted and meant to say was that the finished product should also turn colors on demand, or kill any bothersome insect nearby, or they wanted some other bells or whistles; but there is no more budget or schedule time than before, just a larger project scope. A written specification document with buy-in, or agreement, by all parties up-front will simplify and protect the project from scope creep.

Written specifications are also an excellent document for a myriad of other uses. Early on in many projects we need to identify areas of development which may be novel and benefit from patent protection. The specification document is an excellent tool when conveying information about the concept to the patent agent or attorney. The attorney can get a baseline of information that can then be used to build their understanding of each potential design solution. The relevant art searches, as well as

Project #3001 March 19, 2001

Specifications for Easy Clean Bird Feeder Concept

1. Modular approach to the design - allowing flexibility in manufacture – several potential products all using the same parts from the same production tooling assembled in differing combinations to satisfy various end user needs.
2. All injection molded parts to be dishwasher safe and dishwasher durable. Weekly dishwasher cycle cleaning will be recommended for the product.
3. Design to allow for rapid and intuitive disassembly for cleaning.
4. All individual sub-assemblies to be large enough not to be lost if dropped.
5. Individual parts of sub-assemblies to be child safe size – too large to chew or to swallow.
6. The design must allow for rapid and easy re-assembly after cleaning (dishwasher) to be ready for re-use.
7. Teardown and cleaning will clean every portion of the bird feeder that comes into contact with the feed materials used – thereby eliminating all mold or bacteria growth which occurs within every feeder once the feed become moist.
8. Use of inserts in the mold tooling to provide large or small feeding ports, or thistle slits will reduce overall tooling costs and will provide additional flexibility in the product matrix.
9. Main hemi-spherical feed tube parts will be molded from a transparent material for instant visual feedback for feed volume status.
10. Sliding engagement of the assembly rails over the "Tee" on each end of the hemi-spherical feed tube parts to allow enough tolerance for assembly and disassembly between freezing temperatures and a hot summer day temperature – say 90 F.
11. Sliding assembly features to be manufactured from Stainless Steel to provide a durable non-degrading functional part for use in outdoor environments. This part must be durable and must be extremely safe to handle – absolutely no sharp edges or corners.
12. Top of the feeder to allow rapid filling but to engage tightly to prevent wind or animals from accidental removal or dislodging.
13. Base of feeder to allow for affixing to a vertical 1" outer diameter vertical rod for mounting.
14. Feeder to have means for hanging which will allow top to be removed for refilling feed materials while still in the hanging position.
15. Feeder top and bottom molded parts to be UV stabilized colored parts and might be in several different colors, allowing for several variations in the final product lineup.

Figure 5.4 An example (partial) of a concept specification document as prepared for a proposed new "Easy-Clean" bird feeding apparatus development project.

the strategy used to form the written patent application will be better focused, more effective, and more efficient when the attorney has the best information possible.

Finally, written specifications document and quantify the commitment that is being made for the development project. Every design decision is based upon this document and all final part or product performance attributes are evaluated against this document. The existence of a well-researched and well-documented specification will always benefit the project team and will allow the project team to be more effective and efficient. In addition, this written document becomes a permanent record which may prove useful much later in the product lifecycle, or perhaps might become useful during the development of additional follow-on or related new products.

5.1.3 Quality function deployment (QFD) re; the "Need"

In some organizations and institutions, an alternate philosophy has emerged as an extension of overall organizational quality improvement efforts. This is sometimes referred to as quality function deployment (QFD), and is implemented at the very beginning of a project, much as our discussion earlier regarding understanding the "need" and the "user". In short, QFD provides a connection between the customer or end-user and the solution creator or designer. As an organizational tool, it seeks to connect not only the designer but also the design team, the management, the engineering staff, the manufacturing team, and everyone in the path between conception of the design and the delivery of the finished goods to the end-user. These philosophies are expanded to cover not only production of hard goods from materials like plastics, but also can include services and information goods (internet information, interactive software, smart phone user interfaces, etc.) as well.

According to the QFD Institute:

QFD is:

1. *Understanding Customer Requirements*
2. *Quality Systems Thinking + Psychology + Knowledge/Epistemology*
3. *Maximizing Positive Quality That Adds Value*
4. *Comprehensive Quality System for Customer Satisfaction*
5. *Strategy to Stay Ahead of The Game*

As a quality system that implements elements of Systems Thinking with elements of Psychology and Epistemology (knowledge), QFD provides a system of comprehensive development process for:

- *Understanding 'true' customer needs from the customer's perspective*
- *What 'value' means to the customer, from the customer's perspective*
- *Understanding how customers or end users become interested, choose, and are satisfied*
- *Analyzing how do we know the needs of the customer*
- *Deciding what features to include*
- *Determining what level of performance to deliver*
- *Intelligently linking the needs of the customer with design, development, engineering, manufacturing, and service functions*
- *Intelligently linking Design for Six Sigma (DFSS) with the front end Voice of Customer analysis and the entire design system*

QFD is a comprehensive quality system that systematically links the needs of the customer with various business functions and organizational processes, such as

marketing, design, quality, production, manufacturing, sales, and so on, aligning the entire company toward achieving a common goal.
It does so by seeking both spoken and unspoken needs, identifying positive quality and business opportunities, and translating these into actions and designs by using transparent analytic and prioritization methods, empowering organizations to exceed normal expectations and provide a level of unanticipated excitement that generates value.
The QFD methodology can be used for both tangible products and nontangible services, including manufactured goods, service industry, software products, IT projects, business process development, government, healthcare, environmental initiatives, and many other applications.

This indented and italicized material is quoted directly from the QFD Institute [2].

As each product development project will be different and given that each "need" and "user" are indeed different, and as each provider/developer of the solution for a given "need" is different as well, some will find the QFD methodology suitable and some will not. However, as an overall quality system it is certainly worth a full comprehension and consideration in every project or new product development situation. It is perhaps most often found and used in organizations utilizing Six Sigma or Design for Six Sigma protocols.

5.2 Solution ideation

Every problem or "need" which necessitates some design action to be satisfied has an unlimited number of possible solutions which may be valuable or beneficial to the end-user. However, only a select few are capable of commercial success. Some will just not supply the solution well enough, some may consume too much time when in use, some may be too expensive to manufacture or bring to market, some might be too large or too complex, and the list goes on. It is our job as designers and engineers to make the choices and selections that lead to the *optimum solution* which can lead to success. One method that has been proven to work is expansive thinking followed by review and idea weed-out. This method sets out to identify each and every distinct and remote possibility, from every possible angle, then goes through a review process where valuable nuggets are identified and nonvaluable ideas are tossed out, attempting to reduce the idea pool to a select group from which the designer can draw and create potential conceptual solutions. It is best done with a diverse group and we typically try and complete the session during a defined time, for example, a one or two day contiguous period, where the participants are typically baited with plenty of good food and lively activity to keep their creative juices at a peak.

5.2.1 Expansive creation of solution possibilities

This is where the "think outside the box" thing comes in … but instead of thinking outside the box, we get far more insane and think both inside and outside the box;

including inside the pencil that drew the box or outside the tree that was felled to create the pulp that was used to fabricate the paper the box was drawn upon—we look under every rock and on top of every star in the heavens for ideas that might have even a smidgeon of influence for a valuable solution. Sometimes we use categories to order the chaos, perhaps by material or by materials state (solids, liquids, and gases) or by energy input means (battery, line powered, fossil fuels, solar powered, human motivated, etc.); the goal is to cause us to think differently—not just jump to some conclusion based upon our direct, and perhaps limited, experience. By thinking through all of the possibilities, even the insane ones, we cause our minds to wander and look at the possibilities that might not be direct but may have some key component that might be the single thread to tie the successful solution together.

The ideation team should ideally be multidisciplined and chosen from all parts of the organization, including marketing, sales, design, engineering, manufacturing, quality, management, and if possible include some potential or representative end-users in the mix. An eclectic group mixture will ensure a diverse set of potential concepts and ideas. Of course, depending upon the degree of difficulty the scope of the "need" imparts, the activity can also be done by a small group or by an individual as well. Sometimes the degree of difficulty in creating a solution will determine the size and diversity of the group, other times the scope is so focused that the ideation is best done by an individual. It is important to provide the ideation team participants with the written specification, including the written description of the "need" and the complete definition of the "user" (customer or end-user).

Many times the ideation session is preceded by a group teambuilding session, breaking the group into small teams to solve some silly task in the most outrageous manner, or something similar. This fun initial game or activity can set the tone and allow everyone to get familiar with the others in the group and relax, and let the creative ideas flow freely without concern for ego or emotional self-preservation. Having plenty of props or toys to augment thinking is also a trick used to help stimulate creative thought. Sometimes similar materials, products, parts, or even competitive products might be on hand for touching and investigation. Plenty of fluids (coffee, tea, water, etc.) and nourishment for the group is always a plus for an ideation session as well.

This expansive creative process is a positive space—no negatives are allowed—every idea gets logged into the list—no idea is left out or discounted. As a group or team activity, as mentioned earlier, all of the ideation concepts and ideas need to be captured and recorded for future use. Experience has shown that either using multiple whiteboards or huge sheets of paper on the walls of the room, putting all of the ideas up on the walls allows everyone in the room to see and digest every single idea from every single participant—see Fig. 5.5 as an example of an ideation session space. This visual bombardment of ideas usually causes even more ideas to spring forth, which is the entire essence of the activity; create as many possibilities for the solution as possible. The availability of plenty of colorful markers, as well as whatever other creative materials might be suitable for the particular part or product will help support the creative efforts of the group. At the end of the session, each platform (whiteboard or paper space) can be photographed easily for capturing the ideas in their native form for future reference or archiving.

Figure 5.5 An example of an ideation session room showing the wall covered in ideas created during the session.
Source: Courtesy of Impel Studio, New York, NY.

5.2.2 *Weed-out—focus in and reduce the possibilities to a select few*

The weed-out process should, like ideation, be a group activity. In fact many times the ideation sessions will be arranged with a time defined ideation period, followed by a time defined weed-out session. Usually there will be some other activity scheduled between ideation and weed-out, lunch if the sessions are back-to-back over a full day, or sometimes ideation would be day one, followed by a group dinner and then weed-out would commence the following morning. Scheduling some alternative break between the two sessions allows the participants to come down from expansive thinking "high" and to get back to reality. The weed-out session tends to be a negative activity, finding faults with concepts and ideas, and, as such wants to be isolated and removed from the ideation session by some other distraction. Aside from being a negative activity, the weed-out is extremely important to the whole process. If not enough ideas are removed then there are too many choices, and the process gets fuzzy and nonfocused. On the other hand if the negatives get out of hand no good concepts or ideas may make the cut. The participants can also have an adverse effect, they need to be instructed to evaluate each concept or idea as it pertains to the "need" not to evaluate based upon where, or from whom, it originated. Egos will get in the way unless the participants can separate the concepts and ideas from their originators and understand that the best ideas or concepts that satisfy the "need" will be used and when the best are used, regardless from whence they came, the end solution or product will be the optimum solution. Every contribution, even those cut in the weed-out process were valuable in finding the best or optimum solution. Watch out for the "... we tried that already ..." comments or thinking—often the "... we tried that already ..." attempt was just curiosity and not fully investigated or vetted.

Weed-out can also be an individual process—sometimes even after a team ideation session it is best to accomplish the weed-out session individually or in a much smaller and focused group. The ideas and concepts that survive the weed-out become the base for the solution creation to satisfy the "need" and, as such must be evaluated and analyzed thoroughly and carefully to understand the values which each, or each combination thereof, brings for the final solution.

5.2.3 Alternatives to group ideation

Some projects just don't warrant the time investment of a full ideation and weed-out session; the project is just narrowly focused or too small to justify the resource investment. These "small" design projects are often handled by a single team or perhaps even a single individual designer, but the expansive thinking and then purposeful weed-out process will remain quite the same. I have designed many products on a solo basis, but on each I followed the same process; think of every single possible solution, then weed through in a disciplined manner to pull out the ideas, with possibility and toss the ideas with no change of assisting with a final solution.

Care must be exercised when making the decision to qualify a design project as "too small" to warrant the ideation group method investment. The return from having many brains engaged on a design solution simultaneously is immense and in reality the cost for a couple of meetings with several people in attendance is relatively small, especially considering to impact the design solution could possibly provide for the customer.

5.3 Selection (choices, choices, choices)

The foregoing process of ideation and weed-out can be used as a method to find a solution for any "need" and is not specific to the design of something from plastics. However, since many devices or products are comprised of plastics components entirely or for some vast portion, it is prudent to think broadly. Now that the concepts and ideas have been created and weeded, the survivors form the base of knowledge that will be utilized to form the first tangible efforts to satisfy the "need." It is here that the designer's experience and full knowledge of materials, processes, and tooling (remember the *Four Pillars*) will come into play and be tested. The choices at this point are meaningful and will dictate the outcome of the development project and the inherent quality of the solution once implemented or commercialized. These are important decisions and choices and each will have a sizable impact on the quality and excellence of the resultant part or product. Each choice or decision must be made to be the best decision to satisfy the "need" for the intended "user." If the choice or decision is made to satisfy the selected material provider, or the desired molder, or process machinery, then the product or solution will suffer.

5.3.1 *Initial solution sketching*

Given the results from ideation and weed-out, I will typically sit down and make several crude pencil sketches of ideas and concepts that might provide the solution. These visual hand sketches are a quick way for me to think visually out loud. These sketches are never part of the deliverables or part of the development record; these are simply thinking tools. Once several are completed, I begin to group ideas into more and more refined sketches. My thinking and sketching start to find the way to organize the design into discrete parts formed in discreet ways and assembled with discreet intermating features. Things begin to fall into place, the concept is emerging.

5.3.1.1 *Written form descriptions*

Once the initial pencil sketch thinking is complete, I attempt to reduce the concept to written form. This causes me to think through the design not just visually but in terms of language, so I must think differently. This often will cause me to need to describe something that I had not fully considered in the visual sketches and will cause me to go back and sketch some more to figure it out—see Fig. 5.6 for an example of sketch form and written form descriptions. The written end product becomes another part of the final specification and can be used as a reference as the design moves forward. The goal is to completely and fully define the design so that someone else that is not familiar with the "need" or "user" would understand the concept and understand the design approach.

5.3.1.2 *Pencil sketches and/or "high-level" CAD visualization renderings*

While the initial sketching used for thinking through the concepts and ideas are useful for thinking out loud, they are so rough that no one else would be able to make heads or tails of them. Now with a written description, better sketches or illustrations are needed to document the concepts for evaluation and review by others. In many cases, simple pencil sketches are adequate to assist the written descriptions to convey the concept. In some instances, high-level CAD illustrations may be quicker and more accurate for conveying this information—see the example of high-level CAD illustrations in Fig. 5.7 as produced after the hand sketches and written form description in Fig. 5.6 were produced. Every case is different and every designer has different creative gifts, where one can sketch fast and accurate another may be a CAD wizard and may be able to produce an illustration better and faster with CAD, so to each his/her own. The bottom line is that suitable illustrations must be available to support the written descriptions, such that others (management, sales/marketing, legal, etc.) can fully appreciate the concepts and be able to compare one concept against the other and against competitive products or parts fairly and accurately.

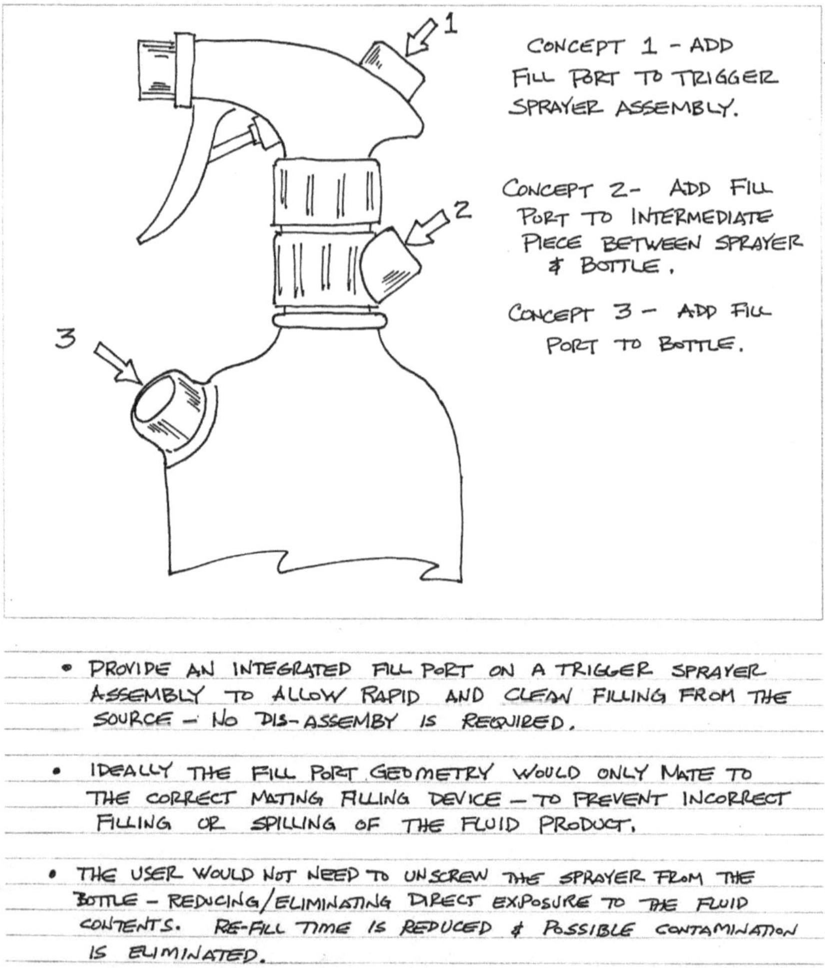

Figure 5.6 A representative hand sketch of a concept with annotations and descriptive notes from a project investigating rapid and safe trigger sprayer refilling options for institutional cleaning and sanitation use.

In addition, both the written description and the sketch illustrations will be excellent for use by the patent attorney in doing research and in preparing the application strategy. They will be useful by the marketing team in preparing their marketing strategy and sales presentation theories. They will be useful by the strategic sourcing professionals for initial research on potential vendors' in particular geographic areas where the potential products might be used in the highest numbers. They can be useful within sales and marketing for market research and analysis studies.

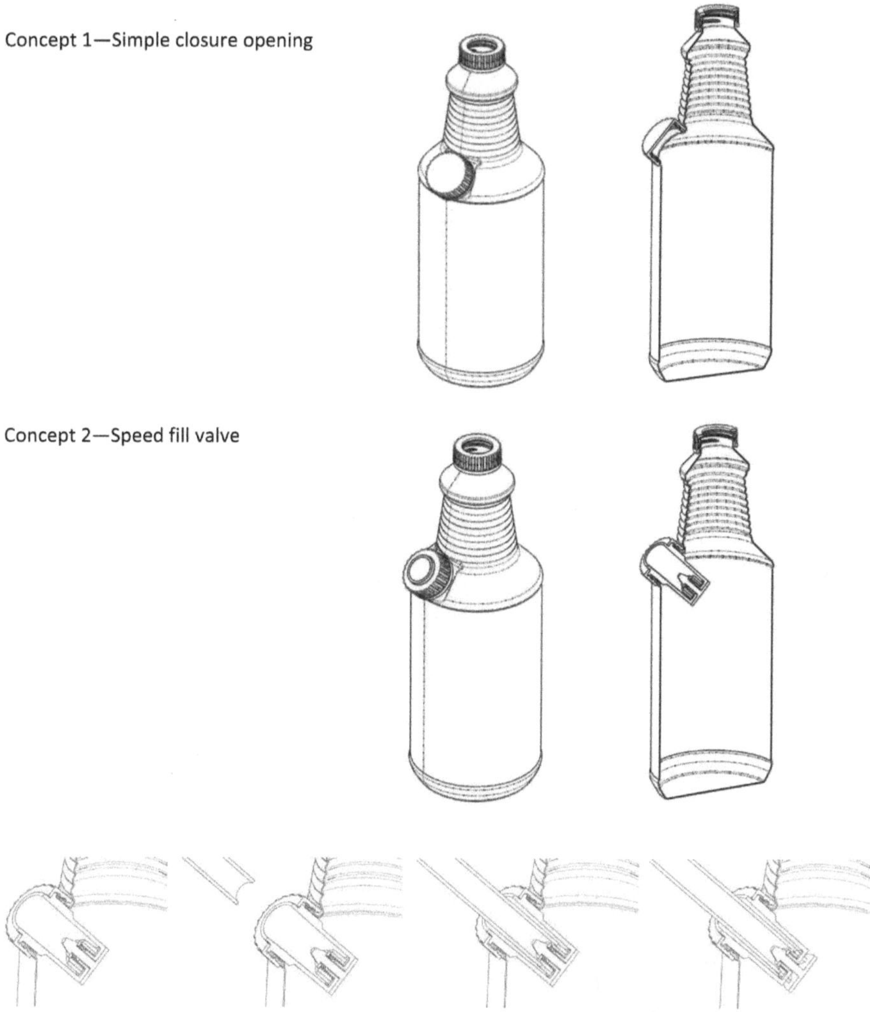

Figure 5.7 Representative high-level CAD illustrations of the trigger sprayer rapid fill concept as shown previously in Fig. 5.6 as a hand sketch concept.

5.3.1.3 Comparative analysis versus competitive products or concepts

Given the top three, four, or five design concepts to be considered for commercialization, it is usually helpful to review them against the current state of the art. Comparing the potential concepts head to head against the current competition will often provide insight into which may be received best in the market. Finding ways and means to differentiate the new concepts from the existing offerings will assist the marketing/sales efforts once the product is available and offered in the market. Comparative analysis of how each of the current competitive products truly satisfies the "need" and the "user" relative to the proposed concepts can sometimes be a useful tool to highlight false assumptions or judgments, further allowing reduction of potential risk.

5.3.1.4 Initial review of intellectual property (IP) and patents

Some organizations will want to use this initial concept description to perform a broad concept patent search, looking for both protection opportunities and infringement risks. While it is still extremely early in the development process, this can prove valuable and can limit the risk of wasted time and resources. Most often any provisional applications filed during the conceptual stages of the project will end up being refiled as the design develops, however, as we are now in a first to file world, it is wise to file early and often when developing valuable new products and intellectual property (IP).

5.3.2 Process selection—conceptual stage

Selection of the process to be used for the part or parts within a given product will usually happen during the concept reduction and review process. The designer will be seeing the part(s) in his or her mind as they conceive the concept and often they will see the part(s) as being from a certain process. Certainly, if the part is a hollow shape used to hold a fluid, they will be thinking blow molding. If the part is structural, they could be thinking machined, thermoformed, structural foam, rotational molding, reaction injection molding, injection molding, or blow molding. However, as stated earlier, given the scope of the "need" and the disposition of the "user" along with anticipated market size, the designer will be already thinking about the economic realities for the product and will be thinking about a suitable process for the particular part or situation.

The conceptual stage process selection serves to set up many of the final design parameters. The presentation of the conceptual design at the concept design review, coupled with the process for each part in the concept, is paramount at the conceptual design review. The detailed design will begin with the concept and process already given preliminary approval from the conceptual design review; and, the designer can proceed to prepare the part(s) design based on the concept and the process as presented and approved. This allows a quick transition from conceptual generation to production level reduction of the concept into the detailed and final ready for production part design. The selection and approval of the process at the conceptual stage allows the designer to begin the detailed design with a clear understanding of which rules to apply as the detailed design begins and with a clear understanding of what kind of tooling will be utilized; again providing insight on design rules which may be critical to final part geometry optimization.

5.3.3 Material selection—conceptual stage

As with process selection, the experienced designer will have an inkling early on which materials may be in play for the part(s) based upon the application requirements and the potential processes which may be used for the part(s). That said, once the final concepts are ready for evaluation, every possible material that can be utilized should be screened for suitability. Often this will reduce the pool of possibilities to two or three ideal candidates. Once these material candidates are identified, it is recommended that the designer contact the material suppliers and ask for their review of the application and recommendations from their offerings. Often they can draw upon experience and find a material that has performed well in similar applications in the past, reducing risk in selecting that particular material. In addition, they will be knowledgeable about new formulations which may be ideal for the application but not yet in the mainstream of their offerings. Most suppliers possess a wealth of knowledge about specific formulations with specific additives or modifiers that they have developed over the years—most of which remains nonpublished!

As with each step of this design process, producing a comparative analysis table, listing each potential material and the values for the most important material characteristics, in relation to specific part or product attributes as defined within the concept specification, is recommended—see an example of this in Fig. 5.8. In this manner, it becomes easy to reference a particular material characteristic during detailed part design and review.

The best approach is to select two or three possible materials, having a back-up if choice number one is disqualified for some reason. For example, for an injection-molded part, choosing materials with similar shrink rates and similar melt flows will ensure the ability to be flexible once in production should some material issues arise. Obviously, tooling features like the gate and runner geometry might need to be tweaked for a material change but that is more cost effective than constructing entirely new molds.

With full comprehension of the design requirements, the processes selected, the anticipated end-user and end-use environments, the designer is positioned as the ideal source to recommend the final material to be used for the part or product. However, experience has taught us to be careful when recommending the material; of all of the choices and decisions made by the designer during the course of development, the selection and recommendation of the material will perhaps carry the largest responsibility as the product moves from the drawing board into the stream of commerce.

Project 3001 - Material Selection Matrix

		ABS	PC	PMMA	PSU	HIPS
Specific Gravity	g/cm^3	1.04 to 1.11	1.18 to 1.22	1.14 to 1.20	1.24 to 1.25	1.03 to 1.05
Shrinkage	10^-3 in/in	4.5 to 7.4	5.3 to 7.0	3.9 to 6.4	6.9 to 8.0	4.5 to 6.1
Tensile Strength	(yield) PSI	4940 to 7420	8120 to 9340	5420 to 10700	8480 to 11100	2480 to 4520
Elongation	(yield) %	2.0 to 12	3.9 to 7.4	2.4 to 5.2	5.0 to 75	1.4 to 61
Impact - Unnotched	(Izod) ft*lb/in	1.1 to 31	59 to 61	4.3 to 15	7.0 to 45	1.0 to 15
CLTE	10^-5 in/in/F	4.2 to 5.8	3.2 to 4.0	3.1 to 6.8	3.1 to 3.2	4.4 to 5.6
Hardness	Rockwell	97 to 115	69 to 121	31 to 105	119 to 120	46 to 113
DTUL	(66 PSI) F	179 to 213	265 to 287	189 to 199	352 to 361	166 to 206
UV Stability		Not Good	Good	Good	Not Good	Not Good
Transparency		No	Yes	Yes	Yes - not clear	Yes
Moldability		Good	Good	Good	High Temp	Good
Cost		Moderate	Moderate	Moderate	High	Low

Figure 5.8 A material selection comparison chart as used for the "Easy-Clean" bird feeding apparatus project.

With over 85,000 commercially available materials (and counting) available [3], it remains best to allow the materials recommendation to come from the material supplier, the processor, or the customer.

5.4 Conceptual design review

Ideally there will be two or three potential concepts to be evaluated during the conceptual design review session, however, in most instances there will be one clear front runner that is really the design being evaluated for detailed design and implementation for commercialization. The conceptual design review serves to notify all stakeholders of the design concept, the process, and the materials which are being considered as primary candidates going into the final detailed design effort. Again, buy-in from all parties is expected at this review session. The investment of resources and schedule time going forward will be formidable and will be nonrecoverable if the product does not go to market or is not successful in the market. If there is any concern at this meeting session that the proposed combination of design concept, material, and process will not be successful in the market, then a serious evaluation must be in order to determine if the project should continue or be halted.

5.4.1 Balanced review of all potential concepts against the "Need"

Given that there is more than one concept on the table for evaluation during the conceptual design review, it is imperative that each concept be given a fair evaluation relative to the specifications for the "need" and relative to the benefits to the end-user. To this end, there are several potential questions to ask regarding the slate of potential design concept candidates. Some teams will have each design review participant keep a running score, from 1 to 10, for each question and do a straw poll to gauge the group's overall assessment of each concept.

5.4.1.1 Which has the best chance of success?

This question might be best restated as which pleases the end-user and satisfies the end-users perception of the need the best. It questions the measure of acceptance by the end-user. If the end-user is satisfied would they purchase it again? Would they recommend it to a friend? Would they gift it to a friend? Great products end up being sold by referrals from person to person; will this solution be so liked that this will occur?

5.4.1.2 Which is the value offering?

End-users and customers are drawn to products or solutions that have real "intrinsic" value. Does on the concepts just ooze with value as compared to the others? Real value sets many products apart from their competition, even at higher price points. Try to determine just how the end-user or customer might see the value in the prospective concepts.

5.4.1.3 *Which provides the economic realities for the "need"?*

How does each of the concepts stack up regarding manufacturing cost and where do they fall regarding estimated cost to the end-user? How do they compare to other solutions already on the market, if any? Is the first cost recoverable in a manageable period of time? Can the first costs be amortized over the first 2 years? Does the prospective end-user perceive the concept as a needed value item or as an expensive luxury item?

5.4.1.4 *Which provides the potential IP protection for the solution for the "need"?*

Do any of the concepts provide exceptional opportunity for IP that can both differentiate the offering and protect it through patents as well? Patent protection is a wonderful opportunity but it does not ensure success. In fact, product differentiation and intrinsic value provide a better measure of potential success, but success brings competition; IP protection helps to keep the competition at bay for at least a short while. However, potential patent protection for concepts which may not be commercialized, but would be suitable alternatives to the concept selected for commercialization, might prove valuable. If the product commercialized is successful then other competitive products will surely follow. Securing IP on any suitable alternative can help to deflect this competition and protect market share erosion from competition over the life of the product line.

5.4.1.5 *Which concept provides a solution that has market appeal?*

Market appeal is a perception of value or need that is distinctly different from user satisfaction. Everyone just knew they needed a pet rock in the early 1970s; that is market appeal. In the early 1980s everyone had a Rubik's cube, but that was because the user experience was intriguing and only one person could be using one at a time—so, everyone had to have one. Perceived value helps to make the sale, often as an impulse sale; real value promotes organic growth in sales by grass-roots word of mouth, user-to-user organic sales.

5.4.1.6 *Do any of the concepts infringe or come close to existing IP?*

Concepts which are close to existing IP or are on the fence will often bring more troubles than the benefits that they might bring. If the best concept is close to some other IP, then seek out that IP owner and attempt to work out an arrangement where both parties can benefit. Unless the IP is a direct competitive product in the exact same market, most times IP holders are glad to license their IP for modest royalties. Don't allow ego or fear to enter the equation, it's just business [4], you want to provide a functional device or product that they have some IP which may (or may not) be close, so work together and create an agreement where both parties can benefit.

5.4.2 What if the proposed concepts fail the conceptual design review?

Not all "needs" can be fulfilled for a given "user" by a given product or solution, from our client or company; so now what do we do? Understand that from my direct experience, the implementation rate from conceptual development is around 50% or so. Sometimes we go back and try to reassess the "need" and "user," and see if we can find alternative concepts. Other times we make the decision to stop and move onto the next opportunity for the client or company. It is better to decide to stop and invest the resources wisely, than to blindly forge ahead "*damn the torpedoes*" style. As a design consultant, one the most difficult discussions to have with a client, especially with a small business, or inventor, is the "*know when to say when*" conversation. Entrepreneurs are ego driven and telling them that the project should stop and not move forward never goes well. I have become the dumbest guy in the room more times than I want to admit, but if the solution does not meet the criteria for success, why go further?

5.4.2.1 Documentation

Even though the project is stopped, either forever or for reevaluation, it pays to document well the results of the Conceptual Design Review session. Market dynamics tend to change as time moves forward and there is no guarantee that in some short-time span, whatever caused the concepts to fail, might change. Full and complete documentation allows the new team, charged with revisiting the failed project, to review with and fully understand the rationale behind stopping the project. Full and complete documentation is just good project housekeeping.

5.5 Functional specifications

Once the conceptual design review is complete and the ideal concept has been selected for development, the functional specification should be prepared for the project. The functional specification is perhaps the most important document produced during the design process which is why I recommend producing it at the end of the conceptual development stage of development. This functional specification is an extension of, or perhaps a refinement of, the initial concept specification, except it can be far more specific now that a concept is defined.

At this point in the development, several concepts have been refined, several discarded, and several chosen as ideal candidates, with one primary ideal concept selected for development; hence, there is a much better sense of the final functional requirements for the product being developed. This is why it is most efficient to wait and create the functional specification at this point. The goal for the functional specification is to capture each and every attribute that will be contained, or possible, with the new product or part being developed and to quantify each of these attributes, such that each can be verified as part of the design verification process. In the example Functional specification shown in Fig. 5.9A–B there is shown the format

(A) **Functional Requirements for Design**

Abstract:

A device for helping a boat operator reach and affix the boat to a mooring post or pylon. This device allows the operator to reach overboard and drop a semi-rigid loop/strap around the mooring post or pylon and has means to allow the device to tighten the loop/strap around said post/pylon. In addition, the device has a rope attached with which can be fastened to the boat via a cleat or other means and has means to provide a tightening of the said rope to provide secure dock tie in. Said rope tightening means has a release for loosening the rope when it is time to un-dock and push-off. Loop/strap tightening means also is provided with a release for loosening the loop/strap when necessary.

DRAFT	Preliminary Draft – In Process	2004-09-29
Revision	**Remarks**	**Date**

MBA
CONSULTANTS & DESIGNERS
MacLean-Blevins & Associates

(B)

Item:	Function:	Type of Value	Value	Verification Method	When
1.0	**Reaching Overboard**				
1.01	How much reach is required?	Length in Feet	3 feet +/-	Physical Testing in field	Concept prototype
1.02	Hand held arm extension helps get more reach.	Desired	Enough reach to grab pylon without extreme bending by user	Physical testing in field	Concept prototype
1.03	Strap or Hoop to have some stiffness to hold a loop shape	Must Have	Hoop must not collapse under its own weight during the reach for the pylon operation	Testing in lab	Concept prototype
1.04	Handle must be easy to grab	Desired	Handle grab to be ergonomic and easy to hold for all user types – it may be wet when in use so features for secure gripping are required.	Testing in lab	Concept prototype
1.05	Must be lightweight to make it easy for entire boating population to use.	Pounds	Maximum device weight should be 2 # or less and the weight distribution needs to account for the hoop weight when extended and the rope weight at the handle end.	Testing in lab	Concept prototype
2.0	**Hooking to Pylon**				
2.01	Pylon Diameter	Inches	11" maximum diameter	Physical testing in field	Concept prototype
2.02	Hoop material must be suitable for pylon material –non-abrasive	Must Have	Material must resist splinters from Pylon and must not damage Pylon in any way.	Physical testing in field	Concept prototype and/or production prototype

Figure 5.9 A functional specification document example, as used for a Boat Mooring device development project; including a title page (A) which is comprised of an abstract and a revision log, and (B) the first page of the 22-page functional specification.

which includes a document management portion for accurate interpretation and revision control (Fig. 5.9A); and the actual specification portion including a breakdown of the functional categories and the information assigned to each allowing the functional attribute to be quantified and verified (Fig. 5.9B). Although only the first page of the functional specification for the device is shown (Fig. 5.9B), the actual document contained some 22 pages of function attributes, including product safety attributes, aesthetic attributes, storage when not in-use attributes, packaging attributes, customer loyalty and branding attributes, and so on. The functional specification can be used as a single repository of an enormous amount of information for the product under development and can be referenced by many throughout an organization.

5.5.1 Functional categories

General organization of the functional specification seems best done by functional categories. For any given product or part there is some end-use function that the device must need to do, for example, if we are creating a bottle for storage and retail display of hair shampoo, the main function would be to hold a liquid. The functional category liquid holding could then have many attributes listed, such as volumetric capacity, permeability, light sensitivity/transmission, closure seal material compatibility, closure seal method, and so on. Other functional categories for this product could include; consumer safety, filling and packing, labeling, shelf-life, and so on. Within each functional category list every attribute that fits within that particular category, then continue by giving each attribute several additional classifications.

5.5.2 Type of value

One of the first classifications would be what type of value we would use to describe the attribute, for example, if an attribute for our shampoo bottle were volumetric capacity, the type of value might be ounces or milliliters. Alternatively if the attribute were an ergonomic grasping surface for dispensing, the type of value might be "desired" or "wanted." If the attribute were sealed nonleaking containment of the liquid product, then the type of value might be "must have."

5.5.3 Value

Certainly once a type of value is defined the next obvious classification would be the magnitude of the value; hence for the volumetric capacity attribute with a type of value of ounces, the value classification would be something like 16 oz. For the ergonomic grasping surface attribute with a type of value of "desired" or "wanted" the value might be defined by some reference, such as "at least as comfortable as our bottle for hand lotion #9." This idea is to have a measurable value, even if it is a subjective measurement, at least have some reference to use as the baseline or comparison. For the nonleak attribute with the "must have" type of value, the value might be defined as no visible product leaking when the container is inverted for 2 weeks at room temperature.

5.5.4 Verification method and timing

This one seems intuitive but is often difficult to define. Once we have a type of value and a value defined for a particular attribute, just how will we determine if our design can deliver the value? And further, when can we determine if the design can deliver the value? For example, our volumetric capacity attribute can be verified during design by creating a CAD solid of the liquid contents and generating a part volume (CAD generated) to verify that the capacity meets the specification. However, the ergonomic grasp attribute with a "wanted" type of value and a "at least as comfortable as …" value cannot be verified until someone can actually compare a physical object; hence the verification method would be hands on comparison and the timing would be at the prototype stage. Our goal is to verify each and every attribute value prior to releasing the design for production of capital tooling. For some attributes, this will not be possible but for most we can limit risk by verifying early and often.

5.6 Action items upon departing the conceptual stage

Once a final concept is determined and all groups are agreed that the concept will move onto detailed design and development, several action items remain active as the design team transitions to develop the chosen concept. Clear definition of each group's responsibilities at the close of the concept design review is recommended.

5.6.1 Design team

The design team will obviously move onto the detailed design and development of the selected concept, however, not before documenting the results of the concept design review and preparing or finalizing the functional specification for the selected concept. Full buy-in from each of the other groups should be secured before investment of any further resources and before entering the detailed design phase of the project.

5.6.2 Marketing and sales

Given the selected concept, the marketing and sales group should begin to create and implement market research studies to determine factors which may influence an end-users choice to purchase and use the intended product. These studies may help to determine things, such as potential retail price points, packaging options, alternative channels to the market, potential marketing or cobranding partners, and the like. Any potential variations, or alternative versions, which may be desired as follow-on products or the like, should be identified and communicated to the design team so they can include this information in their thinking as they move into the detailed design phase. Often tooling can be specified in a manner to leave flexibility for alternate versions of features or attributes, if there is prior knowledge of the desired alternate versions; allowing follow-on products with a fraction of the initial investment that might be required otherwise.

5.6.3 Legal

Given the selected concept, the legal team should be looking for opportunities for protecting any IP which may be involved, as well as looking for potential liability issues that could be mitigated during the detailed design stage. If there are any unknown or unclear elements in the design, the legal team should be verifying their complete understanding so they can evaluate the situation completely and correctly. Often engineering and design types speak and communicate in their own language style and there is sometimes a bit of a barrier between the legal professionals and the design team resulting. Be mindful and ensure the concept has been thoroughly defined and explained in a manner which is understood by all within the organization.

5.6.4 Manufacturing operations and/or strategic sourcing

Those responsible for implementation once ready for production will want to begin thinking about from whom and where they might need resources for implementation. In addition they can begin preparation for estimating any outside resources or capital requirements which might be needed for implementation. Often the vendor selection process can be time intensive; hence if it can be completed during the same schedule time as the detailed design, the overall project schedule efficiency can be enhanced. In addition, if potential vendors can be identified early they can perhaps participate in the design process, offering suggestions or advice as needed by the design team. At the very least they can be familiar with the parts and the project, so that when time comes for estimates they can be produced quickly and accurately.

References

[1] G.L. Beall, Rotational Molding, Design, Materials, Tooling, and Processing, Hanser, Munich, 1998, p. 62 (Section 3.1).
[2] Available from the QFD Institute: http://www.qfdi.org/what_is_qfd/what_is_qfd.htm.
[3] Michael Sepe in UL/IDES Prospector database—as copied from the 3rd edition of Pocket Performance Specs for Thermoplastics published in 2006 by Injection Molding Magazine.
[4] The quote "*it's just business*" comes from the Francis Ford-Coppola movie *The Godfather*, based on the novels of Mario Puzo.

Resources for further study

S. Levy, J.H. DuBois, Plastics Product Design Engineering Handbook, second ed., Chapman & Hall, New York, 1984.
S.E.M. Selke, J.D. Culter, R.J. Hernandez, Plastics Packaging—Properties, Processing, Applications, and Regulations, second ed., Carl Hanser Verlag, Munich, 2004.
J. Rotheiser, Joining of Plastics—Handbook for Designers and Engineers, third ed., Carl Hanser Verlag, Munich, 2009.

G.L. Beall, Rotational Molding—Design, Materials, Tooling, and Processing, first ed., Carl Hanser Verlag, Munich, 1998.

E.R. Larson, Thermoplastic Material Selection—A Practical Guide, first ed., Elsevier, Oxford, 2015.

R.A. Malloy, Plastic Part Design for Injection Molding: An Introduction, second ed., Carl Hanser Verlag, Munich, 2010.

P.A. Tres, Designing Plastic Parts for Assembly, seventh ed., Carl Hanser Verlag, Munich, 2014.

G. Boothroyd, P. Dewhurst, Product Design for Assembly, Boothroyd Dewhurst Inc., Wakefield, RI, 1989.

Web links of note

Great Design information from DuPont can be found here:

http://www.dupont.com/content/dam/dupont/products-and-services/plastics-polymers-and-resins/thermoplastics/documents/General%20Design%20Principles/General%20Design%20Principles%20Mod%201.pdf.

Sabic provides many tools for designers here (you may need to register):

http://www.sabic-ip.com/webtools/toolsHomePage.faces.

The UL Prospector site is a valuable tool for any designer reviewing materials for any application (you will need to register):

https://plastics.ulprospector.com/.

Solvay provides a great design guide for using their Udel polysulfone materials:

http://www.solvay.com/en/binaries/Udel-PSU-Design-Guide_EN-227550.pdf.

Bayer has a nice design guide for RIM systems:

http://reactioninjectionmolding.com/wp-content/uploads/2013/09/RIM-PartMoldDesign-Guide.pdf.

Covestro offers several design related guides, including part and mold design, snap fit design, and joining techniques for plastics:

http://www.plastics.covestro.com/Library/Technology-brochures.aspx.

Preliminary detailed design

6

Now that the concept has been presented, discussed, reviewed, and the green light has been illuminated for project go status, it is the time to get into the details. Yes, there are many details, and, yes, the details make the difference between acceptable parts and optimum parts. Get the details correct and the parts will function and be manufactured with ease. Omit a detail or two and the parts will provide pain for everyone for a long time. The detailed design is really just the natural progression from the conceptual design; and for some, several of the steps outlined in this preliminary detailed design chapter actually are occurring within their thinking processes simultaneously during the conceptual stage of the project. It is sometimes hard to separate the two, once you are adept at thinking through the design, the steps sort of meld together and your brain tends to evaluate each concept from each perspective in a concurrent fashion.

There are rules that we all have been taught or have learned in practice from our mentors and these rules are important to understand and consider as we move through the design, but they are not the last word. They are guidelines that we can and often have to bend, break, or dodge. Breaking these rules is not the end of the world but breaking them without thorough forethought is indeed risky. Hence the method to employ is to *think first*, see the part, see the process, and see the materials flowing or forming into the final shape; all in your mind's eye.

6.1 Seeing the part(s), the assembly, and the process

This is similar to Tiger Woods seeing the putt traverse the green breaking one way then the other speeding up and slowing down just before rolling over the edge and into the cup all before the putter ever strikes the ball. It is no different in the arts; the artist sees the final product before the first chip of stone or the first brush stroke. Designing a plastic part or a product is no different; given the concept as a base, visualization of the entire part, including the process and tooling, aids the competent designer in preparing for starting the detailed design effort. But don't stop there, see the part or product in use, visualize it when it has finished its useful life, visualize the end-user dealing with it in 3, 5, or 10 years. These kinds of mental exercises are used to help and think through the lifespan of the part or product, and will often have a positive effect on design decisions during development.

6.1.1 *Holistic visualization or* think first

Just like the golf analogy, holistic visualization is useful to the designer in many ways. The ability to see the part, to see the part in the mold or tool, to see the material formed

Designing Successful Products with Plastics. http://dx.doi.org/10.1016/B978-0-323-44501-6.00006-4

or flowing into the mold cavity, to see the parts being ejected or removed from the tooling, to see the parts being assembled, packaged, and used is a true gift which many design practitioners do not and will never have. That said, going through the motions, thinking through the process, the tooling, the secondary operations, and so on, will benefit every project whether the design practitioner is gifted or not. This process helps the designer to identify early all sorts of considerations which may be unique to the particular part, the particular application or end-use, the process, or the materials used. While there are any number of methods that can be used to do this, the following steps seem logical from the design chronology point of view.

6.1.1.1 Think and see (in your mind) the structure, form, and shape of the part

Before you can visualize the mold filling with molten polymer and the gas being driven from the cavity, you first must have a solid idea of what the part geometry and structure might look like. What is the form of the part? What are the load bearing points or in-use flexural requirements? What kind of environmental conditions will exist in use? And so on—every part is different and every project or end-use application presents new and unique challenges. As with conceptual design, thinking through and with pencil sketches, or other graphical tools, can sometimes be helpful in achieving a more complete and full visualization of the part structure. Having a clear vision or picture of the structure and geometry in mind first is paramount to the subsequent visualization of everything else required as the designer works and moves through the *think first* process (Fig. 6.1).

6.1.1.2 Simultaneously see the part filling or forming in slow motion

Given a good visual image of the part structure, think about the material filling the cavity from one point out to all the others. Think about the extruded parison in the extrusion blow molding process being pinched-off and then the resulting material being expanded under pressure in all directions simultaneously. Think about the semimolten sheet in the thermoforming equipment draped downward nearly touching the mold then suddenly dragged over the mold and pushed against the mold with atmospheric pressure alone or with some additional pressure assist. Think about the resin rolling around as it becomes molten within the rotational mold and then cooling as it is flowing by gravity and momentum and the tooling is rotated in one direction then the other as the cooling flows over the mold. Think about the material being heated, stretched, and then expanded to form the mold surfaces, the molecular chains being oriented by both the stretching and then the expansion. No matter the process, if you understand the process and can visualize the part structure, form or shape, you can think and visualize the process happening in slow motion. This helps to see where thicker walls may be helpful, where an oval parison might be preferable to a round one, where the gate might be better located in one place than another. All of these "insights" come from this holistic visualization, or the *think first* method.

Figure 6.1 ***Think First*****—visualize and see from each different perspective, before making the first entry or keystroke.** Let your mind "see" before you begin and then carefully follow the insight. Keep the many considerations, as well as the "user" and the "need" in mind simultaneously while thinking through any potential solutions.
Source: Photo from iStock.com/maurusone

6.1.1.3 Simultaneously see the part being cooled and removed from the tooling

Just as the filling or forming portion of the process is visualized, one should continue and visualize the part cooling on the form, or in the mold, and then being removed from the form or mold. Remembering that all plastics processing methods are thermal events and the parts will always cool and change size while doing so. Some parts are naturals, that is their geometry or structure is ideal for the molding and ejecting process; most are not. Some are easy to fill and terrible to eject; others are easy to demold but near impossible to form or fill. Our job as designers is to create a part design that satisfies the end-use functional requirements and that can be filled or formed, and ejected with ease. Sounds easy to me?

6.1.1.4 Simultaneously see the part being assembled in the next stage of the assembly process

Keep expanding the "*think first*" thinking and see the part going through secondary operations or assembly operations. Are there details that might interfere or could details be added to aid in registration or handling? Think through the operator picking the part from the bin placing it on the assembly nest and adding the electronics or the next physical part with fasteners or other assembly means. Visualize the operator removing the assembly, inspecting it, testing it, and moving it to the finished goods staging area. What can be added to the base part(s) design that might assist? What can be removed that might be an inhibiting feature or cause some fatigue or failure? What can be added to assist in orientation of the part with ease or by touch? Visualize and see all of this in your mind, *think first*, then make notes and sketches to remind yourself of the small details that you might want to discard or to include.

6.1.1.5 Simultaneously see the part in use by an end-user performing the functions required to fulfill "the need"

Think through the end-user interacting with the part or assembly, how do they sense it, grasp it, interface with it, rely on it, stow it, play with it, misuse it, abuse it, and so on. Get a feel for the end-users emotions while using the part(s) or assembly, try and evaluate the satisfaction, which might be present once "the need" has been resolved by the end-user. Does the part or assembly convey the perceptions and functional realities that are intended by the design concept? *Think first*, and then use the insight gained to ensure that end-user satisfaction with the design solution is conveyed by the final detailed design.

6.1.1.6 Simultaneously see the part at the end-of-life being recycled or discarded

Finally, think about the end-of-life for the part. What happens when it no longer serves any purpose? How is it disposed of? Does the user intuitively know that the part can be recycled? Can the user dismantle the assembly into recyclable bits and pieces? Is there some natural secondary reuse which can be conveyed to the end-user? Sometimes simple design decisions early on can make the difference for an end-user when this part or assembly end-of-life time arrives. Think about how you as a designer can implement features or intuitive instructions that will assist the end-user in proper disposal and recycling of the part or product when the time comes. Alternatively, think about how the design can be made to last over several lifetimes, being used by more than one end-user. Perhaps a small additional manufacturing cost can be the difference between a product which is discarded without thought and one which is so nice and functions so well the end-user will think twice before discarding it. True sustainability!

6.2 Initial computer-aided design (CAD) layout in 3D

The three-dimensional (3D) computer-aided design (CAD) layout is the design foundation for every part and assembly created by the designer. When I first began designing with plastics materials there was no CAD, only large flat drawing boards where we worked with pencil and paper and produced layouts to scale with many layers of velum or Mylar, each with different design variations and all tied back to a datum set common to all parts in the layout. What used to take us a week of work on the board can now be accomplished in hours in CAD and the data can be copied, mirrored, made into a linear or circular array, and the list of tools continues to grow. The CAD layout is the design starting point, the blank canvas on which the designer will create the part(s) and product assemblies for which there has been so much conceptual work and visualization efforts invested.

This CAD layout can and will be produced from many perspectives and for most projects from many of these perspectives simultaneously. These perspectives are not totally independent, but they are not really dependent on each other either. Each different part or product will demand a certain perspective or combination of perspectives and the designer needs to be flexible enough to understand the requirements of the design in question, and to apply the needed perspective(s) as the CAD layout begins and moves through the design layout journey. The following sections attempt to outline many of these design perspectives which are used during the creation of the 3D CAD design layout (Fig. 6.2).

Figure 6.2 Mindful consideration before and during the creative layout will provide the insight required for optimum part and product design selections.
Source: Illustration courtesy of Caroline MacLean-Blevins.

6.2.1 Design from an end-use application or part function perspective

The design layout begins with any functional information inserted, including any mating parts or bearing areas identified, any assembly points or load points identified, and the major axis of rotation or axis of mating parts locations identified. The functional perspective places a high value on getting the structure to fit within the framework of the functional system in which the part will be used. Several key considerations will be required for the functional perspective, including physical structure, property changes over time, and anticipated misuse or other possible failure modes.

6.2.1.1 Adequate physical structure for anticipated loads in use

Given the placement of the loads and reaction points, the layout will strive to facilitate adequate structure to support the loads and reaction points anticipated. Structure, as well as polymer chain orientation might be reviewed and considered during this physical structure design effort, looking at the flow directions from the anticipated gate and looking at the design of the structure to maximize these effects. In addition, the presence of knit lines around holes or openings in the structure can be analyzed as the design progresses, ensuring that any potential knit areas are explicitly not in areas where loads are anticipated on the part or assembly in question. Features, such as ribs and bosses can affect part warpage during cooling; and, warpage may in turn either enhance or detract from the structural capability of the part. Always consider the final processed part condition, as well as the end-use environmental conditions when designing from a functional perspective. End-use environmental conditions can be game changers, especially in structurally challenged design applications.

6.2.1.2 Additional structure to allow for changes in properties during the part(s) lifespan

Physical properties of parts manufactured from plastics materials tend to be an ever changing set of values. As such, it is wise to anticipate the degradation of properties that might be expected from a particular end-use application or environment. Often we attempt to compensate by overdesigning the part or product, but sometimes additional features can be incorporated to accomplish the same function without additional wall thickness or excessive part weight. For example, in a heavy UV exposure application the designer chose a polymer that was rated excellent for UV exposure and then went on to specify a painted finish, to further block UV penetration. The investment of the small incremental additional manufacturing cost for the paint application per unit, allowed the part to function well beyond its expected 3 year lifespan and the product line was gifted with a reputation for durability; all because the designer went the extra mile to ensure the part was fit for the application. While the example was a secondary operation, sometimes additional structural elements can be employed to facilitate similar life sustaining use for a part or product.

Figure 6.3 Luggage roller assemblies in manufacture.
Source: Photo from iStock.com/yoh4nn.

6.2.1.3 Geometry to support anticipated misuse, or to combat possible failure modes

Sometimes the part or assembly in question might be subject to other loading and mechanical or chemical stresses than the application might suggest. For example, look at the simple roller wheel on any piece of carry-on luggage. All the wheel has to do is roll along—simple and no real extreme loads or conditions, right? Not so fast, that simple little roller wheel sees more impact loads (thank you baggage handlers and airport automation), more chemicals, high temperatures, low temperatures, dirt, debris, biologic materials too disgusting to mention, and so on. The simple roller wheel operates in a battleground of street grime, filthy rain drainage, salted snow covered sidewalks, gas and oil covered roadways, not to mention airport bathrooms—yuk. No material or design can withstand everything which could possibly get thrown at it; however, our job is to anticipate the majority of these "risks" and to provide a design which can survive, if not thrive (Fig. 6.3).

6.2.2 Design from a part or product safety perspective

There is nothing more frightening than to think that some product or part that I have invested my time, talents, and energy into creating would harm some child, some pet, or anyone for that matter. Always think about the possibility that a child could get hold of the part or product. If they did, what could happen? How can my design prevent harm if it were to happen? This can be a materials selection issue, a design feature issue, an overall part size issue, a processing control issue, and the list can go on. Mindfulness by the designer during the creation of the part or product that a child might be touching the part or assembly should come without even thinking about it; however, we hear of tragedy every day and wonder why the designer was not more rigorous in their design efforts.

6.2.3 Design from a part assembly perspective

For the design of a part that is one of many, to be assembled into some mechanism or functioning apparatus, this perspective turns to features that can assist downstream operations. Sometimes we add pins or holes, ribs or bosses, which might interface with other parts or with fixtures to allow positioning or assembly without requiring additional fasteners or additional operator time. Sometimes we add or design in features to assist the operator in orienting the part during the assembly; these orientation features can be tactile so the operator can keep their eyes on the main assembly, reach, grab, and orient the next part rapidly and accurately. This is similar to the design for assembly (DFA) methods but is unique to the particular part in question.

6.2.3.1 Design for assembly methods

Building upon the aforementioned part assembly perspective, the DFA method looks at the whole assembly, not just the individual parts. This method teaches us to look for ways to actually combine functional features on one part rather than multiple parts. Sometimes this can be done, other times not so much, but it is always worth looking at a multiple part assembly from this perspective. In addition, especially in multiple part assemblies, it might be beneficial to look at combining features from the functional perspective, and also from the aesthetic perspective. Sometimes these are mutually exclusive, but other times looking at the assembly from multiple perspectives may prove valuable and provide alternate feature combinations that might not have been see from a solely DFA perspective.

6.2.3.2 Can we add features to reduce assembly time or other physical parts?

One of the main goals of DFA is to reduce overall resource input into the final assembly. This can be accomplished by reducing the number of parts, which in turn reduces the number of handling operations required during assembly. Another way to reduce overall resource input is to add features to the parts to aid in fixturing or to eliminate the need for fixturing during assembly altogether. Part orientation, or the need for orientation, during assembly can also contribute to the resource load for a given assembly or product [1].

As designers, we are tasked to create parts and assemblies to facilitate these DFA goals. Most designers will strive to eliminate as many parts as possible—sometimes producing overly complex parts, which may reduce some assembly operations, but instead produce a part design which is too complex to mold well. My philosophy has always been to keep the parts and the assembly as simple as possible; there is an unspoken and inherent beauty in simplicity.

Case Example

On one tabletop receiver and display unit for a remote thermometer product design, the electronics to be contained within the enclosure included three distinct parts; (1) batteries which needed to be accessed regularly, (2) a control circuit board with a board mounted LCD display connector and three board mounted snap action switches, and (3) a separate LCD display module which was connected via flat cable back to the main control board. The enclosure aesthetics were defined and were to allow access to the batteries via a sliding door with a snap closure on the back side of the unit.

We looked at several ways to arrange the housing parts to contain the electronics and simultaneously provide a means to mount the display module firmly then have access to connect the flat cable back to the main control board. Every arrangement was poor as there was always one item in the way of one other as the assembly went together. Another compounding factor was that once assembled, the radio receiver needed to be fine-tuned as part of the assembly process—the metals found in the battery parts and display could alter the radio signal, hence the tuning needed to be done with them in position. The unit was connected to an instrument and placed into a shielded enclosure and tested at various radio thresholds then the assembler would physically turn a component on the circuit board to fine tune the system. Initial design concepts would require one or more parts to be removed and reassembled during each tuning cycle.

The assembly was reviewed from the aesthetic perspective first; we then attempted to create a solution from a DFA perspective and could not find a suitable arrangement. When we began looking from the functional perspective, challenging the aesthetic criteria, and eliminating parts consolidation we could visualize a potential solution. Now, with the new information gained from looking at the assembly from a different perspective, we offered a new concept; one that actually had additional (yes, more) custom parts.

This may seem directly in opposition of the goals of DFA, however, the new concept was a success. Here is why; the new added part was an internal frame that all of the electronics were mounted into and that the external case halves were mounted onto—see Fig. 6.4 which shows the assembly with the internal frame bead. The batteries were still mounted from the exterior into a battery compartment with electrical contacts mounted to the main circuit board and there was still an exterior mounted slide and snap latch battery door for easy and repeatable access to replace the batteries. We took this internal frame idea and stretched it a bit, providing an external feature on it as well—this was a trim bead which ran completely around the perimeter of the housing, providing a new surface which could be in a contrasting color and texture to the exterior case halves. The industrial designers embraced the idea and altered their style design a bit to include it. The resulting assembly allowed the circuit board to snap into the rear of the new internal frame part which was then snapped into the external housing back half, which allowed batteries to be inserted for testing and fine-tuning. The display module was mounted into a frame molded into the new internal frame part and allowed the flat cable to be easily inserted from the front side into the connector on the control board. This completed electronic assembly could now be handled during testing and fine-tuning with access to the board

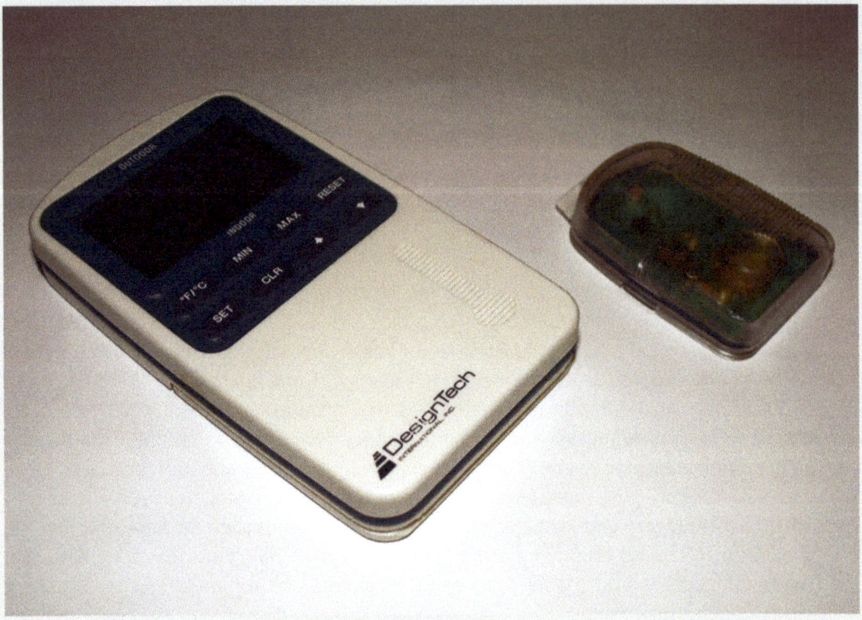

Figure 6.4 Finished manufactured RF thermometer assembly illustrating the gray line at the case separation line which is the added part that allowed the design to function for the end-use and for the in-process assembly, testing, tuning, and quality assurance steps.

mounted components used for the fine-tuning operation. The external housing front was placed into position and four plastite screws were inserted from the back side through the internal frame and into bosses on the external housing front to complete the mechanical assembly.

One of the hidden but nice features of this design was that the snaps used to hold the circuit board to the internal frame part were completely backed up and supported by the external housing back half when it was installed. The display module with a slight press fit into its frame molded into the internal frame part was then trapped completely once the external housing front was installed. This was an extremely robust assembly design made possible by looking at the design from many perspectives and challenging the criteria when it was necessary to do so.

Going against the DFA grain and actually adding the new internal frame part did add parts, but it provided a much simpler set of parts and a much simpler assembly process which allowed for all of the process steps to be completed with good visual sightlines by the assembly technicians and without any disassembly steps required for fine-tuning. Our customer went on to use this electronics housing design to package several other instruments before selling the product line to another manufacturer.

6.2.3.3 Sealing or joining features

Sealing surfaces which might be used as seats for O-rings, or face seals for die cut gaskets, or perhaps even for molded part to molded part seals, need to be identified early and become primary considerations as the part concept is being visualized by the designer. These areas can be complicated easily by mold or tool considerations; hence the designer must anticipate the tool design to be utilized for the feature prior to the first line being drawn for the part in the CAD program. Parting lines, ejector pin marks, vent pin marks, and the like, all resulting from tooling features will absolutely destroy a sealing surface. Design features adjacent to the sealing surface or connected to the wall thickness used to support the sealing surface can also be detrimental, especially intersecting walls in high shrink materials for injection-molded parts. In addition, the designer must be aware of the sealing means to be utilized to achieve the seal. Some sealing means can produce high concentrated loading and this loading will be a continuous static load once the seal is made. No polymers (at least no thermoplastics) do well in high static load applications, the system will be motivated by the high stresses applied by the sealing compressive loads and polymer creep will result. Some polymers will fare better than others, but in general the designer must comprehend the loading and lifespan required for the seal and provide a system of molded part sealing surfaces and sealing methods to accommodate the particular application under development.

Joining features, such as welding beads, adhesive gluing joints, snap beads and recesses, and the like, all must be considered early in the design process as well. Just as with the sealing surfaces mentioned earlier, the potential for tooling design to impact and prevent good joining is very real. Witness lines from parting lines, split lines from cam actions, ejector pin marks, and so on, can make an adhesive joint or ultrasonic weld bead absolutely worthless; hence these need to be considered and visualized before the design is even started in the CAD program. Take for instance the example of a low-pressure fitting used to connect an oxygen-enriched air source to a medical instrument with a side port for flow presence sensing, as shown in the Fig. 6.5A–D. The original designs are shown separated in Fig. 6.5A(a–c) and assembled in Fig. 6.5B(a–c), illustrating the design intent. However, there are several ribs which intersect with sealing surfaces. The original design intent was supported by a robust set of part designs to exhibit strength and durability for the intended application. The fact that ribs intersected with sealing surfaces did not enter into the design thinking whatsoever. The design review session showed these as potential failure points and the design was tweaked just a small amount to remove these potential failure points. The revised design is shown in Fig. 6.5C(a–c) as separate parts and shown assembled in Fig. 6.5D(a–c). This example is way down in the weeds but these are the kinds of "little" details that can make or break a product in the marketplace. Don't underestimate subtle features and be extremely careful of these especially when using semicrystalline materials which tend to shrink more than amorphous materials (Fig. 6.5A–D).

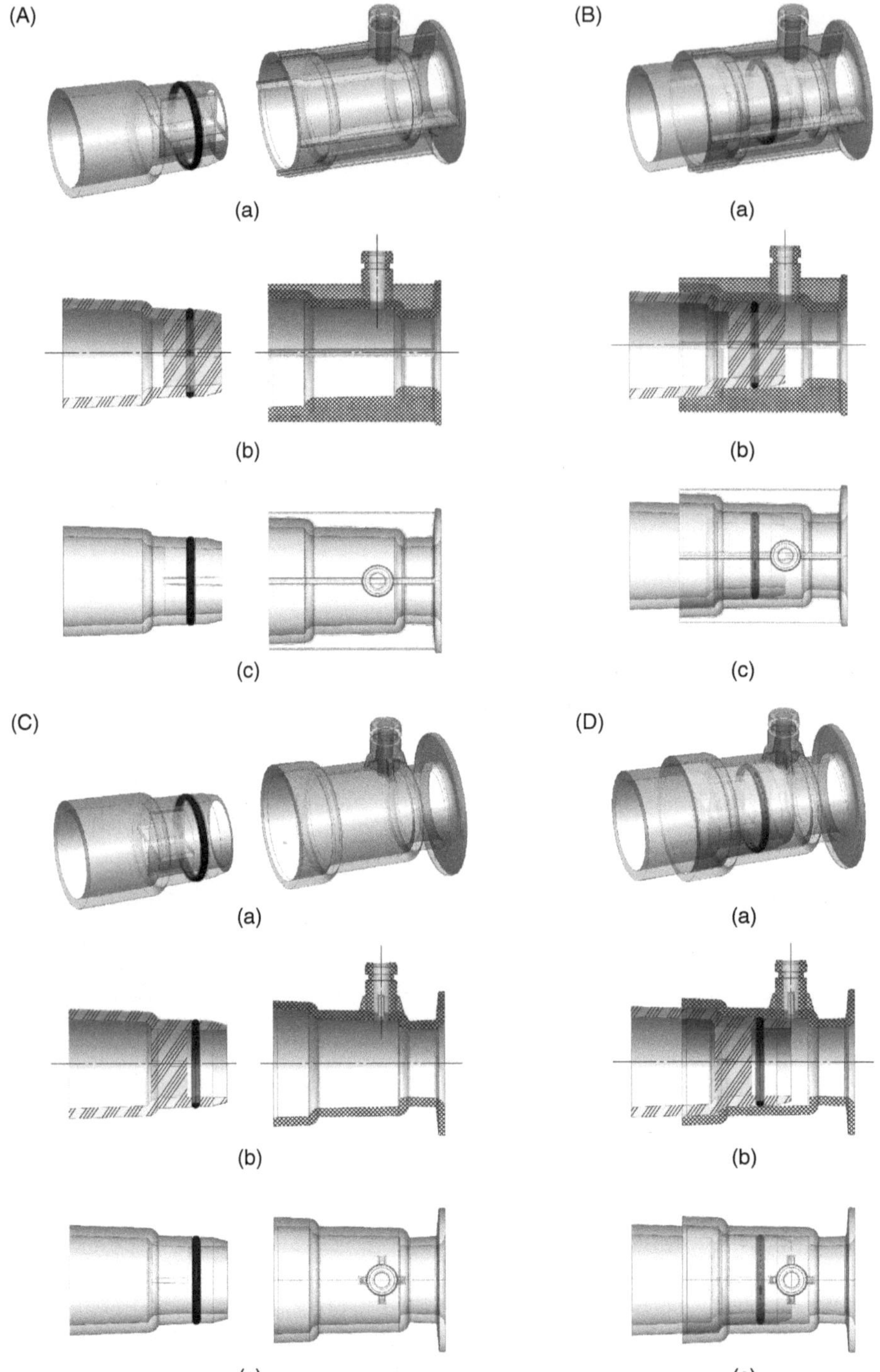

Figure 6.5 (A–B) The original design of a low pressure gas connection showing the robust part design with ribs located intersecting with walls in sealing areas. (C–D) The revised design of the gas connection showing the rib structures revised to avoid intersecting with the sealing areas on the parts. Joining features can make or break a product, so be sure to review and ensure they can be molded well and function as desired.

6.2.3.4 *Snap features*

Assembly of plastics parts utilizing snap features is one of the many benefits we have when we are designing in plastics. These features can reduce fasteners and simplify assembly; resulting in lower cost, lighter weight, and sometimes fully recyclable final assemblies. However, snap features will often demand interesting tooling designs to form the snap or undercut required for the snap. A full comprehension of what kind of snap the designer is intending to use and the tooling requirements for that particular snap prior to the beginning of the CAD design layout will assist the designer in creating the structure in a manner that will allow the tooling to function as needed to provide the feature as designed. As an independent design consultant, I am often called on by molding manufacturers to help them with a customer who comes to them with part(s) from some industrial design (ID) firm that are absolutely beautiful but cannot be molded. Snaps are usually one of the main features that are designed into the parts last by these designers to help in holding everything together. Design the exterior, figure out just how the exterior can be split to get it together, and then throw some snaps inside to hold it together. Fortunately for me, this results in billable work and a chance to interface directly with the customer and often with the industrial designers. These opportunities allow me to forge new relationships and help both the client and their ID firm learn more about producing parts from plastics (Fig. 6.6).

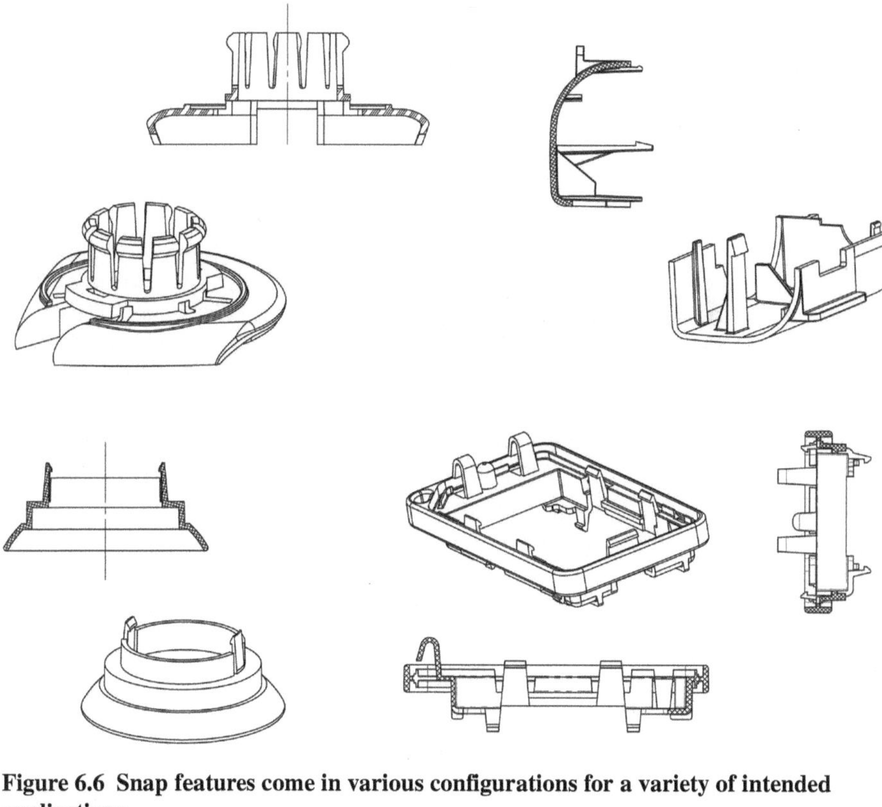

Figure 6.6 Snap features come in various configurations for a variety of intended applications.

6.2.4 Design from an aesthetic perspective

The design of any part or assembly to be used, touched, or interfaced with in any way by humans should be a cause for consideration of the part or assembly aesthetics by the designer. Sometimes we are given the product design by the ID firm or client, other times we are asked to provide the human factors considerations. However, these considerations need to be part of the early design thinking when starting the CAD layout. These considerations impact part design, tooling design, as well as material selection, so early consideration will help reduce the risk of redesign and schedule impact later in the project.

6.2.4.1 Master external style CAD surface file

When dealing with an aesthetic external part or assembly, the external design and style are often defined by industrial designers or artists concerned with the appearance or aesthetics of the finished product. One excellent recommendation from an experienced professional designer and friend, Michael Paloian of Integrated Design Systems, Inc., was to create a master external surface file and to archive it as a separate file to use for checking and reference as the detailed part design is developed. In many cases a product will be made up of many parts, each of which may share some responsibility for the overall external style. As each part design is completed and the parts are set back into a working layout the master external style file can be overlaid to check that all parts conform to the original style intent. A great method for ensuring the exterior aesthetic design intent is preserved throughout the detailed part development and design efforts.

6.2.4.2 Surface finish and color considerations

If the part or assembly has visual requirements, the colors or surface finishes will become a primary requirement and production characteristic. Providing adequate structural design to allow uniform filling and adequate draft for the surface finishes specified will be primary consideration early in the CAD layout process for these kinds of parts. The ability of the design to hide shrink marks must also be high on the designers list as the structure is conceived, keeping intersecting walls to a minimum and paying attention to the rules concerning wall thickness and rib thickness, as well as providing adequate corner radii are a must. Watching the flow paths and looking out for knit line locations will also help the designer to avoid creating a part where there are surface color blemishes that cannot be processed away.

6.2.4.3 Surface touch or feel considerations

Many handheld devices, for example, toothbrushes, shavers, hair brushes, and so on, require a good grip on the device during use. Many of these kinds of devices are using soft-touch or resilient materials layered upon a structurally sound base material, sometimes in an overmolding or two-shot process. This is an amazing process and it can be exploited in many ways for many different functional applications; however, the tooling considerations, and tooling costing, can be an impediment. A thorough understanding of the overmolding process and the tooling required is paramount prior to beginning the CAD design layout of a part for this process. Seal-off locations for the second shot must be considered in the design of the first shot, material compatibility of the two materials with each other and, with the functional application requirements, thermal stability of the first shot during processing of the second shot must be considered, and the list goes on. This is a wonderful process that can provide an extremely simple finished part but these considerations must be visualized and thoroughly vetted early in the CAD design layout process. In addition, consultation with a processor competent in the technology is recommended early in the design process.

6.2.4.4 Weight distribution considerations

Again with any product handled by humans, the designer will be responsible for providing grip or grab features and adequate surface finishes or structural feature for these said grip features. In addition, center of gravity and weight distribution must be considered. In the earlier case example, we discussed the exterior opening for the replaceable batteries in the remote temperature receiver unit. Since this unit was to be wall mounted or to sit upon a desk or other horizontal surface, we were careful to locate the batteries low on the case and to keep the batteries centered on the case centerline, as the batteries represented 75% of the total weight of the unit. Any new part or product will have some consideration for handling, even if it is only during assembly, hence, consideration of how it will be handled and of the weight distribution can be important. This is especially true when designing larger parts, so be vigilant and think through the part handling and weight distribution early in the design stages. In one design project the part under development was intentionally made heavier on one end as a means to aid the assembly operator during production. As each new part was picked and lifted from the bin the operator could feel the weighted end, enabling the operator to orient the part as they were moving it from the bin to the assembly nest, increasing the time efficiency for the overall operation.

6.2.4.5 Auditory considerations

Often overlooked, a finished part can have distinct auditory characteristics. Sometimes these are desired and sometimes they are not, hence a consideration of the possible sounds that might be created in the application or by the chosen materials is recommended. For example, the sound of a snap action used to secure a product access door or the detent position of a selection dial may be desirable for feedback to the user. Alternatively some large surfaces may vibrate and create unwanted sounds, depending upon use location, as well as proximity to the source of the vibration. Some materials are good at providing these auditory feedback sounds, others are not. Some designed features are better than other in creating the impact to impart the audible sound, others are not. Thinking through just how the sound is to be created before the CAD layout begins will save time and limit risk of redesign or poor auditory performance.

6.2.5 Design from a labeling or packaging perspective

Sometimes a part or product is less affected by downstream assembly than it is by filling and labeling operations. In these kinds of parts or assemblies, one has to consider the nature of the downstream operations, for example, is it cold filled or hot filled? Is it sleeve labeled, heat shrink labeled, pressure sensitive roll-on labeled, and so on, or is it to be in-mold labeled? If the label is roll-on or pressure applied, does the part have the required structure to support the labeling forces? These are all considerations which need to be in mind when beginning the CAD layout of any part with these labeling or filling requirements.

6.2.5.1 Is structure required to facilitate the labeling process?

This is primarily a consideration with blow molded bottles or jars used for containing various materials or fluids for packaging. The forces applied to the part during the labeling process can be formidable and the structure or geometry of the part must be able to withstand this operation. Also, in some instances, the label will need to be oriented to a certain position on the part and as such, some feature or features must be designed into the part for orientation on the filling or labeling line.

6.2.5.2 Is structure required to facilitate the packaging method to be used?

As with filling and labeling, sometimes the part will bear some of the loads seen in packaging. Sometimes these are manageable and sometimes these are fairly formidable. Take the case of a bottle of some cleaning agent, packed 12 or 20 to a carton and the cartons are arranged 6 to a layer and 8 layers high to a pallet. The lowest cartons will see the weight of the seven cartons above and will see the shock or momentum load each time the truck hits a bump or pothole of all seven layers. The carton will certainly absorb some of this load but the individual parts in each carton will also see this load. Can they support it? Will they crush? Does the part geometry have the necessary features to assist? These are all considerations for the designer to ponder as they begin the CAD layout for a part used in these kinds of applications.

6.2.6 Design from a tooling perspective

From a process vantage point, injection molding and blow molding will require more consideration of tooling early on. This is not to say that other forming methods require less, rather the tooling complexities for the higher-pressure processes tend to drive first costs higher and, as such the investment payback requires larger volume production to be economically viable.

6.2.6.1 The parting line is the starting line

After visualization of the part and the complete process; the designer should have a fairly good idea in mind of just exactly how this part will be positioned in the tooling. Personally, I most always use the parting line as my starting line and for one of the main datum planes for the part. This gives the part a common plane that will be utilized by not only me during design but by the tool designer and toolmaker as well. Using the parting line as my starting line has helped simplify both the design and the tooling on virtually every design I have completed.

If the parting line is curved, stepped, or shaped in any manner other than a flat plane, then by all means find the most likely parting line for the tool and define the shaped surface that is the parting line on the part from that parting line plane. In reality, the toolmaker will need to have that mold separation plane and the toolmaker will set-up and work from that same plane, so it is an ideal reference for the design to be started from as well. Another suggestion for shaped parting lines is to find some room somewhere within the part or on the part to add three bosses, or surfaces, that are planar and the plane they define should be parallel to this mold opening plane on the tool. Having these three surfaces or bosses allows us to construct a fixture that the part can sit upon for reference measurements. Quite often we struggle to get good measurements on a part that we cannot hold well (lots of plastic parts have shaped parting lines and most have some curved surfaces or features which do not allow us to simply clamp them down to a granite surface for accurate measurements). Providing "on-part" locating features does not take much time and does not cost much to add to the part design or tool, but will assist during debug and for the remainder of the part life cycle.

6.2.6.2 Gate and flow considerations

The flow of material into the mold can be tricky to visualize, but there are a few simple things to think about during the preliminary layout of the part. The desire is to have uniform wall thickness throughout the main structure; however, there will be thicker and thinner walls, ribs, bosses, and so on, in the part. In general, the material wants to flow from thick to thin, that is, to start near the gate in a thick wall and flow throughout the part filling the thin areas last. There are several reasons for this and they are easy to comprehend if you can visualize the molten material entering the cavity and beginning to form a skin at the cooler mold walls and the flow front continues to push into the cavity losing flow area as the skinning process continues. The flow velocity is a function of flow area and if the flow area changes, the velocity (and pressure) will change as well. For this reason we do not want to try and flow from a thin area and try to fill a thick area—the pressure change will slow the flow velocity and the part may well freeze off; otherwise the process may well need to be longer using warmer mold temperatures to prevent the freeze off and then needing extended hold times before ejection can occur. Also, if the structure of the part has openings through the part that the flow needs to wrap around, there will be areas where knit lines will form and the designer needs to anticipate these to ensure they are in areas which are safe.

6.2.6.2.1 Gate location

The location of the gate can be dictated by the part structure or by the tool construction; however, in any design the location of the gate and other flow considerations mentioned earlier will be primary drivers of part performance. Choose a bad gate location and bad things can happen. Plan your design around a good gate location, yielding a good flow during filling, and the part can be processed in a wide process window and perform well over the entire process window.

6.2.6.2.2 Type of gate

Considerations for aesthetics or part structural performance will sometimes drive the gate style to be used for a given part. Other times the gate is chosen for simplicity or efficiency in the molding process. If the gate will be visible or in an area where the user will be touching the part the designer may wish to designate the gate style early on and drive the tool design by specifying both gate style and locations. As shown in Fig. 6.7A–B the tunnel gate will leave a mark on the exterior of the part upon ejection,

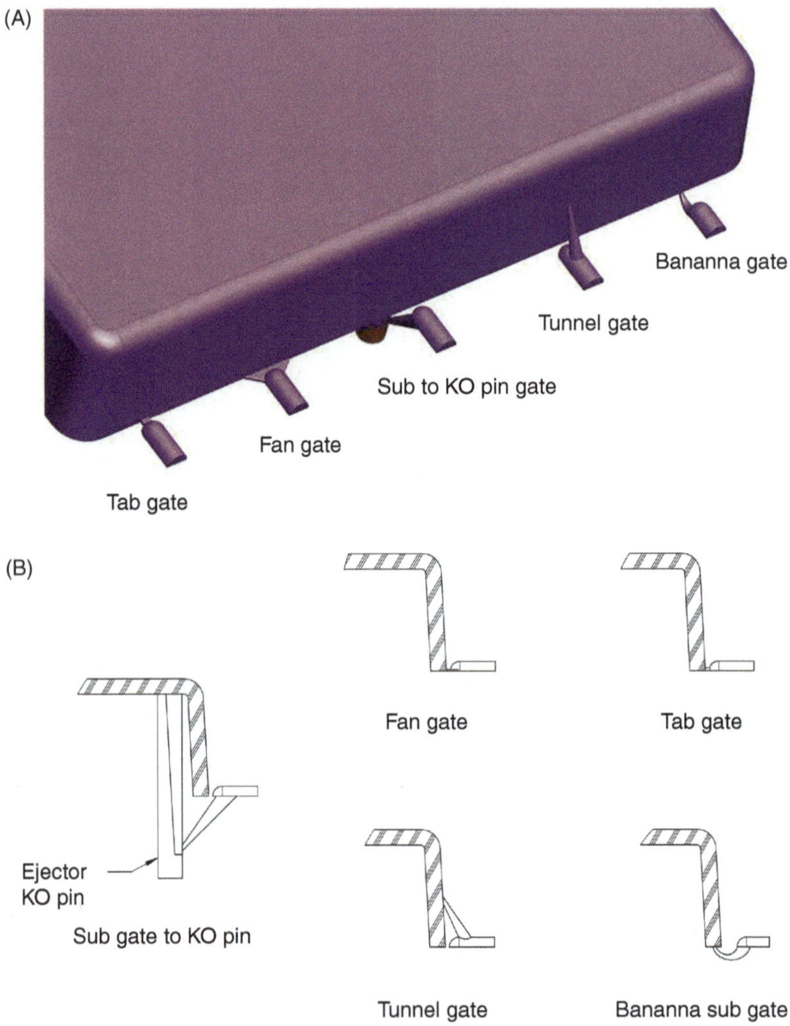

Figure 6.7 Various gate configurations can be utilized depending upon the criteria for the part in question. (A) Shows various gate styles used on a part. (B) Illustrates each gate in a cross-section view—note the relation of the gate to the parting line of the part. Be sure to understand the tooling implications of the part design and how part design can drive tooling costs either up or down.

as will the tab and fan gate. If aesthetics are important, a sub to knock-out (KO) pin gate or a banana gate can be used. Part orientation can be inverted for complex parts and a reverse-side ejection system can be utilized; however, these increase mold complexity and cost.

6.2.6.2.3 Venting considerations

Gate location is indeed one of the most important decisions for any new part; so too is the ability to vent the cavity during the filling portion of the cycle. Remember the cavity is filled with the molten polymer in a second or in fractions of a second and the molten polymer acts as a noncompressible fluid displacing the compressible gas within the cavity. If the gas cannot escape quickly enough then it does indeed compress and cavity burning can occur, not to mention possible part defects and processing challenges. Think through this need for venting during the design and be prepared to discuss any troubled areas with the tooling designers when implementation begins.

6.2.6.3 Cavitation considerations

Multiple cavity molds can be extremely cost effective and can provide enormous capacity for the manufacturer, however, they come with some degree of risk. The tool design must be able to provide uniform processing conditions at each cavity, including mold temperatures, fill times, fill distances from the runner and sprue systems, and so on. Good part design for a multicavity part will anticipate some of these issues and keep the geometries and flow paths just as simple as possible. The combination of good part design and proper tool design is required for successful implementation in multicavity situations. Early consultation with experienced multicavity tool design sources is always recommended for high cavitation part design efforts.

6.2.6.3.1 Runner systems

Generally speaking the part designer will not get too far into the weeds with runner design. However, understanding what runner system could be used will assist the designer in understanding the possibilities for gate locations or for multiple gates. It will also aid in material selection as some materials need to be fully processed in a defined time frame to avoid degradation in the melt. It is also good to understand that different materials process differently even within the same process. In injection molding, for example, different material classes will have different runner geometries and gate requirements as well. Tool and mold designers will routinely depend on the material supplier for design recommendations for runner and gate systems. Early involvement in the design and development process by both the tool designers and the material suppliers can help us (the part designers) better understand and plan for these gate and runner system requirements.

6.2.6.3.2 Three-plate mold

A three-plate mold is perhaps the most common form of multicavity mold arrangement. The runner system is located on the back side of the moving third plate with the cavity on the other side. As illustrated in Fig. 6.8, when the mold opens the ejector half

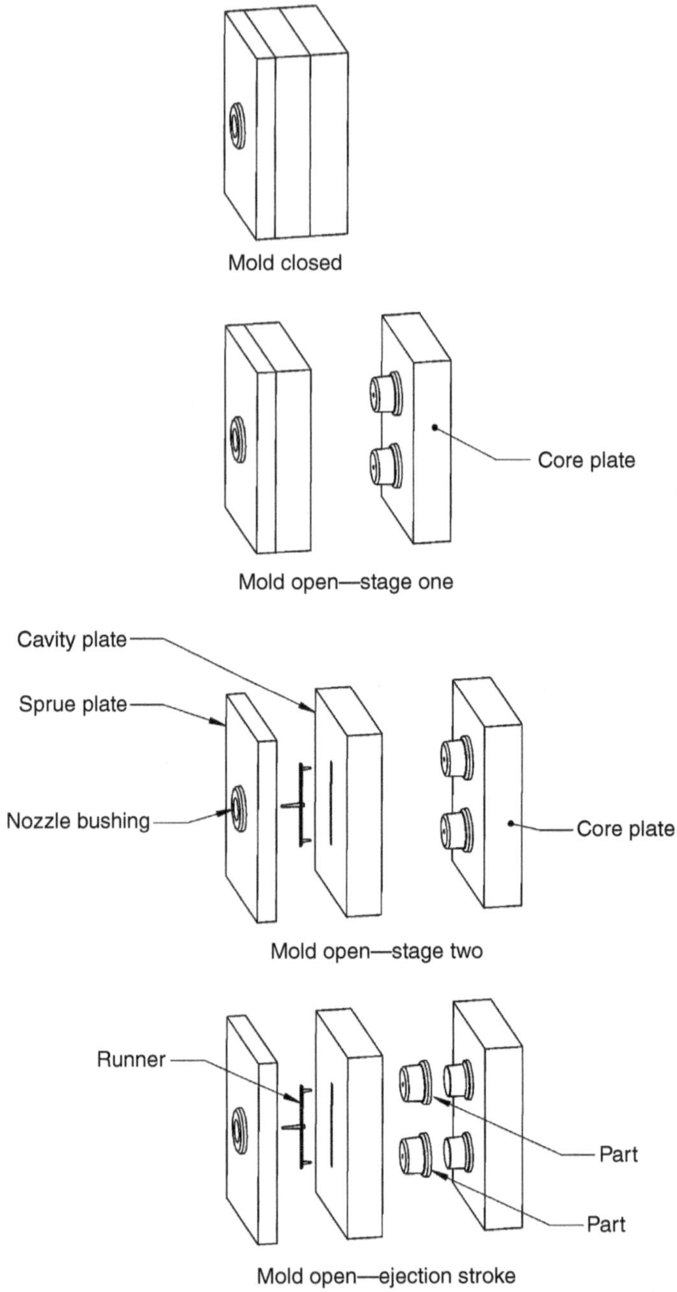

Figure 6.8 The operation of a three-plate mold is illustrated—showing the operation sequence and the runner system on the opposite side of the cavity plate. The three-plate mold is perhaps the most widely used multicavity tooling approach for injection molding of plastic parts.

Figure 6.9 (A) A three-plate mold opened up on the bench with some parts from the mold shown in front of the mold. (B) Shows a side view, clearly illustrating the third plate which opens to allow ejection of the sprue and runner system.
Source: Courtesy of Quality Molding, Inc., Somerset, WI.

opens until some stripper bolts bottom out and then the third plate begins to move with the ejector half exposing the runner system for ejection simultaneously when the parts are ejected. With this system the gates into each cavity can come down to the parting line or they can gate directly into the part from the bottom of the cavity (depending upon part geometry of course). This is especially helpful with round parts as you can fill from the bottom of the part and flow down along the sides of the round part providing uniform flow down around the core to the parting line, reducing the possibility of knit lines and helping to provide even shrinkage and uniform residual stresses around the part (Fig. 6.9A–B).

6.2.6.3.3 Hot runner systems

As the name implies, the hot runner systems keep the material molten within the manifold and are basically a runner-less multicavity filling system. Each hot gate fills each cavity independently and there is no material used for a sprue or for any runners. These systems are expensive to put into operation but once in use can be real cost savers allowing faster cycles and reducing the handling of runners, including the grinding and recirculating of the regrind back into the feed stream. Again, the designer does not need to fully understand the style or brand of hot runner system that will be used but if a hot runner system is a potential, then the designer needs to be prepared to deal with gate locations and gate vestige issues, especially in areas of the parts where the end-user may be touching or gripping the part.

6.2.6.4 Undercuts

Features that can potentially be excellent for simple functional assembly can often be difficult to achieve because of the need for the mold to open and for all of the steel within the cavity to be able to move out of the part as the mold is opened and as the part is ejected. Undercuts, or trapped steel areas, are a fact of life for the part designer and there are many ways to achieve the feature while still allowing the mold to operate. Some are good for structural parts that are not highly visible or aesthetic parts; others are more tooling intensive, but can provide the undercut while still allowing the exterior of the part to be an appearance surface.

6.2.6.4.1 Slide-by sealing

Slide-by methods use steel from both sides of the mold to form the undercut—with one feature telescoping through the part to form the undercut. In this method, the steel from each side of the mold will need to touch to seal-off during mold filling so as not to produce a part with undo flashing evident. In these kinds of situations, it is desired to provide at least 5 degrees of taper at the steel-to-steel seal-off interfaces between the cavity side and core side of the mold (Fig. 6.10).

6.2.6.4.2 Lifters

Lifters are moving mold inserts that are set at an angle and will travel with the part as the mold opens and because they are set at an angle they tend to slide relative to the part during the mold opening motion, releasing from beneath the undercut area. Lifters are typically used when an undercut is located within a cavity and cannot be reached from the outside of the part (where a slide-by or side action slide might be used). The main consideration when planning an undercut formed by a lifter is to understand that the lifter will move some dimension into the part as the motion happens and other features in the geometry of the part cannot block or interfere with the travel required. As mechanical moving parts within the mold, lifters will be a maintenance issue for the mold during the life of the mold, that is, they will need to be lubricated and cared for over the life of the mold, increasing the overhead association with long-term use of the mold (Fig. 6.11).

6.2.6.4.3 Side-action slides

Side-action slides are, as the name suggests, moving elements within the mold where the slide within the mold can form the feature by sliding away from the part, either independently by cam action with the mold opening motion or by hydraulic or servo-motor power. Features requiring slides are typically perpendicular (or, near to perpendicular) to the parting line and either form an undercut or a wall piercing feature like a hole or an opening which could be shaped or have some depth and internal features difficult to form in any other way. Since the side-action slide will move out of the part during the part ejection cycle, draft will be required in direction of slide movement on all features and surfaces formed by the slide, except those face features which are normal to the slide movement direction. Also, the mold will need to close and seal-off

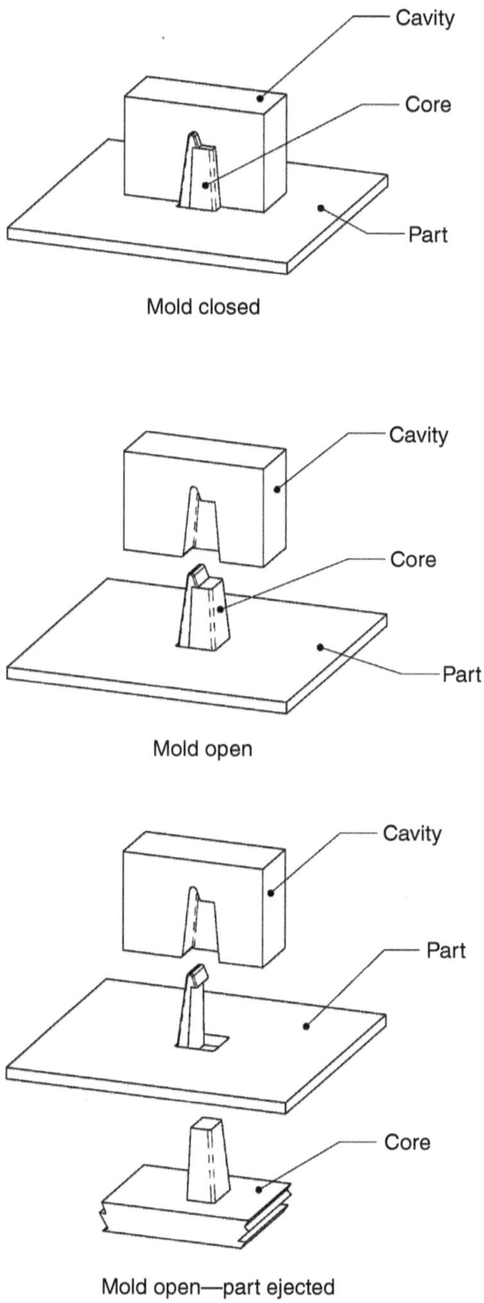

Figure 6.10 Slide-by shut off is illustrated showing the cavity and core sealing off at 5 degrees (steel to steel) forming the cantilevered snap undercut feature.

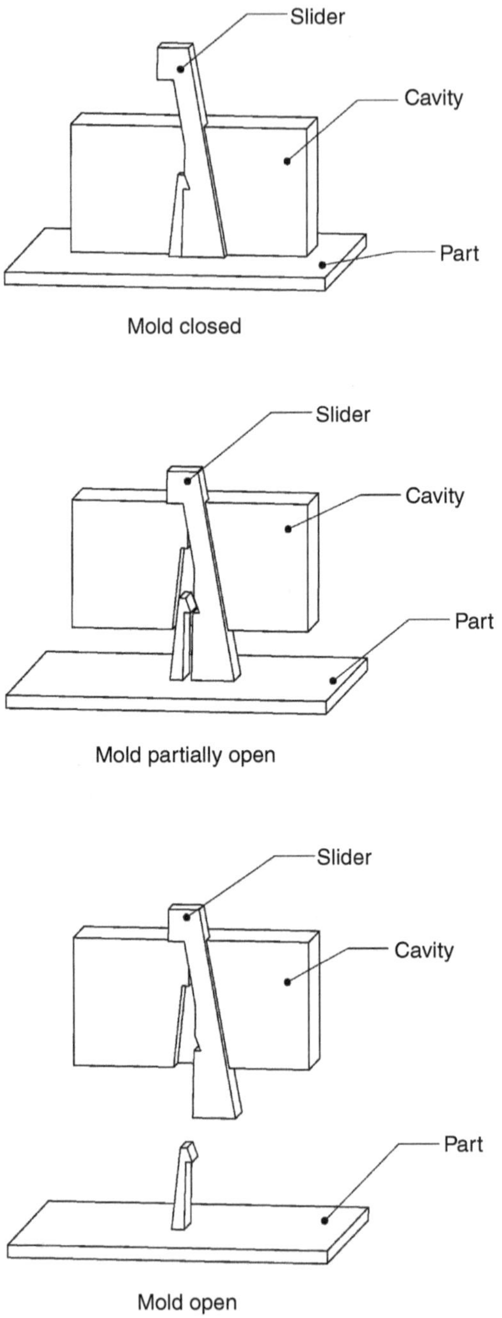

Figure 6.11 A lifter arrangement is shown being used to form a cantilevered snap undercut feature on a part.

around the perimeter of the sliding element to prevent flashing during the injection process, hence the designer needs to keep in mind the mold opening and the slide movement with respect to providing wall sections and surfaces which allow the mold to seal-off easily.

The added costing for the extra steel and mold size required for the slide mechanism in the mold will sometimes cause a designer to make design decisions allowing the avoidance of this tooling method. However, each part and each application need to be evaluated based on their own merits. Sometimes the use of the slide will provide the precise feature required with the best possible geometric arrangement of walls and the best flow path for the ideal gate location, so don't avoid the slide just to save some initial investment. A good part design that has a good tool, even a more expensive tool, specified to be molded in the right material, with a design allowing a good flow path, will save more cost over the life of the tool than the initial extra investment required. Conversely, if the part will be made in a multicavity mold, especially in a high cavitation mold, the extra room required for the side-action slide must be considered right up front. In addition, the slide mechanism will require maintenance over the life of the mold, requiring additional incremental investment, albeit small, over the life of the mold.

On some slide designs the slide is cam driven by the mold opening action; in these kinds of arrangements the stroke of the slide will determine the mold open stroke required to clear the slide feature from the part. Sometimes this creates a much longer mold open stroke than an alternate (nonslide) design might require. This longer mold opening stroke will add time to every mold cycle thereby reducing the overall production capacity of the mold. Hence, there are times when the slide is the perfect answer and there are times when it should be avoided at all cost. The trick for the designer is to do a proper evaluation of the part function and application, and to use that information to correctly evaluate which design method to use to get the desired design feature (Fig. 6.12).

6.2.6.4.4 Strip-off or "Pop-off" features

Sometimes the desired undercut or trapped steel condition occurs on the inside of a part and there is no room for a slide mechanism to operate. In these kinds of conditions and with certain geometries, the part can sometimes be stretched and popped off the undercut feature. Of course, the design of the undercut must be made to allow it to release and the design of the rest of the part geometry must allow sufficient force to be applied to move the part, overcoming the undercut stretch loads without damaging the part in the process. Material selection will play a big part in this decision as well; some materials will allow more stretch than others. For example, a polyetherimide part can stretch about 5% during this event without permanent damage. Alternatively, other materials may not tolerate as much as 2%, some less. Hence, the size of the part as compared to the desired size of the undercut feature for a given material will determine the suitability for this design method (Fig. 6.13).

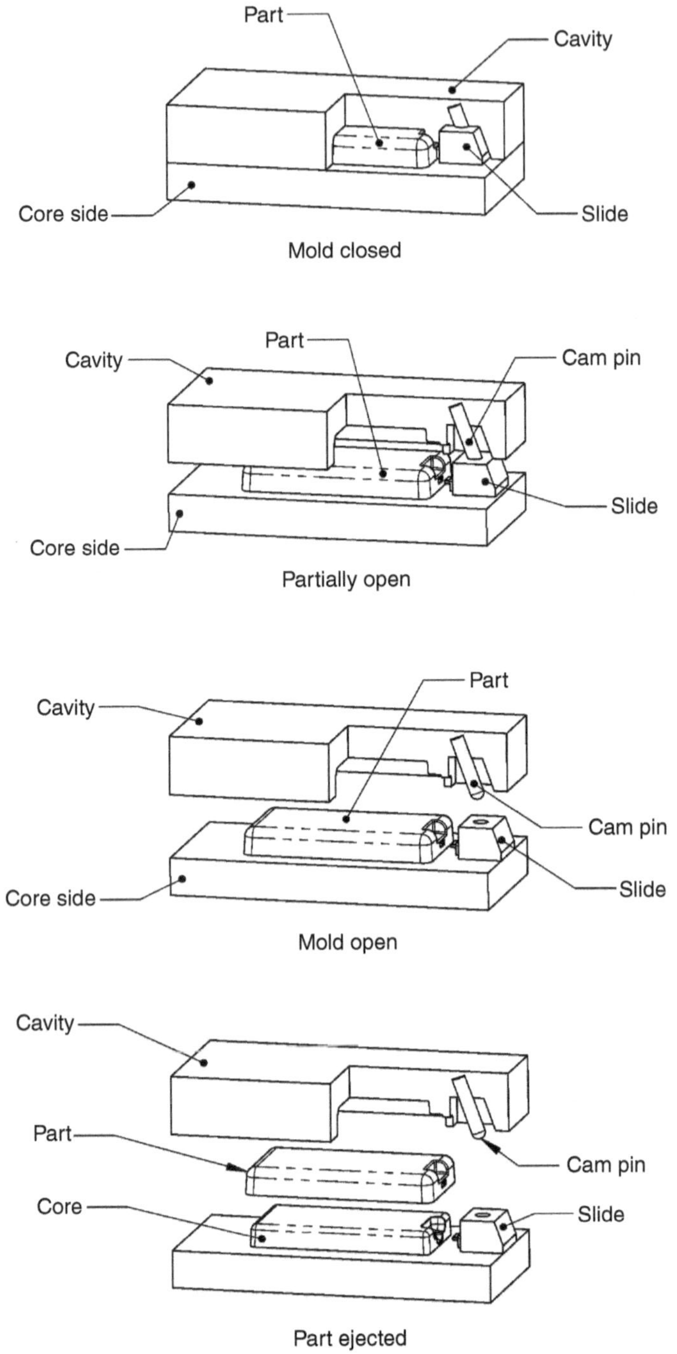

Figure 6.12 A side-action slide illustration showing the slide motion as a function of the mold opening position with the cam pin driving the slide.

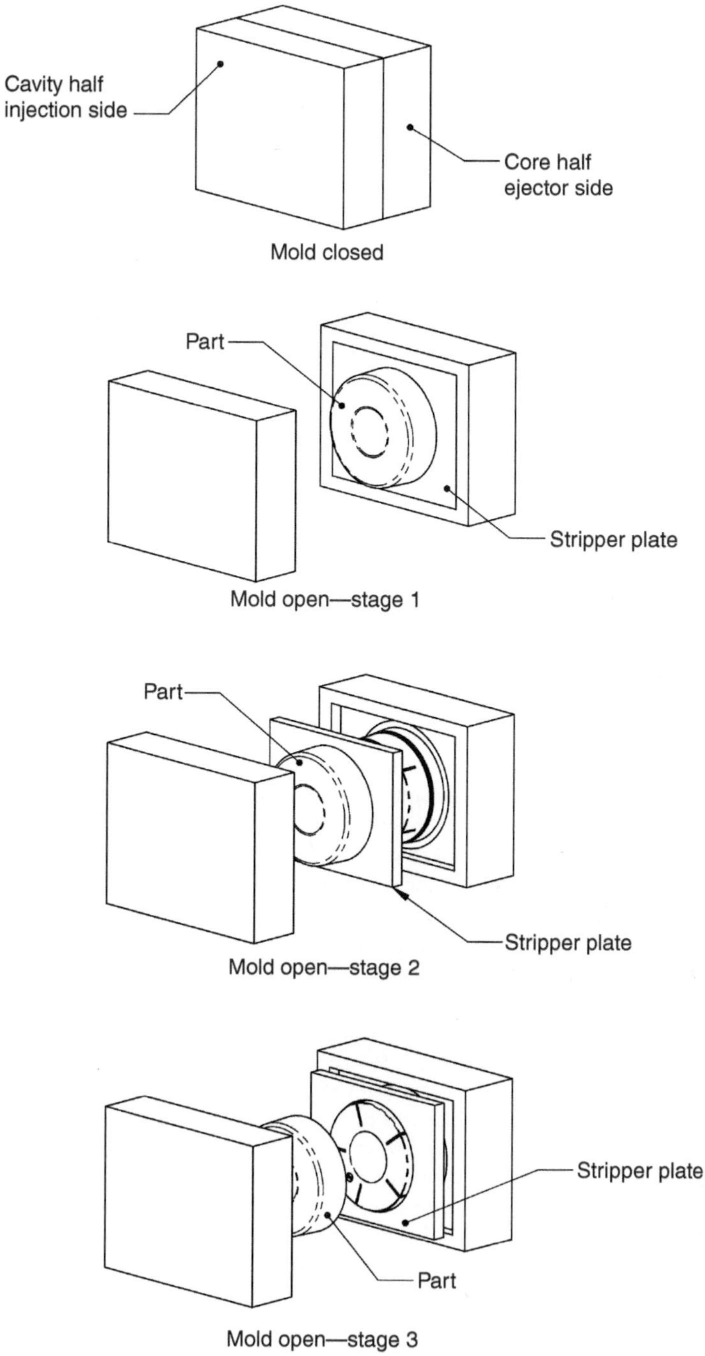

Figure 6.13 An annular undercut, shown as a complete annular feature, ejected by a stripper plate arrangement.

Case Example

One of my first experiences with a strip-off style design was for a small right-circular cylindrical shaped atmospheric gas sensing device. The design was nothing more than a static bucket holding some granular materials, some liquid electrolyte, and had some conductors that needed to make connection with the granular materials and with a precious metal electrode at the gas membrane interface. Pretty boring stuff, right? Well, not when you consider the size of the device, the materials used, and the application where the finished device was used.

The device was fairly small, about 25 mm in diameter and 35 mm long overall; the materials inside were hazardous and toxic, with the electrolyte having a pH of 14, extremely caustic; and finally, the sensor was used in a medical application that was patient central, that is, if the unit failed, the patient could be at risk. Now the design became interesting, even challenging. Our task was to package these hazardous materials in the correct proportions, with the correct conductive path from inside the container to outside the container (without allowing electrolyte leakage over the multiyear life of the device), and to provide external means for connection to both the sensor electronics and to the defined atmosphere to be sensed. In addition, the package had to prevent leaks of the hazardous materials during the use or in the event of a drop or impact to the sensing device. Challenging design criteria after all!

Our design concepts were many and as varied as snowflakes, however, we settled on one particular concept that, in the end, worked out well. We decided upon an inner container structure that was the core container with the electrically conductive feedthroughs. We then inserted this inner container into an outer shell, molded from an impact resistant ABS material with a geometry that supported the inner container and provided the external connection features to be molded in one part. However, we needed a way to secure the inner container in the outer structure without damaging the inner structure and allowing the inner container to expand and contract with thermal gradients. The daily production rates were in the thousands of assemblies per day, hence the assembly operation used to secure the inner container needed to be quick and simple. We decided to use a custom molded snap-in retaining ring and this design would require an annular inner snap undercut in the exterior structural housing part. No problem, except the ID of the exterior outer structure was under 23 mm—less than an inch in diameter; which would severely limit the stretch magnitude and caused the undercut depth to be extremely small, affecting the resulting strength of the snap feature (especially during the drop test impact).

Once installed the custom molded snap retainer part was used to mount a printed circuit board that carried some rudimentary temperature compensation and signal conditioning circuitry and provided a connection means for the conductors from the inner container, allowing electrical communication, through an interconnection means, with the sensing electronics remote from the sensor location. Additionally, we used the rigid circular epoxy fiberglass circuit board OD to slip into a molded feature in the exterior structural housing to add support in a diametrical fashion. The board rigidity supported the structure and aided in keeping the geometry round, even upon impact during the drop test.

The final piece of the puzzle was to be a cover to protect the circuit board and connection means. We specifically used the same material as used for the external structural

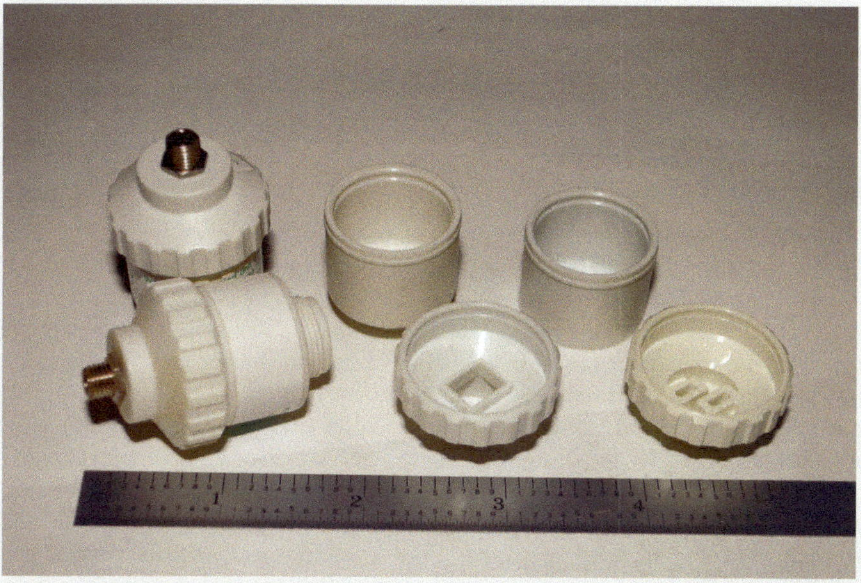

Figure 6.14 Annular snap ABS injection-molded parts and assembly from the case example on annular snap features.

part for this cover so we could keep the colors matched. This cover was designed to be an annular snap fit over the exterior outer structure which was geometrically aligned to overlap the annular snap feature used on the custom molded snap retainer part, thereby providing additional strength to combat disassembly during a drop impact. This was sort of belt and suspenders thinking, but given the risk should the contents leak, the design team felt it only proper to provide every means possible to secure the inner working cell. See the finished device in Fig. 6.14 in several configurations.

The implementation took some degree of testing and iterative learning as to what the best approach and process to assemble the unit would eventually become. However, in the end, the device was successful and the company was able to compete and grow market share handsomely. The base factory cost for the device was reduced by over 75% compared to the previous version and the part-to-part performance range was improved substantially. Costs went down and product performance went up.

An alternative to the "pop-off" undercut design using a stripper plate to "pop" the part from the core is the collapsing core technology. This utilizes a split core with a moveable inner core to allow the external core features to reduce in diameter when the inner core is moved; thus allowing the part to clear the core and be ejected. Collapsing core systems need room for the core segments to move inward upon the ejection stroke, so internal part features need to allow this movement. This collapsing core technology is used with parts with little or no internal geometry, like jar lids or bottle closures. One thing to remember, when the part is "popped" of the core, there is always going to be some smearing or drag of the undercut feature during the ejection stroke and this smearing can sometimes lead to snap retention performance degradation. The collapsing core feature will be clean and crisp and should perform as desired.

6.2.6.5 Ejection systems

When does the designer consider the ejection system to be used in the tooling? I think you will find this answer to be almost never in the early or even in the detailed stages of the part design. Most often the first conversations and consideration about the ejection system comes when the tooling source is estimating the project and makes some remarks on the estimate regarding the ejection system to be used. In some parts it just does not matter; but for others, the ejection system may be a key element in getting the part performance and features required (as in the previous case example). Hence, it remains important for the designer to understand the costs and complexities of the ejection system required for the part design created. There are several methods of ejecting a part from a mold that should be understood by the part designer and considered relative to the part complexity and to the required features for the particular part being designed.

6.2.6.5.1 Knock out ejector pins

In the injection-molding process, the most common form of ejection system is the use of ejector, or KO pins. These are standard off-the-shelf mold components that are designed into the mold to push against a surface of the part to push the part off of the ejector half of the mold—commonly called the core side, sometimes referred to as the ejector half, or the moving half, or the "B" side of the mold. These pins will be located to apply an even push force around the part and especially around any feature that will require additional force to move off the coring in the mold. As the designer is preparing the part design geometry it is wise to consider ejection pin locations and leave some features or room for adequate ejection pin surface area. Larger-diameter ejector pins (larger area) will provide some benefit to reduction in overall process cycle time. Smaller-diameter pins (think higher unit force) can deform the part surfaces if the part is not cooled sufficiently, so the process hold time is increased until the part temperatures are low enough to allow ejection. Total ejector pin surface area can be increased by either using more pins or larger pins; however, the design of the part must provide surfaces for the pins to push against. Overall process optimization will depend on the part design allowing for good tool design for the given material, hence the ejection system should always be one of the considerations in mind as the designer creates the part geometry in the layout.

6.2.6.5.2 Stripper plate

Another common ejection method is the use of a stripper plate system. This is the ejection system used in the previous annular undercut snap feature case example. In this system a plate is connected back to the ejector mechanism tie in components and the entire plate is moved forward to push the part from the coring. Typically this is used for round parts and the stripper plate forms one side of the parting line for the part. This is a more expensive ejection system to build into a mold but it has the benefit of pushing on the entire surface at the perimeter of the part on the parting line. This ejection system will typically leave no ejection system marks whatsoever on the finished molded part. In addition, another benefit of using a stripper plate is uniform ejection force distribution, less opportunity for the part to be deformed during ejection and absolutely square ejection; the part cannot twist, flex, pivot, or warp during the ejection stroke. Sometimes this can provide much superior handoff positional accuracy for a robotic pick-out system as well.

6.2.7 Design from a material flow perspective

Comprehension of the mold filling during the creation of the initial 3D layout seems premature at first. However, when you consider that the polymer will align in the direction of flow and final part strength is sometimes vastly different in the flow direction, when compared to cross-flow direction, the opportunity to control the parts final characteristics by creating the part with material flow in mind seems not only reasonable but perhaps even a primary requirement. That said, most parts are designed and sent over to the tooling source and the gates are placed by the tooling source without any consideration for the functional end-use or processing considerations. On every part that I create, the data package will include the 3D file used by the tooling source to create the core and cavity geometry within the tool and the data package will include a two-dimensional specification drawing that will contain critical dimensional requirements, as well as notes to the tooling source and to the molding source. One note in particular that I use says, "*All gate locations, ejector system locations, parting lines, vents, and other part critical tooling details must be reviewed and approved by the design team prior to implementation in the tooling*." This note is, of course, ignored sometimes; however, the majority of tooling sources will provide an initial tool layout showing all of these tooling details. This allows the design team to understand if the tooling source will create a tool that will produce the part as intended.

6.2.7.1 Part fill

Thinking through part fill and designing the part for uniform filling is a basic task but an important one. Comprehension of the fill path will assist in understanding where the knit lines may form around features in the geometry, or where the wall thickness changes might be possible or, more importantly, impossible. Placement of supporting structures, extending perpendicular from the main part walls, such as ribs or fastening features like bosses can create gas traps, hence understanding fill allows the designer to orient these kinds of features to allow flow and gas movement to enhance flow and to avoid gas entrapment. There are also some features that require a specific

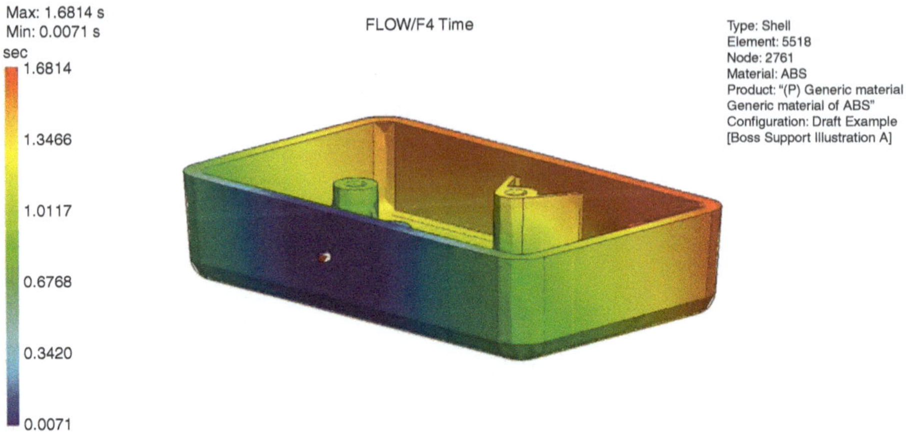

Figure 6.15 A fill time report from a flow analysis run within SolidWorks Plastics showing a single gate on a side wall from a tunnel gate.

flow direction to function as intended; living hinges, for example, work when the polymer chains are oriented perpendicular to the hinge line, hence the designer must completely understand the flow while designing the part geometry around the living hinge details, otherwise the hinge will live for only a few cycles. Some materials are anisotropic, and will shrink differently in the flow direction as opposed to the cross-flow direction. A good part design with these materials will consider the fill and flow directions and orient critical or close tolerance features to be in the direction with the least shrink variance shot to shot (Fig. 6.15).

6.2.7.2 Residual stresses from molding

Material flow, combined with a good part design that allows uniform cavity fill, will most often provide a part, which has a fairly even distribution of residual stresses. These residual stresses are the internal part stresses, formed by the phase change from a molten to a rigid state, while being held under pressure within the mold. Process parameters used during filling and cooling will also have an impact on residual stresses; but, the more uniform the fill throughout the cavity, the easier it will be to fill, cool and remove the part. Postmold warpage of parts is a direct result of these residual stresses and the warpage in effect is attempting to relieve some of these stresses. Sometimes a part will look and perform well until it goes through a thermal cycle large enough to allow some internal residual stresses to be relieved and the part then either shrinks some additional incremental amount or the part warps. Residual stresses can be especially troublesome when a given part is subjected to external environmental effects, such as contact with certain chemicals which when combined with the residual stresses, or with load induced stresses, can cause polymer degradation and can lead to part failure. We call this phenomenon environmental stress cracking or ESC.

6.2.7.3 Flow knit line locations versus structural features

As mentioned earlier, comprehension of the fill during the initial 3D creation of the part can assist the designer in visualizing the knit line locations around features being created within the part geometry. Good designers can think through and visualize how to create geometry to shape the flow front during fill to place the knit line in areas of the part where they will do the least possible damage. Sometimes this is not possible, however, other times it is manageable and good design can exploit geometry to avoid knit line placement in critical areas.

6.2.7.4 Venting the cavity

When designing from a material flow perspective remember that when the material enters the mold cavity, the gas in the cavity must escape. Anytime there are features where the gas can get trapped during the material filling cycle, be sure that the design of the feature will allow the toolmaker to add venting. This can be around core pins, at standing bosses, at standing ribs, and so on. Venting is not necessarily the responsibility of the part designer, but if the part designer can add features that allow the tool designer to vent properly, then we approach the optimum part.

6.2.8 Design from a disassembly perspective

For some products, the end-user should never have the need to disassemble the product. For other products, it is sometimes necessary to allow the user to disassemble the product. A case in point is where the product is comprised of recyclable materials but has nonrecyclable content as well. From a designers view during the creation of the parts and product, why not consider and provide some means for the end-user to dismember the unit enabling reclamation of the base materials? Other thoughts and considerations along these same lines should be useful to the designer when creating part which must be altered by the end-user during installation or use. For example, electrical termination boxes, used in buildings to house receptacles and switch devices are required to be closed devices with no access or exposure to humans or to combustible building materials; however, some number of openings are required for cabling and wire to enter and exit the box and this number of openings will vary from one up to four or more. The plastic box components are designed with the hole there but filled in ready to be removed. The design feature is a purposeful thin section with a sharp corner that provides a stress concentration point when loaded and fails easily when acted upon by the electrician. Another example might be a molded closure need to close and seal a container, with a feature that will disallow opening the container without tearing or breaking a portion away; commonly called tamper evident closures. The designer needs to envision these features and create the initial 3D layout from the perspective of allowing the tooling to produce the feature and allowing the material fill and flow create the part within the cavity. Repeated disassembly, such as a battery door, or a contact lens case lid operation require other thinking on the part of the designer, fatigue, and wear factor into the thinking, as does over stressing the feature by the end-user.

6.2.9 Design from an end-of-life perspective

As mentioned in the previous examples, disassembly at end-of-life for recycling and reclamation of materials is always something that the designer should be thinking about during the initial 3D layout of the part geometry. Thermoplastics are unique materials that provide us this ability for reprocessing unlike no other. As designers and good citizens of the planet we all wish to use materials and energy as wisely and efficiently as possible. Hence, considering how the end-user will deal, with the product at the end-of-life as we just begin to create the part geometry makes sense. For some designs this will be obvious; for example, if I am designing a new disposable water container from PET, we know it will end up in the recycling stream. On the other hand, if we are designing a new electronics enclosure from ABS, do we think much about recycling or disposal? After all, it is durable and it will last for years before disposal. Well, what about all those mobile phones, millions and millions of them, could the polymer from those be reclaimed? Absolutely it could, but with labor input and at great expense. At what point does cost of reclamation deter the activity and can we design the part or product to minimize this cost? Certainly not every part will require this level of thinking regarding end-of-life, however, these concerns can provide insight which just might influence the creative results for a given part or product.

6.3 Design rules review

While in the 3D layout creation process, and while considering the applicable functional, material, process, and tooling requirements, the designer will follow some basic rules for the chosen materials and processes. There are several extremely detailed reference books dedicated to these rules and why they are so important, delving into both the historical and the physical reasons, and rationales for each. For the purposes of this text, we will review these in brief, not in detail. However, the reference texts where more information can be found are listed at the end of this and each chapter of this text with other reference materials and resources. These design rules are fairly process dependent, hence we will review them as categorized by the three main processes used for structural parts, injection molding, blow molding, and vacuum or thermoforming.

6.3.1 Injection-molded parts

For this high-pressure process, the rules are somewhat more important than for other lower-pressure processes. With that said, these rules need to be considered with every new wall, rib, boss, and so on, created as the layout is progressing. Each feature developed will be guided by one rule or another and when features are combined or built adjacent to each other, the rules will sometimes compete. As the design layout progresses, the designer must be always thinking about the material flow, about potential air traps, about how the part will be ejected, and so on. Hence these rules need to be followed but often they will be compromised and the designer will need to assess how far to bend the particular rule in an effort to satisfy the needed feature on the part.

6.3.1.1 Uniform wall thickness

Once the general geometric shape of the part is visualized by the designer, the first thing that should come to mind before even beginning the CAD layout for an injection-molded part is just how this part will fit within the mold. Which side is the cavity? Which side is the core? Where is the parting line? Where is the gate? Given a good visualization of these tool elements for the part in question, the layout begins with the nominal wall, which will be located near or at the parting line. Remember, *the parting line is the starting line*. Most likely the gate will be situated into or adjacent to this wall. Given that this wall is one of the main structural elements in the design and provided that it is the wall that the gate is located adjacent to, it is imperative that this wall be as uniform in thickness as possible and that when we need to have transitions, they need to be gradual. In addition, it is always best to flow from thicker wall sections to thinner wall sections and not to transition from a thin section to a vastly thicker section. These flow considerations makes sense if you think about the molten polymer flowing into the mold cavity at some flow rate and pressure, then hitting a transition where the wall section abruptly gets larger, what happens? The pressure drops and the flow rate slows; now the flow front (which is ever cooling during the filling process) becomes slower yet. Depending upon the initial wall section dimensions and the materials being molded, the thinner section could freeze off prior to getting good fill and pressure in the downstream thicker section. This condition will, in most materials, result in extreme shrinking in the thicker sections, providing surface blemishes, sink marks, and nonhomogeneous residual stresses within the part walls in that area. Likewise, too large a transition from a thicker wall to a thinner wall will increase pressure, increase flow velocity and may induce shear thinning of the material which could lead to material degradation, or at minimum surface blemishes on the finished part. In addition the differential cooling time required from the thin to the thick cross section, especially if the thickness change is not gradual, will cause some warpage and molded-in residual stresses that may impact part performance over the life of the part.

The uniform wall section rule is one of those rules handed down from years of experience and experimentation that demands the attention of every designer creating an injection-molded part. It is truly one of those rules of nature that you don't want to mess with (Fig. 6.16).

6.3.1.2 Intersecting ribs

As the design layout progresses, there will undoubtedly be a need for a wall which is perpendicular to the main wall, either as a side wall or as a rib for stiffening or for supporting some other feature or design element. These perpendicular walls intersect in two ways, either as a corner or as a true intersection. The uniform wall rule earlier needs to be modified just a little when it comes to these perpendicular wall elements in the design. Intersecting walls will create a heavy cross section just at the intersection, which will not cool as fast as the adjacent walls in contact with the mold, often resulting in an area that will leave surface imperfections due to shrink. In most cases this may not be an issue; however, there are two particular cases where the designer needs

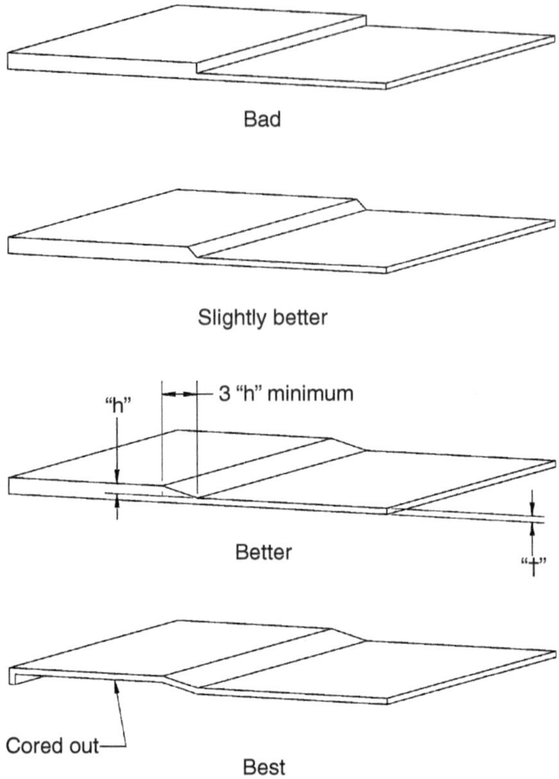

Figure 6.16 Maintaining a uniform wall thickness is one of those rules of nature that just must be obeyed.

to be careful when placing these intersecting walls. One is when the exterior surface opposite the intersecting wall is a cosmetic surface for the part or product. The other is when the surface opposite the intersecting wall is a joining or sealing surface; especially a surface that might be used for an O-ring seal or perhaps for an adhesive glue joint. Sink surfaces will produce a nonfunctioning seal or weld joint, and worse yet, the variation in material properties and process conditions can produce parts where the seal functions perfectly for some percentage of the time and fails at some small percentage of the time. Now we are producing some quantity of parts and over the process variation cycle (or within the process window) some parts will be produced which are marginal or outright nonsealable parts. Designers be warned, avoid intersecting walls on parts where seals, weld joints, glue joints, and so on, need to be made correctly part to part. Always think about what the final part requirements demand during the design creation process.

The case where the intersecting wall is a corner or a simple wall turn is somewhat easier to work with. It is important to try and keep the uniform wall through the turn, that is, the corner radii on the inside and on the outside should be specified to keep the wall as uniform as possible through the turn (Fig. 6.17).

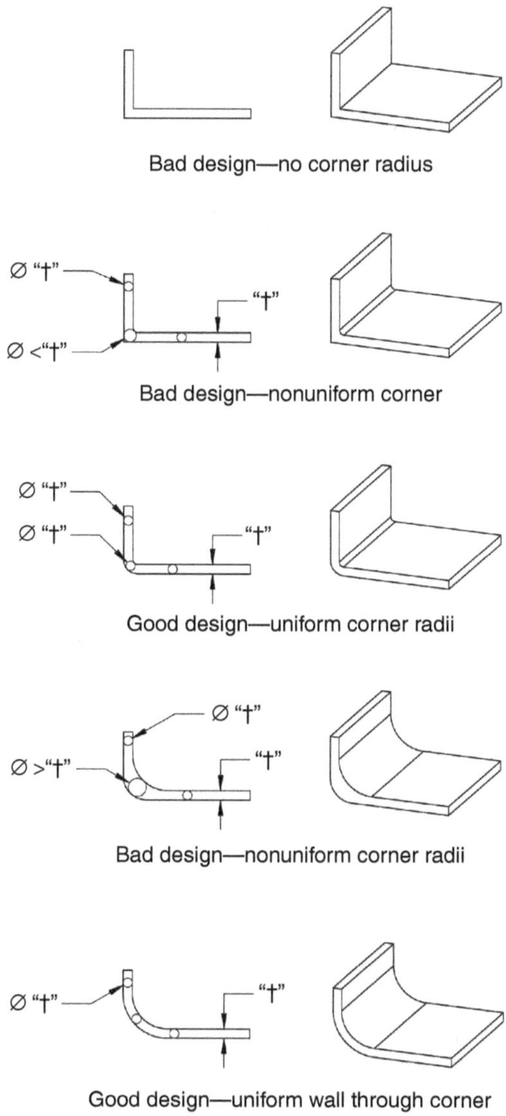

Figure 6.17 Uniform wall thickness through the corner radii illustration.

For the intersecting wall case we tend to reduce the intersecting wall thickness at the base by some factor. The rule of thumb is that the rib or intersecting wall needs to be somewhere between 60 and 80% of the nominal wall thickness. However, this is material dependent and most material suppliers will publish data specific to their materials for these factors (Fig. 6.18).

Another consideration when creating an intersecting wall within an injection-molded part is material flow and venting. Intersecting walls in the direction of flow

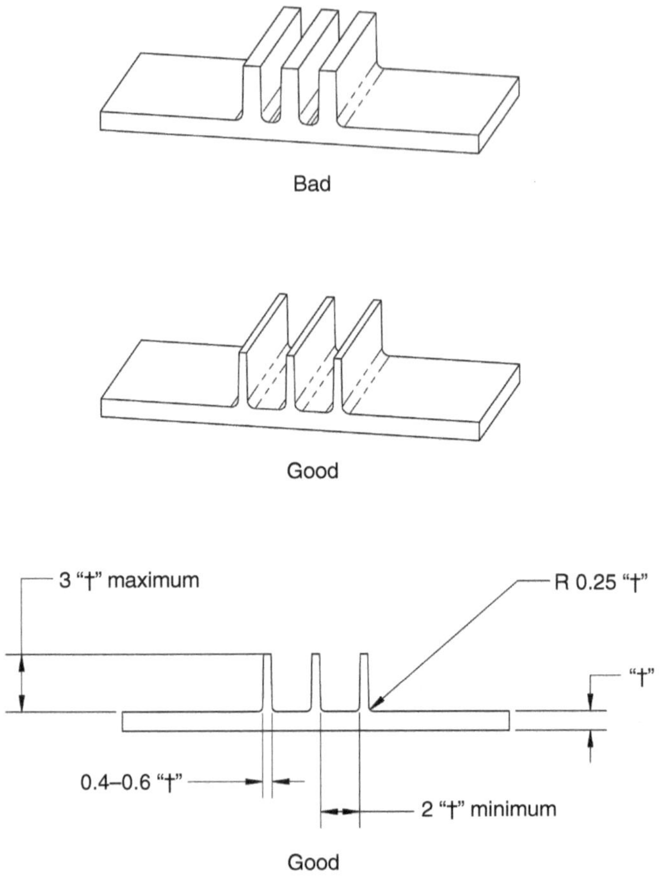

Figure 6.18 Ribs intersecting a uniform thickness base wall illustration.

will fill differently than those situated across or perpendicular to the flow. Intersecting walls and ribs that also intersect with a side wall fill and vent easier than free standing wall features. Any intersecting wall that is in the direction of the mold opening will need to have adequate draft on both sides of the wall. Free standing wall features that have termination ends will require similar draft on the termination ends as well. Larger draft or outright tapered ends to these freestanding walls will aid in filling and venting, depending of course on the position of the freestanding wall relative to the flow direction (Fig. 6.19).

6.3.1.3 Bosses

Bosses are nothing but intersecting freestanding walls in a circular form and as such they represent a hard to cool feature, which must be specified well for good results. Bosses are used for many different part features and functions; mostly they

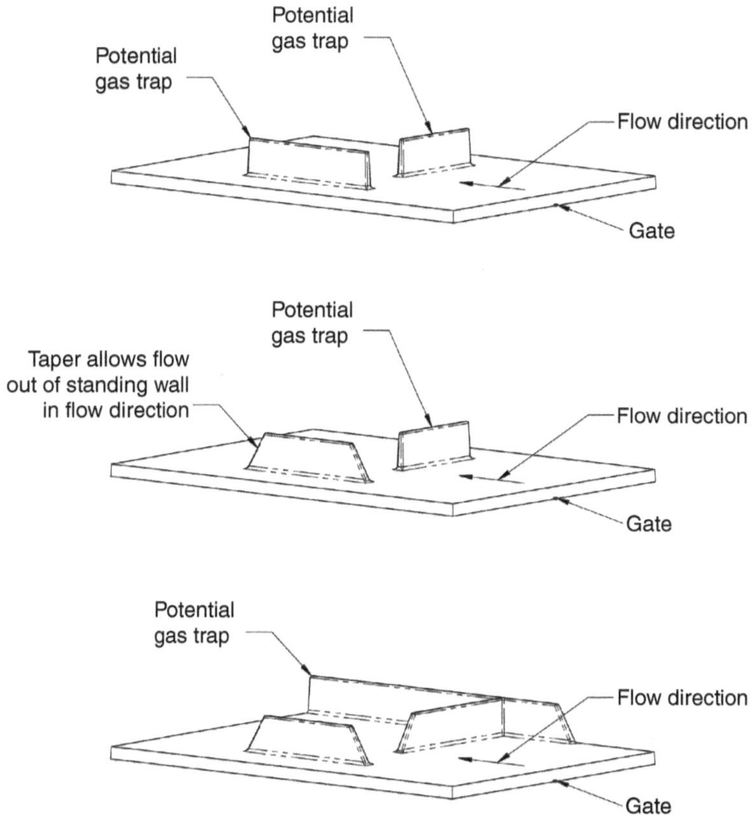

Figure 6.19 Freestanding and intersecting walls shown with relationship to material flow during cavity filling illustration.

are used for joining or alignment features. Sometimes they are used, with a through hole for allowing a screwed fastener to pass to an adjacent part. Sometimes they are used as the material into which a thread cutting or forming fastener will engage with to affix the adjacent part. Sometimes they are used to hold a dissimilar material insert which may be threaded to accept a machined threaded fastener. As they are most often used for these joining functions, they are subject to loads that need to be considered as the design layout is being created. The loads get transferred to the base wall section that the boss is filled through. Sometimes additional ribs or gussets can be created in the design to assist in supporting the fastening and holding loads generated at the boss feature. Just as with intersecting walls, bosses will cause difficulty when they are opposite a sealing or joining surface, or if they are under a cosmetic surface. These considerations do vary with some materials, but in general, a boss creates a heavy section that is difficult to cool at the same rate as the surrounding walls, hence it is always an area where shrink will manifest for most materials (Figs. 6.20 and 6.21).

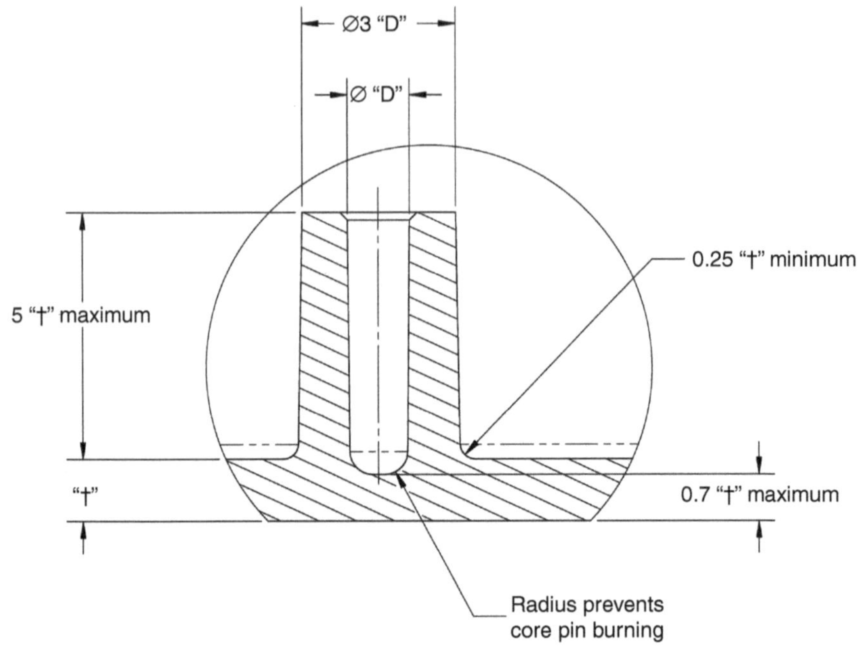

Figure 6.20 Generic boss design rule illustration.

Just as with free-standing walls, bosses can sometimes cause fill and venting issues. However, with a round boss detail, the tool will most often be designed, with a removable round core pin insert. Venting for most boss features can be accomplished by venting to atmosphere alongside this removable core pin element in the mold. Most mold designers will understand and secure venting at each feature or boss during mold design and they understand the process impacts if they do not.

6.3.1.4 Draft

Taper on all surfaces perpendicular to the parting line is a requirement; another law of nature like the uniform wall rule. In order for the mold steel to release from the cooled molded part without damaging the part a taper is required. The degree or angle of the taper does indeed vary with the design of the feature, the materials used, and the surface finish required. This taper is commonly called draft or draft angle and we will typically see a 1 degree draft per side on most injection-molded parts as a minimum draft. In some cases with smooth surfaces we will see 0.5 degree draft specified—always specify draft direction plus or minus. With a heavy textured surface we may see as much as 5 degrees or more specified. This draft is important to keep in mind during initial design layout creation, which is why I like to always start my layouts at the parting line.

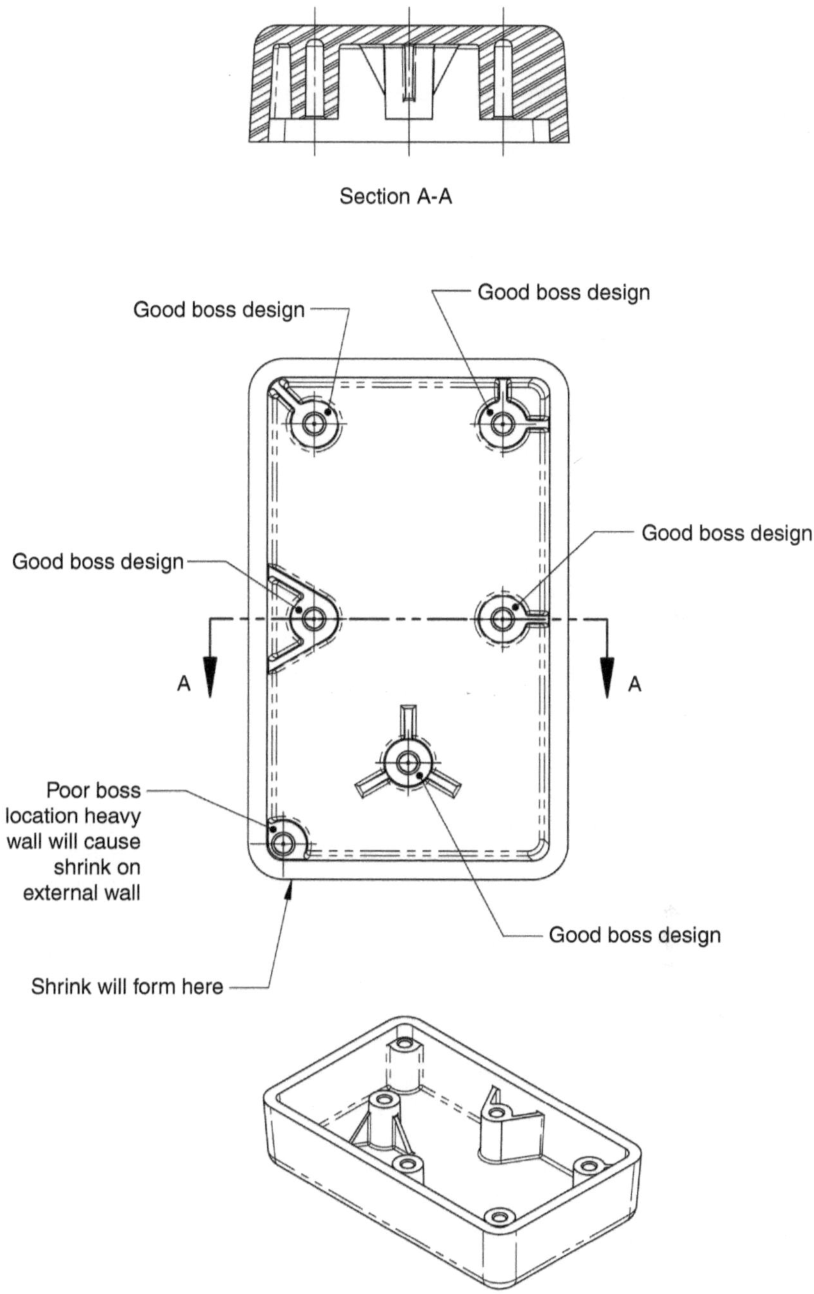

Figure 6.21 Other boss design and support illustrations.

Case Example

On one particular project the client had defined an open ended tank element to hold a liquid, 5 L of the liquid to be exact. All of their initial specification drawings were done by an architect and were quite lovely, artistically nice. However, while the design was specified with gloss surfaces, textured surfaces, opaque areas and transparent areas, and so on, the design did not show any draft whatsoever. We took their exterior styling and adapted it to parts designed for the injection-molding process. We were careful to keep the style as exact as possible and in proportion to the original design concept; however, when the client saw the design with proper draft applied they were dismayed. This looked nothing like their architectural beauty they had brought to us. As the part was fairly large and tall, the draft was extremely visible. In the end, we worked with the original designer to educate him about the molding process and the need for the draft. We then modified the design somewhat to have portions of other parts possess a similar taper, so the different parts all tended to look similar and the final assembly ended up being acceptable to the client. Extremely small details sometimes have a huge impact and presentation with draft at the onset would have changed the course of the project in a positive direction.

Draft must be applied from both sides of the molded part; both cavity and core need to have adequate draft. This becomes tricky when the core will sometimes telescope into the cavity to form a hole or some other feature (Fig. 6.22).

6.3.1.5 Corner radii

Corner radii at every inside corner, and at most outside corners, is another rule which gets classified in that "must have" category. Molten polymer flowing into the mold cavity wants to flow in and fill as quickly as possible before freezing solid. The strongest parts are attained with the highest final molecular weight material; meaning the longest possible polymer chains. Molten polymer flowing over and across sharp edges causes shearing, which in turn causes breakage in the polymer chains. These shorter polymer chains result in a lower average molecular weight in the finished part; hence a part with potentially compromised strength and properties. Adding corner radii, even a small one, will aid flow and help prevent this shearing phenomenon. Be careful to check with the material supplier to see what their recommendations might be for corner radii. Also do not be too caviler with corner radii size, radii that are too large can produce heavier sections that become hard to cool and can produce shrink marks and can sometimes cause part warpage. Be careful not to specify corner radii where it would cause undue difficulty in the tooling, unless it is required for functional reasons; for example, at the parting line a corner radius would require the cavity be transferred over the parting line into the other half of the mold, adding cost in the production of the tool. Any time the cavity and the core steel meet to form a hole or a feature in the part, corner radii cannot be placed where these steel surfaces meet for obvious reasons. However, in today's CAD programs, one can just point and click to add a radius to every edge on the part, so *think first* then add the radii where it will help flow and

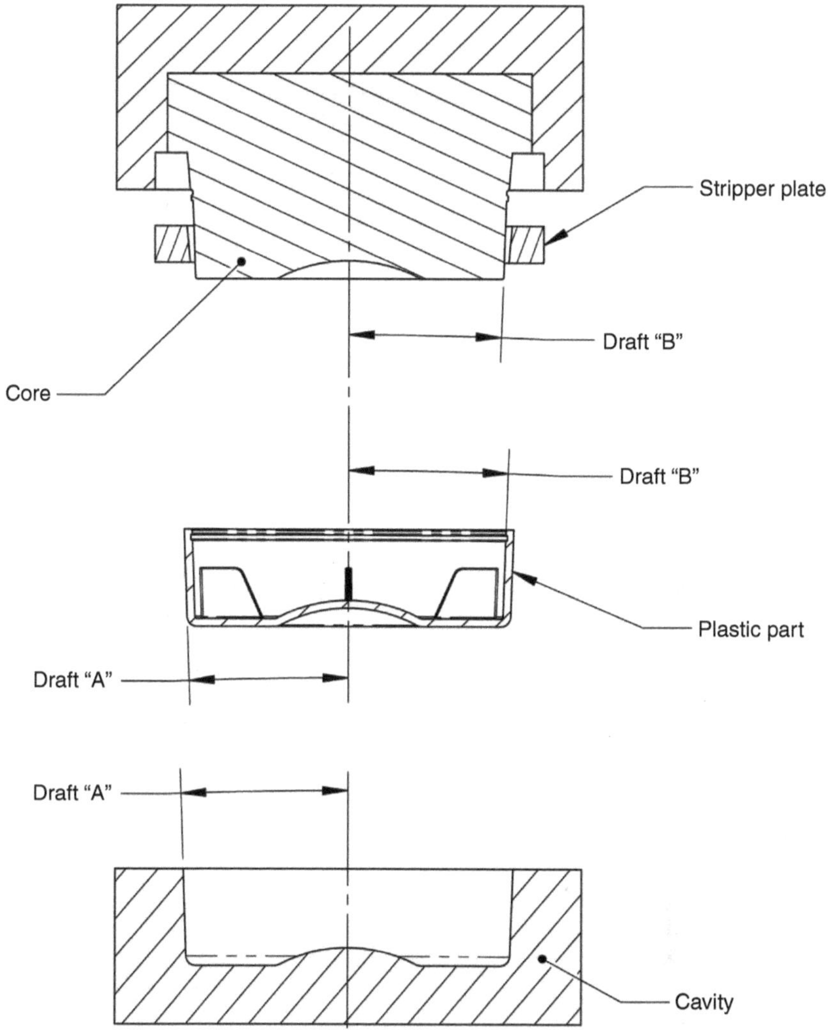

Figure 6.22 Draft is needed on all surfaces normal to the parting line, especially those surfaces which will shrink around a core, to aid in ejection of the part from the mold.

not where it will make tooling difficult. As a general rule, the corner radii should be sized to be 25% of nominal wall minimum, with 75% of nominal wall being ideal for most injection-molding applications (Fig. 6.23).

6.3.2 Blow-molded parts

Blow-molded part design is similar in many aspects to injection-molded part design but there are a few extra considerations to be aware of during the design layout process. Of course, there are several blow molding processes and each requires a bit of

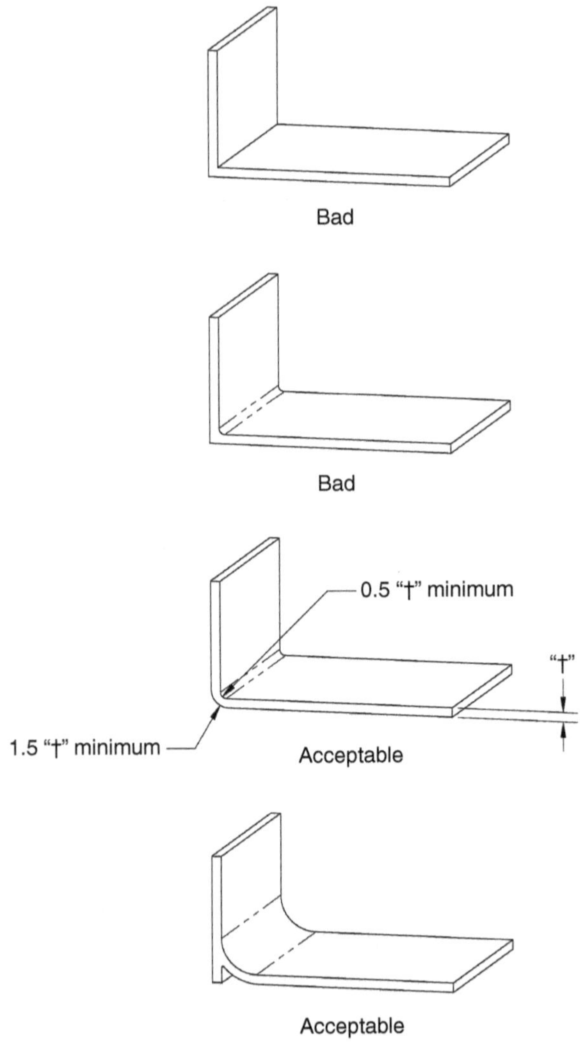

Figure 6.23 Generous radii in all corners allowing a uniform wall through the corner is the best situation when filling an injection-molded part.

specialized understanding and each has slightly different parameters which need consideration. For this review of the design rules overview, I am considering generic rules for all blow-molded part design.

6.3.2.1 Neck area transitions

Given that the packaging market segment for blow-molded products is one of the largest applications and since virtually all blow-molded parts for this packaging market segment will be hollow vessels for containing some liquid or granular substance, most every part will have a neck area on which will typically be found a thread or other closure capture feature. For other markets and for structural parts, there are many examples of blow molded hollow parts that are not containers, but since great majorities are, it is prudent to cover the transition from the neck to the body of the part. Typically the neck area will be the thickest portion of the part and, as such needs to be designed to allow for adequate filling during clamp-up and then for stretch transition to the extent of the body during the blowing operation. This is mainly an extrusion blow molding criteria but it also applies to the design of the injection molded preform for stretch blow molding and for injection blow molding. Providing a generous radius from the neck into the transition area and providing adequate taper in this area will allow the extruded parison material the ability to stretch and flow, filling the mold when pressurized.

6.3.2.2 Pinch-off areas

Specific to extrusion blow molding, where a parison is extruded between the mold halves and the mold is closed on this molten hollow tube of material, are pinch-off areas; where, as the name suggests the parison is pinched in between the mold halves when they close. These areas are defined by the parison shape and perimeter or diameter which is defined by the final part shape and desired minimum wall thickness. These pinch-off areas will cause rapid parison cooling during the pinch-off process and will result in a thicker wall nearest to the pinch-off area. In certain materials this will also cause a noticeable color difference because of the thickness change relative to the remainder of the blown part walls, especially in translucent materials like HDPE and PP. This pinched off material is a solid and remains connected to the blown part until after the part is removed from the mold, when the external pinched off solids (sometimes called flash even though it is really just pinched off material—not flash in the typical sense of the word) are removed. This secondary process is known as deflashing (even though it is not really flashed material) and can be accomplished by several methods depending upon the materials and the requirements for the part in question. However, no matter the deflashing method there will always remain a witness line where the excess solid material was removed from the part. Cosmetic parts designed for this process will need to be oriented so the pinch-off can be in the most noncosmetic, or hidden, position possible.

Injection blow molding and stretch blow molding do not suffer from this pinch-off consideration (Fig. 6.24).

6.3.2.3 Bottom push-up features

Again, specific to extrusion blow molding, there are many instances where the part would require a slight indentation at the bottom or base of the part. Usually this is at the opposite end of the part from the neck. Often this is simply a way to allow the recessing of the pinch off area so that the resulting witness line/mark does not cause the hollow part to rock or teeter when sitting on a flat surface; common to most every bottle shape used for holding liquids (Fig. 6.25).

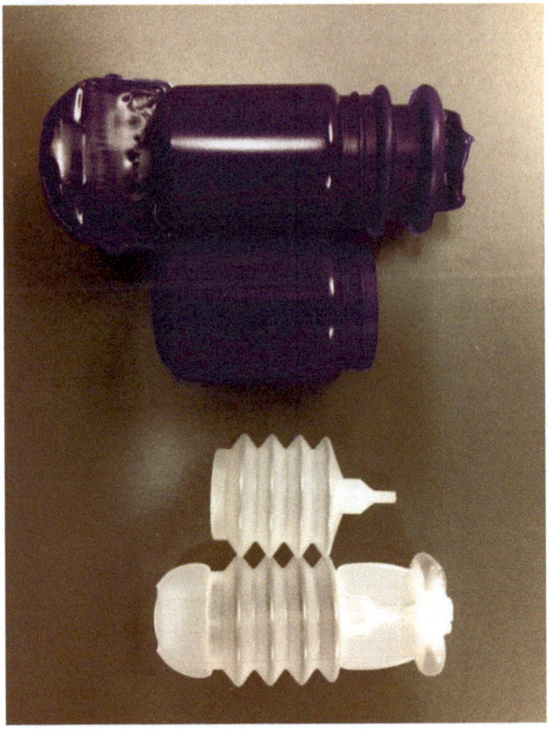

Figure 6.24 Two blow-molded parts shown with pinch-off and with the pinch-off solid material removed.
Source: Courtesy of Blow Molded Specialties, Pawtucket, RI.

Figure 6.25 Three blow-molded parts are shown each having a bottom push-up feature.
Source: Courtesy of Blow Molded Specialties, Pawtucket, RI.

6.3.2.4 Draft

Just as in the injection-molded part design efforts, draft is required in the blow mold cavities to help and release the parts as well; however, there are instances, since we are dealing with much thinner walls and with no core tooling on the inside walls, where zero draft is acceptable, even reverse draft or undercuts can be molded and the demolded by flexing the part walls during the ejection process. Every different geometry and wall thickness combination will yield different degrees of flexibility for the given part so each instance where an undercut or negative draft is required will need to be analyzed carefully prior to implementation. Molded in handles common to liquid bottles for retail sales, as in detergent bottles, milk bottles, and so on, are one area where draft must be considered carefully since the resulting walls are typically shorter and thicker, and since the material will be shrinking onto the core not away from the cavity. Discussion with an experienced process technician or toolmaker, will help to understand the design constraints which might be faced for these kinds of features. As with injection molding, surface finish will also determine the magnitude of the draft required for certain materials in the blow molding process (Fig. 6.26).

6.3.2.5 Corner radii

Since blow molding produces hollow parts from the inside out, the corner radii most important are the external corner radii which allow the materials to flow

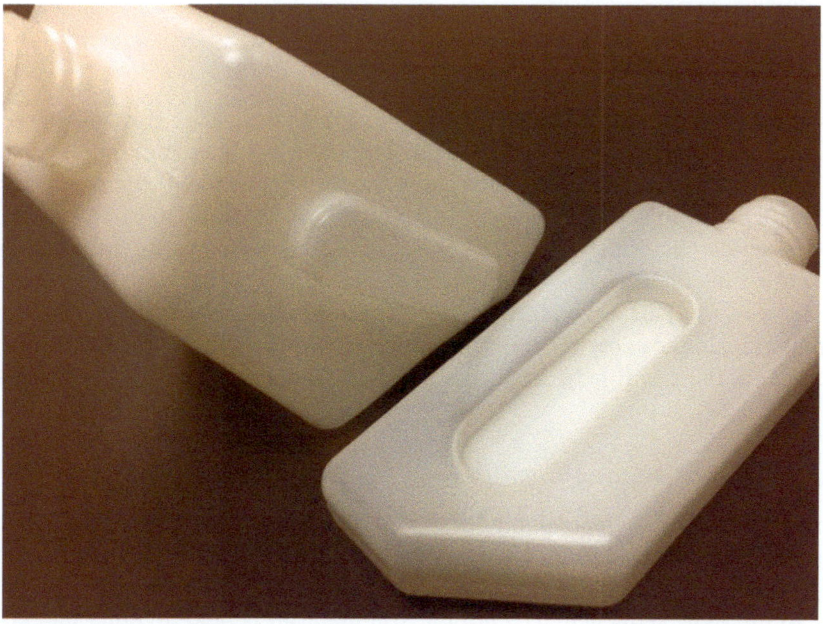

Figure 6.26 A set of blow-molded parts shown with a male and female attachment feature, with reversed taper (undercut) that is flexed to eject from the mold during ejection.
Source: Courtesy of Blow Molded Specialties, Pawtucket, RI.

Figure 6.27 A close-up view of a blow-molded part showing the corner radii at the neck transition and at the blown extreme wall corner.
Source: Courtesy of Blow Molded Specialties, Pawtucket, RI.

outward during the blow process. Sharp radii tend to restrict flow, while generous radii allow the material to flow well. The object in blow molding a part is to achieve the most uniform wall throughout the blown portion of the part possible; hence adequate material flow is paramount to good finished part consistency. Good practice has established the use of a 1/32 in. radius as a minimum, and a nominal radius of ¼ in. is normally acceptable. Sharp corner radii at the extreme perimeter of the part are also difficult to achieve and cause extreme thinning of the materials in those corner areas. Again, a generous radius in these extreme part perimeter corners will help achieve a good uniform nominal wall in the finished part (Fig. 6.27).

6.3.2.6 Ribs and tack-off features

When blow molding a part that needs structure it is sometimes desired to add in features, such as ribs and tack-offs. A tack off is a set of rib features from each side of the part wherein the parison is forced to touch itself by the mold halves prior to the blow sequence. The hot parison material is held together and cooled so the tack off area is thicker than the surrounding material stretched from the blow operation.

6.3.3 Thermoformed (vacuum forming, pressure forming, and twin sheet forming)

Thermoformed parts are made when an extruded sheet of material is heated and then formed into or over a mold and pressurized by pulling a vacuum from beneath the mold and sometimes by dropping a die over and adding additional pressure from the topside of the part (normally atmospheric pressure). Hence, the terms vacuum forming and pressure forming are used interchangeably with thermoforming, each having a bit of a special process twist compared to the other. Since these parts, in a similar fashion to blow molding, have tooling on just one side, the rules for the design of thermoformed parts are somewhat less complicated than those for injection molding (Figs. 6.28 and 6.29).

6.3.3.1 Sheet thickness versus draw depth

There is some rule of thumb for the draw depth or form depth to determine the degree of thinning that will occur during the forming process. This is often called the "draw ratio" and is really expressed as the ratio of the part surface area divided by the part footprint surface area. So for a part with a finished size of 10 × 15 × 6 in. deep the total surface area would be 450 in.^{2} and if the footprint is 150 in.^{2} the "draw ratio" works out to 450/150 = 3.0 which becomes the factor used to determine the starting sheet thickness. For this draw ratio of 3.0 and with a desired finished wall thickness of 0.125 in., the starting sheet thickness should be 3.0 × 0.125 or 0.375 in. thick sheet. If the geometry were changed to a broader base or to a thinned form—say the depth was only 3 in., then the "draw ratio" would be 1.5 and the starting sheet thickness would become 0.188—thinner raw materials equals lower materials cost and faster overall part cycle times, resulting in lower part costs.

In general it is desired to attempt to keep the draw depth as uniform as possible throughout the part, this aids in getting a positive form over and into each of the lowest formed features on the part. Parts having extremes at various locations will need to be processed for the most demanding extreme and the others will follow but cannot be controlled further; hence an extremely tall feature may require a thicker sheet to start with and all of the shallow features will be thicker as a result even if not required. A good design will allow the thinnest sheet to be used to gain the structure required at the minimum overall part weight.

6.3.3.2 Draft

As in other molding processes, draft is certainly required, but perhaps more for material flow during the forming process rather than for the demolding cycle; although it certainly is helpful during the demolding cycle, especially in thick wall parts which are not easily flexed away from tight features. Thin wall thermoformed and in-line die cut high volume parts (think disposable drink lidding) do require draft for rapid forming and rapid ejection, or blow off, from the high-speed process. Disposable food service items, such as cup, plates, and trays are formed using this in-line process method.

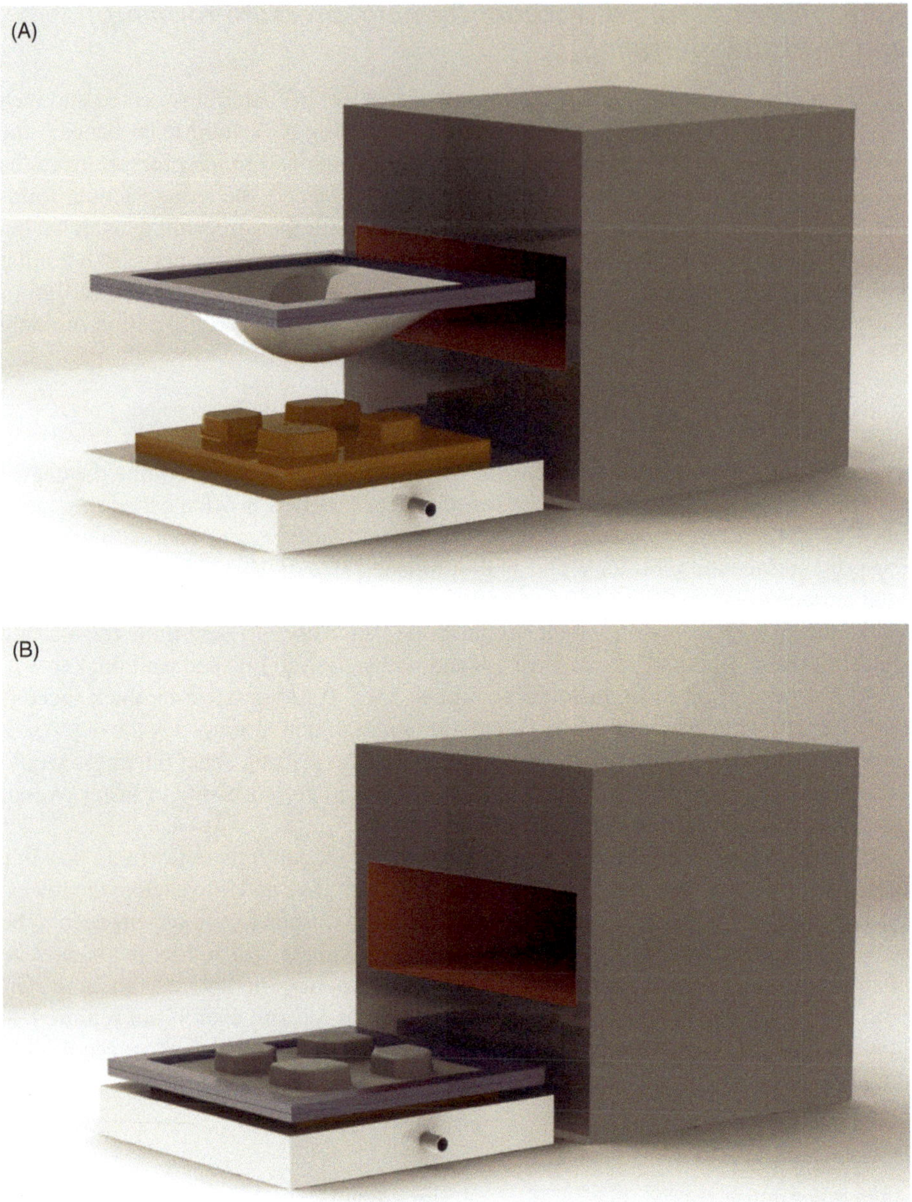

Figure 6.28 A representative illustration (A) showing the cut sheet thermoforming process with the sheet heated and removed from the oven ready to be drawn down over the mold form. (B) Shows the process after the hot sheet has been drawn down and the differential pressure has been applied.

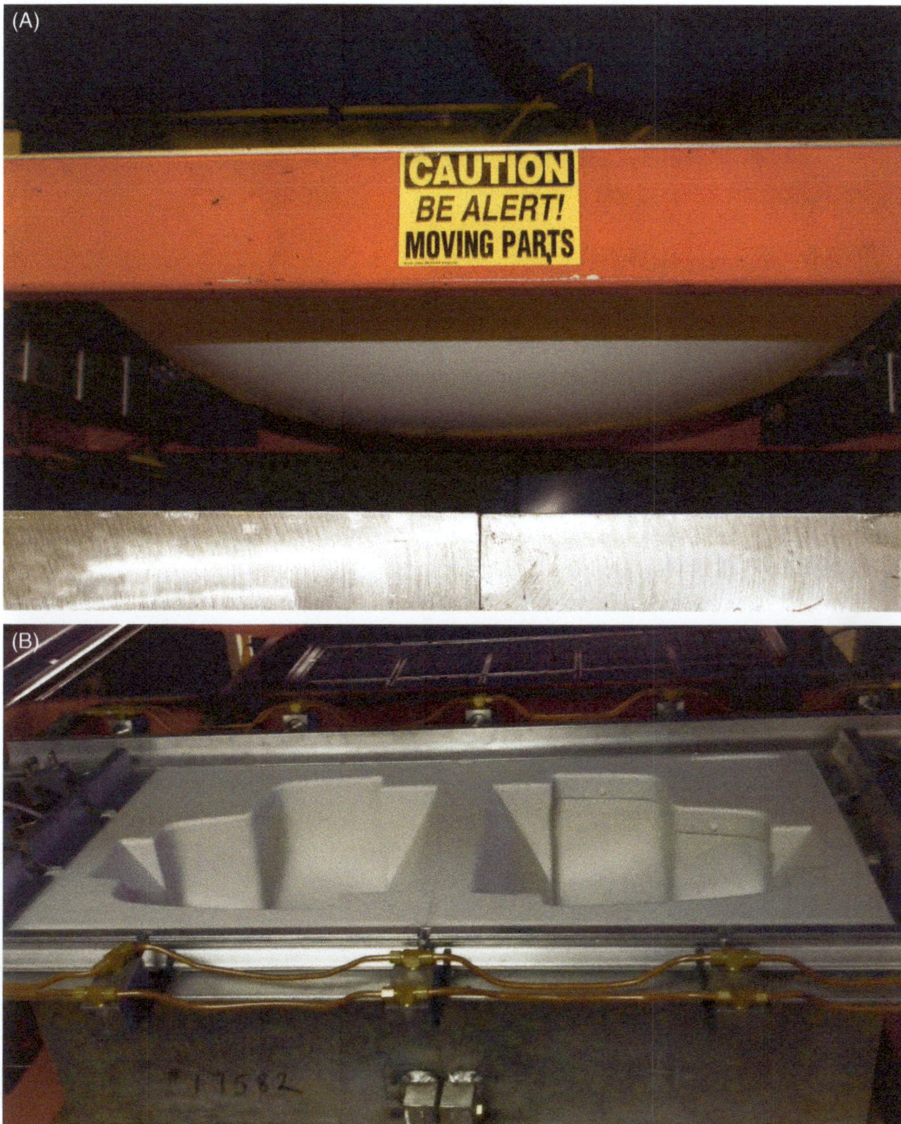

Figure 6.29 (A) This photo show a cut sheet just removed from the oven portion of the equipment, and ready for forming by the tooling. (B) The cut sheet, shown in this photo still held in the clamp frame, after forming into the tooling and before removal from the form tool. On this particular part, a male "plug" form die was used from the top as the tooling closed to complete the process.
Source: Photos from Courtesy of Roger C. Kipp and Associates, and the Plastics Innovation & Resource Center at Penn College.

6.3.3.3 Undercuts

Undercuts can be formed in the thermoforming process; however, there are some limits to what is possible. Sometimes pick-out cores can be used to form various undercut like features or reverse taper areas allowing the part and the insert to be removed during demolding. Thin wall thermoforming allows flex and "pop-off" ejection with undercuts shaped to assist the flex release.

6.3.3.4 Corner radii

As with blow molding, corner radii on features to assist the material in flowing during the forming process are good to consider as the part design is in process. Understanding the flow of the sheet over or into the mold during forming, and visualizing this during the design of the part, will help the designer to understand where generous radii will be required. Normal practice dictates corner radii of 4× the nominal wall; however, with pressure assist (as in pressure forming) the corner radii can get downward to 0.010–0.015 in. In general, deeper features should have larger corner radii to avoid excessive thinning—another way to visualize this is to consider as the draw ratio increases so too does the need for larger corner radii.

6.3.3.5 Ribs

Standing ribs used as stiffening features on thermoformed parts are generally hollow shapes of a pair of drafted but closely spaced near to parallel walls. Single wall standing ribs as are used in injection molding, for instance, are sometimes possible but can be extremely problematic, hence best to be avoided. For thermoformed rib design, the distance between adjacent ribs should be greater than or equal to the rib feature depth.

6.3.3.6 Trimming/routing secondary operations

Since the thermoforming process is a sheet forming process, the demolded formed sheet will need to be trimmed to final net shape. This includes the perimeter of the formed sheet, as well as any openings or holes through the part. Good design will provide a clear path for the trimming router and will provide access for any openings required in vertical or angled walls. Think through this trimming operation during the part design to help and avoid a part design that will require multiple set-ups for the trimming operation.

6.3.4 Rotationally molded parts

Rotationally molded parts are made by introducing a metered volume of powdered or granular polymer within a hollow mold which is closed and sealed from the outside. The powdered or granular polymeric material must have the ability to flow like a fluid as the mold is rotated; some plastics will do this well, others will not. The closed mold is then heated and rotated through multiaxis motions to allow the powdered polymer to become molten and completely coat the inside walls of the mold. The mold is then cooled, while still rotating, to freeze the polymer and form the part. Once cooled, the mold is opened and the part is removed for secondary clean-up of the parting line and/or machining of openings and other design features.

6.3.4.1 Corner radii

Finished part quality and strength are both affected by the designer's selection of corner radii for a rotationally molded part. In general terms, any inside or outside corner on the part will form a more uniform wall, with larger radii and will cool and set quicker with less shrink and residual stress. Accepted normal practice for rotationally molded corner radii is 1/4 in. with a minimum inside radii of 1/4 in. and a minimum outside corner radii of 0.060 in. It is important to remember that both thermoforming and blow molding are stretching process, while rotational molding is an accumulating process and external corners will fill.

6.3.4.2 Draft

Draft for part removal purposes is, as with other plastic processes, material dependent but should be present on all surfaces normal to the mold opening line or parting line. Outside surfaces on the part will shrink away from the hollow mold and will not require the draft that an internal surface might, where the material will shrink onto the core during cooling.

Minimum internal surfaces draft for materials like HDPE or PP can be as little as 1–1.5 degrees, but will allow part removal easier at 2–3 degrees per side. Other more rigid material like polycarbonate, for example, may require up to 4 degrees on internal (shrink to) surfaces. For all materials, additional draft will be needed if the surfaces are textured.

6.3.4.3 Ribs

Since the rotational molding process works best with an inner mold geometry that is smooth and allows the materials to flow freely, ribs want to be open channels of spaced parallel or near to parallel walls; no single standing walls like might be used in injection-molded parts.

6.3.4.4 Closely spaced features or parallel walls

Minimum spacing will be required when rotationally molding parts with closely spaced parallel walls. This helps to prevent the materials from bridging during the molding and cooling process. Bridging will cause warpage and visible shrink will be evident at the bridging site.

6.3.4.5 Undercuts

Undercuts can be easily molded into rotationally molded parts, so long as they are on external (shrink away) surfaces. Undercuts in interior surfaces (shrink to) should be avoided. Each material will demold differently so discussion with the rotational

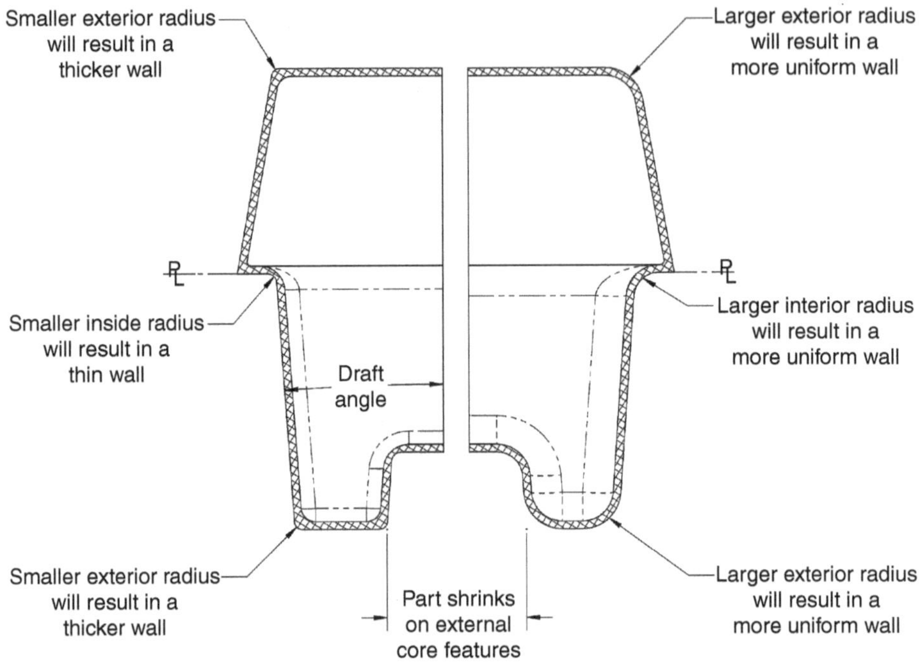

Figure 6.30 Given the physics of the rotational molding process, the materials are melting as they flow over the heated mold surfaces and tend to build up or thin out depending upon the degree of sharpness at a corner or edge transition.

molding vendor early in the design process is recommended when an undercut will be designed into the part (Fig. 6.30).

6.3.5 Profile extrusion

Plastic profile extrusions are some of the most efficient parts made from plastics. The process is inherently efficient and accounts for a large segment of the overall use of plastics. Profile extrusions are used in most every market segment from building and architectural products, to appliances, to health care, to scientific research apparatus, to automotive and transportation uses, and high-tech medical devices. However, from a capital tooling and design standpoint they are far simpler than most other processes, as they are essentially a two-dimensional process. Yes, some equipment allows for downstream secondary operations which sort-of makes them a 3D process, but the majority are extruded as a profile and then secondary operation are added well after the primary processing.

6.3.5.1 Wall thickness

Since the extrusion process is a pressure process, wall thickness throughout the cross section as it is extruded, does not have a great impact on the process.

However, as it cools after extrusion, wall thickness, or the change in wall thickness from one portion of the profile to another may well have an impact on final part geometry and quality—remember this is a thermal event and the thicker sections cool slower, and will shrink more and warp relative to thinner faster cooling portions of the profile. Of course, this is material dependent and some materials are more susceptible than others to this differential wall thickness cooling phenomenon. Uniform walls throughout the cross section of the profile are recommended for optimum processing windows. Be sure to consult with experienced extruder processors and material suppliers when dealing with large differences in profile cross-section thicknesses.

6.3.5.2 Profile interior detail

For profiles with hollow sections, minimize or eliminate detail within the hollow section whenever possible. In fact if possible, create the design with standing features or ribs to attempt to avoid a hollow section all together. Hollow tube like shapes with interior detail containing secondary hollow geometry (a hollow within a hollow) should be avoided.

6.3.5.3 Corner radii

No sharp corners—all corners should have a radius, both inner and corner intersections must have a radius. Corresponding interior and exterior corners should be designed to keep the wall as uniform as possible through the corner. Sharp corners, especially sharp interior corners can become stress concentration areas when in use leading to part failure.

6.3.6 Machined from stock parts

Parts machined from available plastics stock shapes can be as varied as the raw stock from which they can be fabricated, hence there are no general one rule fits all for these fabricated parts, however, there are some things to be mindful of when producing parts from stock plastics sheet, rod, tubes, and bars.

6.3.6.1 Stress concentration features

Sharp corners and chips from machining operations, even microsized chips, can lead to crack initiation points when stressed or under load. Always try and specify features with generous corner radii or fillet radii. Look for any feature which when loaded might create a stress concentration. Remember to let the machine operators know any details which might be material specific and note these on the part specification drawings so that every potential vendor or operator will have the information. These might include information regarding tool sharpness requirements, tool speeds and feeds, special coolant requirements or cautions, and the like.

6.3.6.2 Annealing

Parts machined from stock are perhaps the most likely place that an annealing step will be used in plastic part production. Be sure to experiment with the annealing process profile to achieve the most efficient process for the particular part and materials, and then be sure to annotate the process completely on the part specification drawing.

6.3.7 FDM additive manufactured parts

Fused deposition modeling (FDM) has been known and used for prototyping for a number of years. Now commonly referred to as 3D printing, recent developments for FDM in both the process and the materials used in the process have increased the efficiencies to the point that the process is now being considered as a true manufacturing technique. Not just for prototyping any more—but a real manufacturing process, referred to now in manufacturing circles as *additive manufacturing* which is differentiated from machining from stock processes considered subtractive manufacturing (material removal).

Rules which apply to part design from FDM are completely contrary to the rules for any other process, since the process is a bit-by-bit thermal event, not an entire part thermal event. The orientation of the bit-by-bit layers relative to applied loading or strength requirements is but one of many process specifications that the designer should be annotating on the production specification drawing. Specification of secondary operations for clean-up of external surfaces or functional surfaces is another aspect which the designer must note and describe on the part drawing. Understanding of joining techniques used for FDM and proper specification of threaded inserts, and so on, is a must for parts produced by this process. The current best source for design rules comes from practitioners in the field, using their knowledge of the design capabilities of the process and materials to assist clients in solving part issues. Another resource are the technical papers presented each year at the Society of Plastic Engineers ANTEC conference, which are available through the conference proceedings as published by the Society of Plastic Engineers. Many of these research papers focus on part geometric orientation with respect to the layering process and the resulting strengths attainable with respect to said orientation. Good information when considering the process for any part design where strength on a given axis or direction might be critical (Fig. 6.31).

6.4 Prepare a preliminary bill of materials

On most every design project I will prepare a bill of materials (BOM) as I create the design layout. On most projects there will be several parts which are the custom plastics parts, but there will also be hardware, or gaskets, or labels, or fittings, or electronics, or stamping parts, and the list goes on. Starting a BOM side-by-side with the design layout helps to identify all of the required bits and pieces for the product but it also serves as the central information store.

Figure 6.31 FDM parts were traditionally used for prototype use and are now becoming a mainstream manufacturing method—commonly called additive manufacturing.

6.4.1 *Identify and list each part, quantity required, tentative material, and proposed process*

The preliminary BOM should identify each part in the final product or assembly being developed. These do not all need to be listed at once, rather I tend to add them as I add them into the design layout. The quantity required per finished assembled unit should be identified. The initial selection for material should be identified as should the proposed process for making the part. Once the part is near completion in the design layout I tend to record the part volume and sometimes calculate part weight if I am secure in the material selection for the part. Any other special attribute for each part in the BOM should be listed as well; these might include surface finish, color, durometer, flammability rating, and the like (Fig. 6.32).

6.5 Prototyping during active design

Once the design layout is nearing a final form, prototypes are often desired by the design team for initial "touch and feel" feedback. Sometimes these are requested by the marketing team or the management team so they can now see exactly what the designers and engineers have been talking about with so much enthusiasm over the past weeks. Prototyping has many befits for a creative design project, but there are also risks when in a development mode.

Project 5125 - PARTS LIST - Bill of Materials									
Part #	Rev	Description	Req'd per Unit	Material	Color	Process	Mold I.D.	Part Volume (cc)	Data File
5125-1001	B	Handle	2	PVC	White	Injection Molded	1	130.25	5125-1001 Rev B - Handle.SLDPRT
5125-1002	C	Sleeve Body	1	ST Nylon - Zytel	Natural	Injection Molded	2	43.32	5125-1002 Rev C - Sleeve Body.SLDPRT
5125-1003	D	Nose Collar	1	PVC	Blue	Injection Molded	3	29.96	5125-1003 Rev D - Nose Collar.SLDPRT
5125-1004	N/A		0	n/a	n/a	n/a	n/a		
5125-1005	A	Loop Strap	1	PVC BELTING	Blue/Black	COTS	n/a		
5125-1006	B	Ratchet Lock Spring	2	POM	Yellow	Injection Molded	4	11.56	5125-1006 Rev B - Ratchet Lock Spring.SLDPRT
5125-1007	B	End Bumper Grip	1	EPDM or TPE	Blue (to match Nose Collar)	Compression Molded	5	53.94	5125-1007 Rev A - End Bumper Grip.SLDPRT
5125-1008	F	Rope Lock Cam	1	PC 25%GF	Red	Injection Molded	6	8.49	5125-1008 Rev F - Rope Lock Cam.SLDPRT
5125-1009	C	Cam Support	1	ST Nylon - Zytel	Natural	Injection Molded	7	40.64	5125-1009 Rev C - Cam Support.SLDPRT
5125-1010	B	Housing Top	1	PVC	Blue	Injection Molded	8	44.76	5125-1010 Rev B - Housing Top.SLDPRT
5125-1011	B	Housing Bottom	1	PVC	Blue	Injection Molded	8	42.1	5125-1011 Rev B - Housing Bottom.SLDPRT
5125-1012	A	Rope Cam Spring	1	SS Compression Spring	n/a	COTS	n/a		
5125-1013	A	Belt Clamp Plate	4	SS Sheet	n/a	Stamping	n/a		5125-1013 Rev A - Belt Clamp Plate.SLDPRT

Figure 6.32 A bill of materials (BOM) example from Project 5125 comprising a Boat Docking Aid product, showing the part number, the current part revision, the number of each part required per assembly, the material, the desired color, the selected process, the mold ID, the part volume, and the data file designation for each part.

Fortunately, prototyping is now a relatively fast and painless effort. In days gone by, we hired experienced model makers who fashioned the prototypes from wood, or from multiple thin sheets of laminated styrene. Sometimes these could only be exterior models for style and feel but the models were solid, so their weight and densities were not true. These "hand-made" models would take weeks and weeks of schedule time and would be as costly as some of the capital tooling. Then came rapid prototyping ushered in by a process called stereolithography which when given a 3D solid data file, would process the file into layers, and then used a high intensity UV laser to selectively cure (in a bit by bit fashion) an epoxy resin. The results were amazing; we could get a representative model in a few days' time. Since that time, the rapid prototyping world has progressed to the point where nearly every design office has some form of a 3D printer or has access to one.

6.5.1 Visualization benefits

More than just a means to let the management team or the marketing team see the future part or product, prototyping can help the designers and engineers thinking through the design as well. Even with the advanced graphics of today's CAD programs, nothing beats holding a part in your hand. I would venture to say that in a majority of cases, there will be at least one, if not many, small iterative changes that will be made after the designer of a part sees the first 3D prototype of a part they have designed. In addition, if there are several parts in an assembly which fit together, the designer can visualize better holding the part and assembling them. The ease (or not so much) of assembly, of holding, of nesting for subsequent operations, and so on, can all be comprehended easier with a solid object in your hand.

To be certain, it is wise to prepare a final design prototype or two for archiving, and for use within the design review sessions. Prototypes can help to convey information within the design review saving time and efforts. There is some element of visualization that designers and engineers possess, that many within an organization simply do not. The prototype helps those "visualization challenged" individuals in the design review sessions to get up to speed with the group quickly and effortlessly. A wise investment.

6.5.2 Prototyping cautions

Prototypes are nothing more than tools for the design team to use to help them think through all of the facets of the design project, right? Well, that is true; however, be careful when making and using prototypes in a product development scenario. Prototypes convey different meanings to different people. For instance, to the marketing team, once they see it, that's it, they want it … then, later in the project when first parts come off the tool, which are slightly different (because the design progressed) the marketing team is upset because their expectations were set by the prototype. Now to be true, sometime just the opposite happens, so this can work both ways. Another example is management, when they see the prototype, guess what they are thinking, "The end is in sight, we can turn off the funding faucet, the project is almost over, I saw a unit in engineering, it must be near time to get this puppy into production."

In addition to the "expectations" caution, there is one additional caution to mention, that is the reality caution. These 3D solid objects are prototypes, NOT REAL PARTS. Yes, they may be tightly tolerance and fit just as real part would, and yes, they might be made from similar materials just as the real parts will be made, but they are not the same. They cannot, and will not, provide one smidgeon of functional performance information, except perhaps the ease of grasping the device, or the aesthetic feeling one gets from viewing the shape, and so on. However, the physical properties of the prototype will be vastly different from the real as processed (injection molded, thermoformed, rotationally molded, extruded, etc.) part and don't let anyone within the organization believe otherwise.

6.6 Review of the preliminary design solution

Once the preliminary design layout is complete the design must be vetted against the concept specifications and, more importantly, the functional specifications. This is best accomplished in a multidisciplinary team approach, where the team is made up of individuals from different departments within the organization. Ideally, each group within the organization that will have some responsibility in bringing the product to market should be represented. Often, only marketing and engineering are represented, but I like to encourage early involvement of the anyone who will have any part in bringing the part or product to market, including, but not limited to, material suppliers, tooling sources, the molding or part processing resources, assembly and packaging resources, sales and marketing, strategic sourcing and purchasing, and quality assurance.

6.6.1 Gather the stakeholders

There are no better critics for a design review than those individuals who will be tasked with implementing and caring for the ongoing production of a new part or product. Each of the participants has skin in the game and wants to influence the design so they can make their portion of the project better or easier for them or for their department.

6.6.1.1 Engineering team

The design and engineering team which has created the design to be reviewed will be required for the design review of course. However, if possible invite another team of engineers or designers from another project within the organization. Let them listen and participate as outsiders, perhaps they will identify something that may have been overlooked or underappreciated by the design team that created the design under review. Fresh eyes and brains are always desired for a thorough design review session.

6.6.1.2 Tool designers

I always like having tool designers in the room during a design review. They are always looking for ways to improve the design to make the tool more robust or just plain simpler and often small changes in the part or product design can simplify the tool, providing a large impact on tool life and capacity. Since many tooling sources are independent and not inside the organization resources, this will often cause some grief with the purchasing or strategic sourcing professionals. They do not want to allow an outside vendor to feel that the project is theirs prior to estimating and actual release of an order. Given the relative importance and the value of the risk mitigation achieved by completing a thorough design review, simply hiring the tool designer as a consultant for the day seems a good option, which I have seen benefit projects time and again.

6.6.1.3 Part manufacturing (molding, forming, etc.)

As with having the tool designer, having the part processing resource represented seems natural; these are the individuals who are going to live with the part or product for years to come. If there is an issue, they will fight it year after year. If they see something in the proposed design that experience tells them should be modified or improved they will speak up. Having buy-in from this group will motivate them to add extra effort once the tools are in their hands for processing and part manufacturing.

6.6.1.4 Product assembly manufacturing

Whether this is a true assembly, or a container to be filled, or just a single part to be packaged, someone has to handle the part after processing and before it leaves the factory for its final destination. These operations professionals will be reviewing the design for things that might make their operations easier or less risky, as well as looking for things which might go wrong that could be remedied in the part design.

Case Example

At one review meeting everyone was on-board with a change to an existing part. We were adding a small fin, which was only on the ejector side of the mold, with reverse draft to help and hold the part firmly to the ejector side of the mold. The part has been sticking to the injection side of the mold and flexing during the mold opening sequence, causing slight but nearly unnoticeable defects. Everyone at the meeting was convinced that this design modification would resolve the issue and remedy the root cause of the defect.

Just prior to completing the meeting and sending the change to the tool shop, the assembly operations guy spoke up, "… um, I think we should look at this a bit more. What if one of the guys in assembly should drop a tool or otherwise hit one of these fins? If the fin cracks, could that crack propagate and cause the part to eventually fail?" In a flash, we knew he was spot on, while the thin fin feature solved one problem, it could, without question, cause a larger one.

After several additional iterations, the team was able to come up with a feature to solve the root cause in processing the part and satisfy the robustness needed for this particular part and application. If the assembly operations team had not been invited to that meeting we would have made two mold changes and possibly had units fail in the field, a far worse condition than the original processing defect problem that were all handled and caught by the quality assurance program in-place at this organization.

6.6.1.5 *Quality assurance*

As discussed earlier in this chapter, having the quality assurance team partnered into the project early and often is a good thing. Make sure they are represented at any and all design review sessions and ask for their input early on during validation and verification discussions.

6.6.1.6 *Material Suppliers*

Depending on the material, the supplier, and the manufacturers processing experience with the chosen material, the material supplier's technical service engineer may be invited for the meeting. Especially when the part is calling for a new or novel material and the processor's experience with the material is limited. This is usually a call made by the design and engineering team, although they could confer with the processor to see if they feel there would be value in having the material supplier present.

6.6.1.7 *Marketing/sales*

Representation and buy-in from marketing and sales is a must. They should be looking for the features and benefits that they can demonstrate to their potential customer base and they should be providing feedback as the voice of the end-user. Marketing should be ready to identify any variations of the product or part that might be desired for a follow-on or compatible "sister" product in the future.

6.6.2 End-users feedback—user studies

Prior to allocating funds to being the design project, someone saw the "need" and that someone was most likely marketing. Since identifying the "need" and since completing the conceptual design selection, the marketing team should have been working to lock down some of their requirements, including retail price points, packaging and graphics, product color schemes, and the like. Having this end-user feedback available for the preliminary design solution review is indispensable.

6.6.3 Critique the concepts and the design

The purpose of this review session is to communicate the design to the various stakeholders and to get their critique relative to their portion of implementation. It is a chance to discuss the proposed design and to flush out all questions and concerns before any further investment of time or resources. Once completed successfully, additional resources and time will be invested to push the design toward full implementation.

6.6.3.1 *Against the written specification*

From the functional specification there will be several attributes which are verified by the design layout; these should be highlighted and reviewed. Each functional category should be discussed and the design team should describe how the preliminary design satisfies each. In addition, if prototypes were built and used for design verification of any functional attributes, these findings should be presented and discussed.

6.6.3.2 Against the budget and timeline

The preliminary design should be detailed enough that preliminary estimates can be made as to the projected capital requirements and implementation timeline can be determined and discussed. Engineering and design team investment until completion should be discussed, as well as any budgeted prototype requirements which may be needed.

6.6.3.3 With respect to end-user/marketing "wants"

The design should be vetted thoroughly regarding the end-users experience and the end-user desires, relative to the "need" as perceived by the potential end-user. Will functional prototypes be required for in depth end-user studies to verify any of these assumptions?

6.6.3.4 Alternate materials and/or processes?

Have any and all alternate materials or processes been discussed and reviewed for potential opportunities? Sometimes a material or process gets overlooked or is not considered because of current infrastructure within an organization. It is always prudent to open the table for discussion of any alternates which may be considered. Most times the engineering team will have vetted their decisions well, but with all stakeholders present, this question can sometimes promote some valuable discussion.

6.6.3.5 Cost and lifespan targets?

Do the preliminary cost estimates match up and align with marketing's retail price point findings? If not, then some discussion and work may need to be done to figure out the benefit of developing the project further. Can the preliminary design solution meet the anticipated lifespan from the marketing research findings? Should the product carry some warranty or guarantee? Is the preliminary design solution robust enough to support the desired marketing claims?

6.6.3.6 Compliance issues, hazards, and risks?

Is the product used in an application where there are regulations which require compliance? If so, does the preliminary design solution satisfy those regulations completely? Are there any potential hazards associated with the preliminary design solution, in manufacture, in storage, and/or in use? Are there other risks associated with the preliminary design solution once implemented and put into the stream of commerce? Could the part or product fail in some manner that could be life threatening? Could the part or product pose a risk to an unsuspecting user, say a child or pet? Could the part or product be misused in some manner that could be dangerous?

Observation for Thought

Packaging washing chemistries in single use "pods" was a wonderful creation, assisting users in simplifying the chore of metering out the product at each use. However, the brightly colored small packages looked deceivingly like something good to eat, if you are a child. They looked like so many single wrapped candy products. Was this a risk? You bet it was. Not too unlike small mint flavored candies all too often shaped exactly like a pharmaceutical pill. It is a great shape for swallowing, or for placing in the mouth for enjoyment of the mint flavor, but at what risk? How many pharmaceutical pills were ingested by children thinking they were eating a candy? Perhaps none, but these illustrate potential risks that designers should be considering and should be open and on the table for discussion during the design review.

6.6.3.7 *Is there a plan B?*

What if? If something goes terribly wrong, is there a parachute? The team contributing to the design review and making the decision to proceed to invest resources in the project needs to thoroughly determine what, if any, backup plan they can put in place.

6.6.4 *Preliminary design FMEA*

Failure modes and effects analysis (FMEA) is a powerful analytical tool that we can exploit to assist in limiting risk for the preliminary design solution. Applied at this preliminary stage it can help direct some of the effort invested during the detailed design implementation and then can be reused once the final design is ready for evaluation.

Design efforts, in general, are positive thinking experiences; that is, we are in creating mode, generating new and beneficial things that can change the world. FMEA is not one of those positive thinking exercises, in fact it is just the opposite; you have to cross over to the dark side when working on a FMEA. It is sometimes recommended that the Design FMEA be conducted by a group other than the actual design team for this very reason. It is unlikely that a design team that has invested time, intellectual resources and creative resources, will be able to then think and feel that the failures can actually occur. In many instances the failures categorized will not occur, they are theoretical. However, the exercise will provide insight and a means to provide remedial design effort to attack the potential failures and provide the greatest potential benefit.

The FMEA requires one to think about all of the possible failures which could occur, all of them. It causes the team to think about what could go wrong, what the causes might be, and what the effect of the failure will be. Once each possible failure is listed, each is assigned a theoretical rating, or score, in three classifications, including; the relative magnitude, or severity of the failure, the likelihood of the failure, and the ability to detect the failure. These three theoretical ratings are assigned a score on a relative scale of 1–10, with each classification having a different comparative metric for the range of what is a score of 1 and what is a score of 10. The three ratings once recorded are then multiplied together to produce a dimensionless number called a risk priority number. The higher this risk priority number becomes more likely the failure it represents will occur and the higher priority the design team should place on ensuring the failure cannot occur.

6.6.4.1 *Scoring, or theoretical rating, for severity*

Minimal	1	Trivial fault that the end-user will not be aware of.
Minor	2–3	Minor failure, cosmetic, no performance degradation.
Moderate	4–5	Annoying/upsetting failure—small degradation of performance.
High	6–7	Noticeable degradation of performance—may disturb the end-user.
Very high	8	high severity failure—significant performance degradation.
Extremely high	9	Failure with no advance notice to the end-user—major degradation of performance—total system failure.
Most severe	10	Failure without notice to the end-user—failure is catastrophic and constitutes a danger to the end-user.

6.6.4.2 *Scoring, or theoretical rating, for likelihood*

Remote	1	Highly unlikely that the failure will occur—a history of similar designs with no problems experienced.
Low	2–4	A low probability that this failure will occur—few problems experienced in similar designs.
Moderate	5–6	Some failure likely but not in major proportions—similar designs exhibit regular, although occasional, failures.
High	7–9	High probability of this failure—similar designs traditionally cause problems leading to failures.
Very high	10	Failure is certain in significant proportions.

6.6.4.3 *Scoring, or theoretical rating, for detection (process)*

Remote	1–2	Assembly/manufacturing process/technique virtually precludes defective manufacture—means of inspection/test screen out virtually all failures.
Low	3–4	Probability that the process can yield some defective parts/assemblies is low—inspection/test procedure detects most failures.
Moderate	5–6	Assembly process and/or inspection procedures have a moderate probability of detecting this failure.
High	7–9	Low probability that process and/or inspection procedure will detect this failure.
Very high	10	Remote possibility that the process will avoid this failure mode or that inspection will detect it.

The value in performing the FMEA is the resulting list of RMA values, providing a priority list for the design team to attack. The obvious goal is to reduce all failure modes to the lowest RMA number possible, often the first and easiest means is to attack the likelihood classification by modifying the design to reduce the likelihood score to 1. Detection would then become the next target for remedial design efforts; creating features where the failure can be detected early in the process or can be inspected early in the process. The FMEA score sheet will be revisited and updated at the end of the detailed design to provide a metric of the remedial design effort during the detailed design stage of the project, prior to production implementation (Fig. 6.33).

The FMEA can be accomplished prior to the preliminary solution design review but I like to recommend that it be accomplished afterward. It is truly a separate event and the results are valuable for the detailed design efforts not necessarily for the preliminary solution design review. In fact, if the preliminary solution design review sends the design team back to prepare an alternate design solution and does not approve the initial solution, then any time invested in FMEA work may be wasted effort.

6.7 Action items upon departing the preliminary detailed design stage

Once the review of the preliminary detailed design solution is completed with success, the focus is now directed to the due diligence of finalizing the design and detailing all aspects to be ready for implementation in a production context. Each team will need to work independently on the portion for which they have responsibility, but should stay in communication with all of the other teams so that the project can move forward efficiently and promptly.

6.7.1 Design team

Gather all notes from the review sessions and implement any required remedial changes and updates back into the master CAD layout and archive for safety. Provide information as needed to support the operations or strategic sourcing team for gathering budgetary estimates from potential outside vendors and/or from in-house manufacturing groups. Provide information and/or prototypes or mock-ups as might be requested by marketing for preparation of sales and marketing literature and studies. Prepare and provide information regarding the detailed implementation of the preferred concept for use in patent research and application preparation as needed. Make preparations for detailed piece part development and validations, including verifications of materials and processes selected for each piece part.

6.7.2 Marketing/sales team

Refine and continue to develop marketing literature and concepts, including research into best options for packaging and messaging. Develop color schemes and product marking requirements, if any. Begin to lock down any and all variables needed in the final packaged product. Develop and lock-in a price structure, ready for roll-out.

FMEA Score Sheet	Project: 5094	Author: MTM-B
Document File: 5094-1011	Title: Bottle Cap	Team: Design
Revision / Iteration: C	Part/Assembly Number: 5094-1011	Sheet: N/A

Title Part No.	Function and/or Operation	Potential Failure Mode	Potential Cause of the Failure	Potential effect of the Failure	A Severity	B Likelihood	C Detection	RPN AxBxC	Recommendations and comments	Actions
Bottle Cap 5094-1011	Retains device to bottle	Stripping off the bottle threads	Improper specification of the tread form	Leakage of contents from the bottle during storage, shipping, and handling.	8	1	5	40	Double check specification at Design Review and during Qualification	Design Team to follow-up
		Stripping off the bottle threads	Improper molding of the thread form during production	Leakage of contents from the bottle during storage, shipping, and handling.	8	1	5	40	Supply thread checking fixture and specify inspection dimensions and min/max conditions.	QA to follow-up
		Disconnects from device	Snap feature on device is out of specification	Catastrophic leakage failure	10	1	5	40	Double check specification and validate during prototype build.	Design Team
		Disconnects from device	Snap ring on cap is out of specification	Catastrophic leakage failure	10	1	5	40	Specify inspection dimensions and validate during FAI – incorporate into in-process QA criteria.	Design Team and QA to follow-up
		Disconnects from device	Retainer ring feature on cap shears off	Catastrophic leakage failure	10	1	8	80	Verify maximum shear load and determine adequate in-process load test to qualify after each mold set-up.	Design Team to QA and production team
		Disconnects from device	Snap ring on cap shears off	Catastrophic leakage failure	10	1	8	80	Verify maximum shear load and determine adequate in-process load test to qualify after each mold set-up	Design Team to QA and production team
	Anti-Back-off feature to prevent cap from unscrewing	Doesn't hold cap at torqued closed position.	Retainer features not specified properly	Cap can loosen during storage, shipping, and use allowing leakage of fluid contents.	8	1	2	16	Double check specification and validate during prototype build.	Design Team
		Doesn't hold cap at torqued closed position.	Retainer features out of specification on molded cap	Cap can loosen during storage, shipping, and use allowing leakage of fluid contents.	8	1	5	40	Specify inspection dimensions and validate during FAI – incorporate into in-process QA criteria.	Design Team and QA to follow-up
		Doesn't hold cap at torqued closed position.	Lug features not specified properly on Bottle	Cap can loosen during storage, shipping, and use allowing leakage of fluid contents.	8	1	2	16	Double check specification and validate during prototype build.	Design Team
		Doesn't hold cap at torqued closed position.	Lug features out of specification on molded bottles	Cap can loosen during storage, shipping, and use allowing leakage of fluid contents.	8	1	5	40	Specify inspection dimensions and validate during FAI – incorporate into in-process QA criteria.	Design Team and QA to follow-up

Figure 6.33 An example FMEA score sheet used on a closure design for a bottle mounted dispenser device. The score sheet shows how the RMA values are determined and the notes used to focus the efforts for each potential failure mode or cause.

6.7.3 Legal

Review the preliminary design concept for patent potential and against any current art which may be close in the art. Discuss and review product marking requirements relative to any regulations which might encompass the product or market arena in which the product is sold or used. Complete a thorough review of the project/program from a liability standpoint. Working with management and marketing, the product warranty and any disclaimers to be included with product use and care instructions should be crafted by the legal professionals.

6.7.4 Manufacturing operations and/or strategic sourcing

Continue to evaluate potential outside vendors and gather budgetary estimates from both outside and internal resources for implementation costs and timelines. Ensure that quality assurance is on-board and has an understanding of the implementation plans.

6.7.5 Quality assurance

Review the manufacturing implementation plans as they become available and look for opportunities to include quality system check points within the implementation plan. Work with the engineering team to fully identify and document any and all quality specific notes and instructions required on specification drawings used for primary and secondary manufacturing of the part, product, or assembly. Including identification of specific control dimensions or attributes on each custom manufactured/processed plastic part, product, or assembly.

Reference

[1] From lessons within G. Boothroyd and P. Dewhurst, Product Design for Assembly, Boothroyd Dewhurst Inc., Wakefield, RI, 1989, pp. 2–12 (through 2–23).

Resources for further study

S. Levy, J.H. DuBois, Plastics Product Design Engineering Handbook, second ed., Chapman & Hall, New York, 1984.

S.E.M. Selke, J.D. Culter, R.J. Hernandez, Plastics Packaging—Properties, Processing, Applications, and Regulations, second ed., Carl Hanser Verlag, Munich, 2004.

J. Rotheiser, Joining of Plastics—Handbook for Designers and Engineers, third ed., Carl Hanser Verlag, Munich, 2009.

G.L. Beall, Rotational Molding—Design, Materials, Tooling, and Processing, first ed., Carl Hanser Verlag, Munich, 1998.

E.R. Larson, Thermoplastic Material Selection—A Practical Guide, first ed., 2015, Elsevier; Oxford.

R.A. Malloy, Plastic Part Design for Injection Molding: An Introduction, second ed., Carl Hanser Verlag, Munich, 2010.
P.A. Tres, Designing Plastic Parts for Assembly, seventh ed., Carl Hanser Verlag, Munich, 2014.

Web links of note

Great Design information from DuPont can be found here:
http://www.dupont.com/content/dam/dupont/products-and-services/plastics-polymers-and-resins/thermoplastics/documents/General%20Design%20Principles/General%20Design%20Principles%20Mod%201.pdf.
Sabic provides many tools for designers here (you may need to register):
http://www.sabic-ip.com/webtools/toolsHomePage.faces.
The UL Prospector site is a valuable tool for any designer reviewing materials for any application (you will need to register):
https://plastics.ulprospector.com/.
Solvay provides a great design guide for using their Udel polysulfone materials:
http://www.solvay.com/en/binaries/Udel-PSU-Design-Guide_EN-227550.pdf.
Bayer has a nice design guide for RIM systems:
http://reactioninjectionmolding.com/wp-content/uploads/2013/09/RIM-PartMoldDesign-Guide.pdf.
Covestro offers several design related guides, including part and mold design, snap fit design, and joining techniques for plastics:
http://www.plastics.covestro.com/Library/Technology-brochures.aspx.

Final detailed design

7

Given input from the preliminary solution design review and from the design failure modes and effects analysis (FMEA) session, the design team now is armed and ready to move into final locked-in detailed design. This is where every "i" gets dotted and every "t" gets crossed, a major investment of time and resources to fully specify, annotate, and document every part in the design, ready for production implementation. Precision and thorough complete focus is required while preparing and documenting the final detailed design specifications and data files. In addition, thorough checking and a full design review are completed during this phase to minimize any and all potential risks as the project moves into production implementation. Once the master layout is updated and checked, each of the individual piece part files is separated and broken out from the main layout for final detailing and checking.

7.1 Iterative updates to the CAD 3D layout and BOM

The final master 3D layout computer aided design (CAD) data files from the preliminary design stage and the bill of materials (BOM) files should be archived for safekeeping. A copy of each should be then used to begin the first iteration for detailed completion. These efforts will include completion of all detailed features to every part, including all required draft surfaces, all corner and fillet radii, all fit and tolerance analysis, and the like. However, before we can begin the details it is imperative that the master layout files be updated as completely as possible from the information gained during the preliminary design review processes.

7.1.1 Updates to the 3D CAD layout and BOM as required from FMEA and design review findings

Any and all major remedial design changes to the master 3D CAD layout should be completed first, including everything from the design review and the FMEA. In this manner, the fine details are not added until the major efforts are completed and checked, thereby reducing risk, and avoiding any wasted effort or time. In fact, depending upon the degree of remedial changes, sometimes an interim design review is in order to ensure that the remedial design changes are adequate relative to the potential failure or other rationale for the change in the first place.

Likewise, the BOM is revised and updated to reflect any and all changes that may be required from the design review or from the FMEA session, including any materials changes or any special notes for a given part or process.

Designing Successful Products with Plastics. http://dx.doi.org/10.1016/B978-0-323-44501-6.00007-6

7.1.2 Adding in details to the 3D CAD master layout for matching and mating parts

Prior to breaking out all of the individual parts for detailing, it is prudent to add in the detail where the detail factors into more than one part in the final assembly. For example, if two parts meet and the corner radii to be added on one must mate with or match the corner radii or some other feature on another adjacent part, then it is wise to add these details in while still in the master layout where the mating relationships can still be seen and measured. Likewise, any press fit surfaces, or mate lines between parts should be completely detailed in the layout prior to individual part breakout. This provides the best time efficiency when adding relating details to mating parts.

In addition, reference features, such as a centerline or a plane can be added to both (or more) mating parts so that measurements and alterations can be made once the part files are indeed separated. I tend to name the reference feature identically on each of the mating parts to help avoid confusion during the detailing process. For example, if I were to add in a reference centerline, which might be an axis (direction) of flow for a fluid that is found on multiple parts within the assembly, I might call the reference centerline something like "flow-axis" so as I am working on any of the parts, I always have the same reference feature name. This is simple housekeeping, but on a multipart assembly with dozens of planes and centerlines shared between multiple parts, the time invested in naming the reference features will reduce errors and allow the designer to be more efficient during the detailed part design phase.

7.2 Breakout for detailed design of features and parts

Once the 3D master layout files are completely finished and all matching or mating part details have been added to the master layout, the team moves to a detail and documentation mode. Sometimes the documentation tasks are handed down to others and the design team can move to the next project, but most likely the design team will continue and complete the detailing and documentation, including production ready specifications.

7.2.1 Breakout all parts from the initial 3D CAD layout for detailed part design

Given a product design with multiple parts, each part will need to be broken out from the assembly layout and archived in its own 3D CAD data file for safekeeping. Each part can then be assigned a part number and a 2D specification drawing document can be made. Now with each individual part removed from the assembly file, as an individual 3D part data file, the fine details can now be implemented and defined much easier. In most every CAD system in use today these "individual" part files remain parametrically connected and referenced in the main (assembly) layout file; hence, changes made in the "individual" part files are simultaneously updated in the main (assembly) layout file (Fig. 7.1).

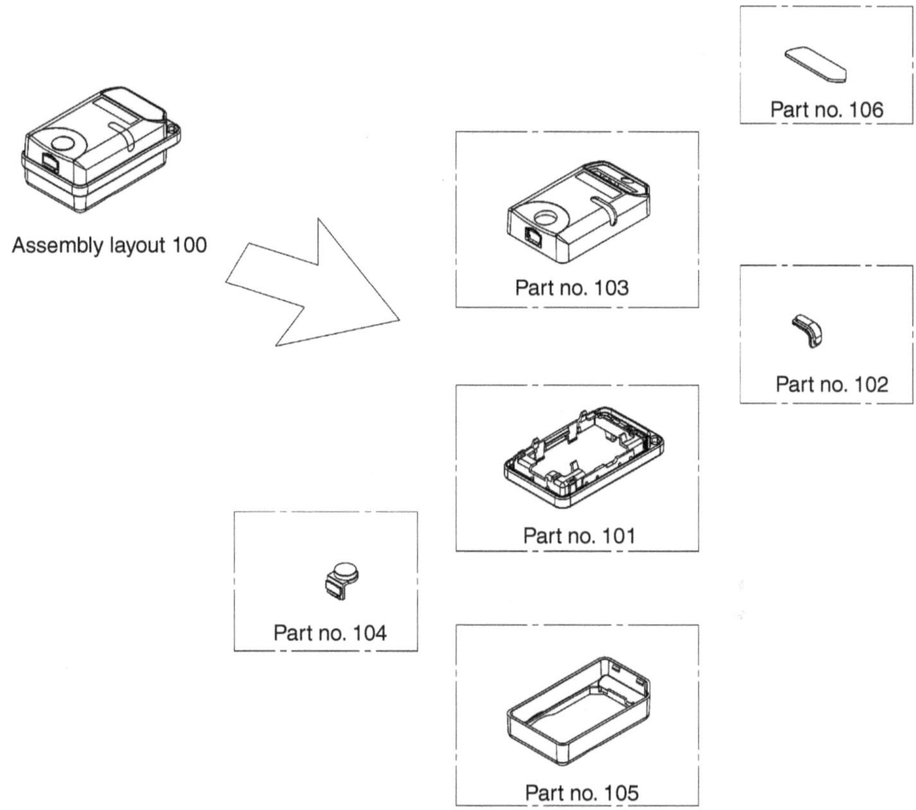

Figure 7.1 A representation of the removing of each individual part file from the layout for individual part feature and attribute detailing. The addition of draft, corner radii, split lines at the parting line, and so on are added in the individual part file once separated from the master layout.

7.2.1.1 Breakout features from parts for detailed feature design and analysis

Sometimes a feature (cantilevered snap, complex pass-through shut off, etc.) is complex and may be easier to develop as an individual feature file, rather than part of a larger more complex part. This break-out allows for not only design of the part, but also for simulation analysis of the feature by itself, prior to integration back in the full finished part. It also allows the designer to document any analysis or calculations made at the feature level for future reference and complete archiving of the design, or for use on another similar design.

7.2.1.2 Add all features to each part, including all corner radii, all draft, split lines, and so on

As an individual part file, check to ensure all features (draft, corner and fillet radii, gate position markers, etc.) are included and completely specified. Add in split lines where a surface may transition a parting line being sure to add draft as appropriate. Insertion of areas where text or graphics may be desired—such as cavity ID, recycling marks, material ID, part number ID, shift ID, date code, lot ID, and so on, can be added and identified so the information can be transferred to the tooling and production processing resources.

7.2.1.3 Follow the rules for the chosen materials and processes

For each given individual part design, review the material and the process selected, and make sure that the rules for that particular process and material selection have been followed as closely as possible. Check the part design for intersecting walls where seals might be located, look for adequate draft, look for proper seal-off angles where a tooling pass-through seal is required, watch for flow from thicker section to thinner sections, and so on. If design changes are incorporated, for example, to better satisfy some process or material rule, make sure to review the impact the changes might have on any mating part or feature on an adjacent part in the final product assembly. Since the part is now separated from the layout, it is easy to overlook some mating relation that might have importance, hence always double check when making part design changes outside of the main layout file.

Case Example

On a recent project, there was a simple tubular part, which functionally was a conduit for a fluid mixture to be dispensed. The prototype functioned but was not optimal with respect to the fluidics for the device. Once we did some testing we found that opening the inner diameter would improve the fluidic functionality for the device, so we did what any red-blooded engineer would do, we increased the inner diameter. The fluidics now worked as desired, however, the processor was complaining that the now thin wall was hard to fill and wanted to change the outer diameter to relieve the processing challenges. A 0.50 mm change to the wall thickness to help process the part was all that was needed; however, when checked in the master (assembly) layout that 0.50 mm change would create an interference with a mating part in one extreme condition. The 0.50 mm wall thickness change caused us to modify the mating part to relive the interfering condition—however, the modification to the second part changed a feature formed by steel from both the core and the cavity, necessitating changes to both. Hence this 0.50 mm slight change has now caused more than twice the cost of the original tool. The moral of this long tale is to be careful and to always check the assembly whenever a change is made to one part.

Once the part file is completed, I will often send the file to the material supplier and ask for a review and opinion based on the preliminary material choice for the part. Likewise if a part processing and a tooling source have been selected for the part, I will ask for a review and opinion of the design from each. Gathering this preliminary feedback, sort of an in-process design review, can sometimes help find small subtle improvements that might be overlooked in a large scale group design review, where many parts or an assembly is being reviewed. The individual part is much less complex and the tooling, processing, and material supply resources can focus their review with much less distraction.

7.2.1.4 Update the BOM with all "new" and updated information

As each detailed part file is completed it is a great time to go back and continuously refine and update the BOM; updating the materials, the part volume, the colors, and any special requirements, which may have changed since the original BOM was created.

7.3 Integration of all parts back into a new 3D CAD master checking assembly layout

Once each and all of the parts are complete as individual files, they can be useful in an assembly checking method, which uses the individual 3D CAD part files to construct a "new" assembly layout built just as the assembly might be built in production. This is sort of soft or virtual prototyping that can show you within the CAD program where issues may arise

7.3.1 Prepare a separate "new" assembly checking 3D CAD layout file

As a means for checking fit and function, I will often use the individual part files to build a "new" assembly checking layout, built using complete parts, and using the assembly order and mating conditions that would be used during production part assembly (Fig. 7.2). If there are components which are meant to deflect or compress, I will often prepare a separate file of each with the solid model in the deflected or compressed condition to use in this checking layout. For example, gaskets are often designed from materials that are specified from some density and are expected to be compressed by some percentage; hence the part file might be produced at nominal manufacturing thickness but a second assembly model might be produced to simulate the gasket when assembly and compressed to X% of nominal thickness. Another example that can be specific to plastics are molded in feature to add a biasing force to some element when assembled; for example, a cantilevered leaf spring feature on an injection-molded part, which may be used to keep some other part in place but allows some variation in part size. On a feature, such as this, I will produce a second solid model with the cantilevered feature in the deflected condition. In most CAD programs these variations of solid models can be integrated within the same solid model by utilizing multiple configurations of the same part, each configuration defining a specific end-use condition or position.

Item no.	Part number	Description	Qty.
1	101	Center band	1
2	Battery		1
3	Sensor		1
4	Circuit board		1
5	105	Case top	1
6	103	Case bottom	1
7	104	Cover	1
8	102	Lens	1
9	106	Nameplate	1
10	Label	-	1
11	Screw, plastite 1.2 mm × 6 mm		1

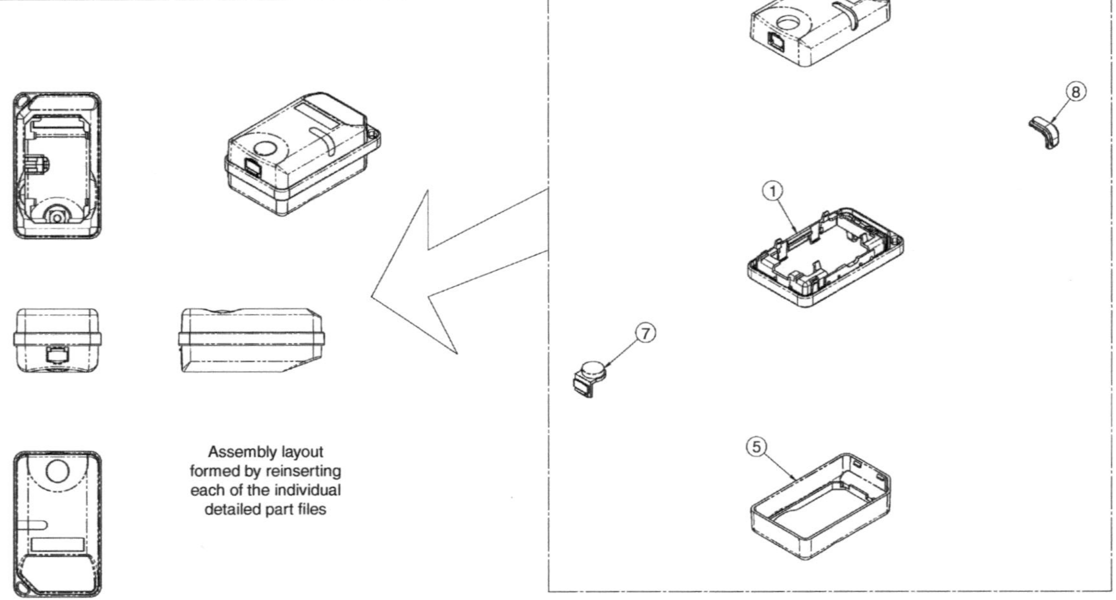

Figure 7.2 Once the individual detailed parts are completed, the recommended checking method is to start a new assembly file and add in each individual part file, as if it was being assembled. Using the same mating arrangements that would be used in production; pins go in holes, surfaces touch surfaces, corners are tangent to surfaces, and so on. In this manner, the clearances and interferences can be seen within the CAD assembly, helping to check each part and the overall layout.

7.3.2 Add parts to the checking layout using physical features for real-world mating conditions

In this manner, the pin feature on part one would fit concentrically to the hole feature on part two and the stepped boss on part one would be coincident, with the flat surface at the top of the hole on part two and part one could rotate relative to part two until the snap feature on part three engaged and snapped into place behind the catch feature molded on part one, and so on. This method of checking simulates the real-world assembly conditions and can sometimes find interferences that otherwise might be missed during the design layout where everything is locked down relative to the X, Y, and Z planes, reference centerlines, or reference planes used during feature and part design creation.

7.3.3 Review for fit, interference, and clearance issues—CAD analysis

Once the assembly checking layout is complete it can be used for further analysis. Remove a concentric mating condition and add in a tangent mating condition to simulate real-world parts fitting at minimum and maximum tolerance conditions. Some CAD programs will analyze fits and given proper tolerance input will provide the range of fits that might be encountered for the fit being analyzed. Other simulation tools including draft analysis can be utilized on individual parts or on the assembly checking layout to ensure that the assembly or part can be made to work at all conditions for the application. With large parts that tend to have large manufacturing tolerances, separate files, or configurations, can be produced at the maximum material condition and at the minimum material condition. The tolerance condition specific files can be inserted into the assembly layout and the resulting final assembly conditions can be observed and measured—sort of a virtual tolerance analysis means. If there are wide temperature extremes expected in the end-use conditions for the part or product, the resulting part sizes (at the temperature extremes) can be modeled and fit, and clearances can be analyzed.

7.4 Detailed part drawings and specifications

Once the part 3D solid data files are complete, the specification process begins. There was much bravado in years past regarding the “Art to Part” promise of 3D CAD; however, there is so much more detail that is required beyond just the part geometry that a corresponding 2D print will almost always accompany the 3D solid file. In fact, I advise many clients to list the solid file name on the 2D specification drawing and to always issue a 2D print with each and every purchase order or shop work order; in this manner the manufacturer will have all of the information required to perform the job correctly. The purchasing originator will have recourse if the wrong materials or files are used, since part drawing files with the correct part number and revision level are provided with each order. In too many cases, production mistakes are manifested from incorrect assumptions because the source vendor is given too little specification with the order.

7.4.1 Individual 3D data file for each part

First and foremost, each individual custom part to be manufactured will require a 3D part file that can be used for creating the tooling and can be used for preparation of inspection fixtures, secondary operation or assembly fixtures, and so on. In some instances there will be two 3D files for a part, for example, if ultrasonic inserts are added to a part post injection molding, then there would be a 3D file for the injection-molded part and then a second 3D file showing the part with the inserts in position. In some CAD programs this may be one file with multiple configurations; however, the point is that the bare injection-molded part must have its own discreet data file for adequate specification, tooling development, and future reference when modifications must be incorporated. The subassembly 3D file will be needed for development of tooling for the secondary operation and for downstream assembly fixture development. If the part is sourced from a subcontract provider, then the subassembly 3D file becomes the controlling document, along with the 2D specification drawing, for the purchasing and quality assurance activities for the part or product.

7.4.2 2D part drawings with specifications

From the 3D part file a 2D part definition and specification file can be produced. This specification drawing can be used for conveying the requirements when attached with a purchase order (as mentioned previously) and can be used to convey quality assurance information as well as cautions and recommendations. This 2D specification drawing can also be used to callout secondary operations, assembly notes, and the like—such as the addition of ultrasonic threaded inserts, as in the previous example (Fig. 7.3).

7.4.2.1 *Specify materials, colors, and tolerances*

A major component which should be specified on the 2D specification drawing is the material that will be used. Full specification is required; not just plastic, or thermoplastic. Be specific, call out the exact material the application requires and be specific if alternates are acceptable. If there are alternates that can be used, list them as alternates to be used with approval of purchasing or engineering. Be careful when specifying a material and adding phrases to the end like "...or equivalent"—equivalent in which manner, with respect to what criteria? Experience has shown that best practices are to call out the exact material, manufacturer, grade, ID number, and so on, be as specific as possible. In addition, be extremely specific when calling for other additives and color. Do not assume that when you specify and exterior grade material that the processor will use the best UV stabilizer on the market. Better to over specify and cause some direct discussion between the processor and the customer than to make any assumptions.

Color specifications can get tricky, many individuals will callout a color relative to some published specification like the Pantone® Matching System (PMS); however, a color paper specification will render itself totally different in a PC resin and different yet in a PP resin. If possible, it is best to get a set of color chips in the specified material and have a unique color identification number assigned by the colorant supplier or compounder; then reference that chip and number on the 2D specification drawing (Fig. 7.4).

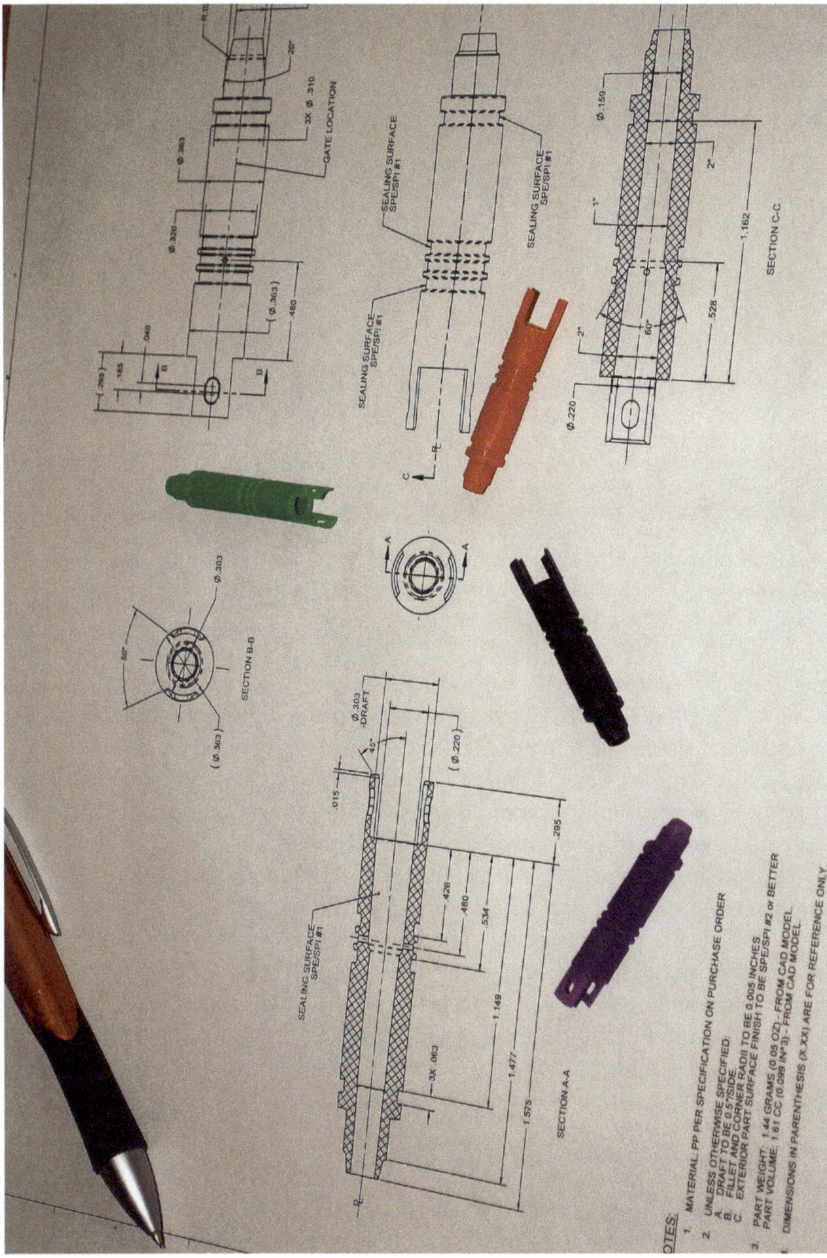

Figure 7.3 A representative 2D part specification drawing, with notes and the finished injection-molded part.

Figure 7.4 Sample color chips are used to specify the colorant, along with a color house color serial number (which should be called out on the 2D specification drawing).
Source: Photo Courtesy of Holland Colours.

Part and feature tolerances will vary from material to material and from process to process, one set does not fit all. Hence, it is recommended that each 2D specification drawing specifically note the tolerances used for the part and dimensions should be inserted for features that are critical and require inspection during first piece inspection or as in-process inspection dimension for use during production operations. In some instances a part may be manufactured from multiple colors. The specification drawing may have a table showing each color version and a dash number for each to identify it discreet from the others. Color variations will be produced, with differing colorant materials at different color concentrate loadings each of which will produce a different final part. Yes each is a different color, but each will have different properties as well. Higher colorant loadings will reduce some properties and enhance others. Part postmold shrinkage will be different as well; hence the tolerance analysis for a part that will be molded in multiple colors need to be reviewed including the net difference in shrinkage resulting from the different colorants.

7.4.2.2 Specify surface finishes

Typically, a general overall surface finish note can be used on the 2D specification drawing for all surfaces, which are not critical to function or cosmetics. Best practices dictate using a reference panel as produced by some national or international organization. Specific to surface finish, the Society of the Plastics Industry (SPI) produced a standard set of plaques showing a series of different surface finishes and outlined how each finish is attained on tool steel use for plastics molding. In general, a note something like "all surfaces to be SPI #3 or better" might be sufficient for this general note, then specifically annotate or callout areas that might be cosmetic, or that might be optical, or sealing surfaces to ensure that the tool designer and toolmaker understand completely that something special is required and to which surfaces these special finishes need to be applied. There are several reference samples, which can be used in addition to the SPI standard; some private companies like Rawal Engraving, and Mold-Tech® have published standards and provide reference plaques (Fig. 7.5) in various plastics materials, in various transparencies, with various grades and examples

Figure 7.5 Reference plaque showing surface finishes by Mold-Tech®, with each specified surface finish showing the depth of the finish, which can indicate the necessary draft required to release without dragging the surface.

of embossed and debossed engraved text, and so on. Many reference plaques will also specify the texture depth, which is directly related to the draft required for releasing the part without surface finish damage. If the draft is not sufficient, then the texture modified surface will drag through the part upon ejection causing an aesthetically unpleasing result. Be sure to specify a texture or surface finish, which corresponds with the draft used for the particular surface in question.

7.4.2.3 Note any sealing surfaces or areas that cannot have ejectors or witness lines

In addition to specifying desired or required surface finishes, it is wise to note areas that have specific functional requirements; especially sealing surfaces, weld joint surfaces, and/or adhesive joint surfaces. Ejector pin placement or cavity split witness lines on these surfaces could be a disaster, so annotate them on the 2D specification drawing to provide the tool designer, with clear direction as to the requirements for the end-use of the part. Tool designers and toolmakers understand sealing and welding; and, provided they are given which surfaces are functional, they will ensure the tool design can produce the part to function as intended. The lesson here is to adequately annotate the specification drawing for the tool designer and toolmaker, the tooling and the part will be better if you do.

7.4.2.4 Note recommended gate locations and other processing information

As the part designer, you know and understand how the part will function in its end-use application and what the part structural loading directions and magnitudes will be. In as far as flow direction will help determine part strength in certain conditions, it is wise to annotate the 2D specification drawing, when deemed appropriate with the desired flow direction and/or recommended gate location to achieve whatever properties are important for that particular part. This is not required on every part, but when the part is critical, the more information you can provide the tool designer the better the results that can be achieved. Once the initial tool or cavity is proven and is acceptable, it is never wrong to go back and update the part files and drawings with knock out ejector pin locations, gate locations, slide surface areas on the part, and so on. This information will help anyone tasked, with implementing remedial changes to the part in the future.

7.4.2.5 Note location for mold ID and/or cavity ID if required

Markings on parts for internal tracking and identification are useful and customary. Add these and specify the desired location on the 2D specification drawing. Be sure to note if these need to be interchangeable, to accommodate revision numbers, cavity identification (Fig. 7.6A) or date lot ID codes, and so on. Other marking for material type and recycling may be required depending upon the application and the end-use customer, so be sure to include all required markings and their desired locations on the 2D specification drawing. If lot ID or date code information (Fig. 7.6B) is used at the part level, it is recommended that the purchase order or shop order for the part be the control document instructing, which lot ID or date code to use for each work order. In this manner there is a direct tie in between the shop floor and the customer which

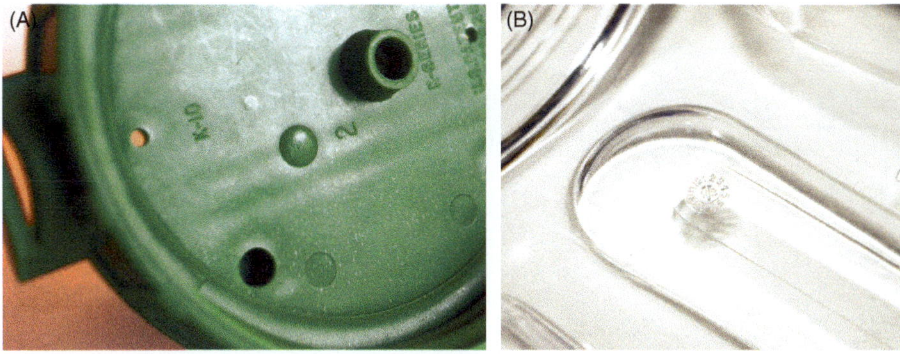

Figure 7.6 Cavity number and sometimes mold numbers are included on an inner surface of a part for accurate tool traceability (A) and changeable date lot ID inserts (B) can provide traceability to date of manufacture, which can be traced to material lot ID, and so on.

will include quality assurance, since all QA or QC activities will typically reference the work order or purchase order as a reference.

7.4.2.6 Ensure all dimensions specify draft where applicable

On most part drawings, a general note will be placed, something like "...unless otherwise noted maximum draft is to be 1 degree per side," which is perfectly acceptable and customary. However, there will be certain features, especially features with tight mating requirements to other parts, where precise specification is needed. I will typically dimension a feature in a side profile (so the tapered surface can be seen) and use the convention calling for plus draft when the feature is getting larger as it moves from the dimensioned surface; and, calling for minus draft when the feature is getting smaller as it moves away from the dimensioned surface. In 30 years of using this convention, I have never had a tool designer question the dimension or make a mistake (Fig. 7.7).

7.5 Detailed part design checking

Every project and every part design are different, some are just plain, simple, and easy; others not so much so. One thing remains constant; checking the part design prior to sending it for tool construction is the prudent thing to do. As the 2D specification drawing is being created the designer has the perfect opportunity to check the part design. In addition, the use of the reconstructed 3D CAD master assembly layout will help to identify alignment and interference conditions.

7.5.1 Tolerance analysis studies—from 2D specified dimensions and tolerances

Again, some assemblies are more complex than others, some require a bit more due diligence, but they all deserve to be reviewed and checked. Tolerance analysis is not a

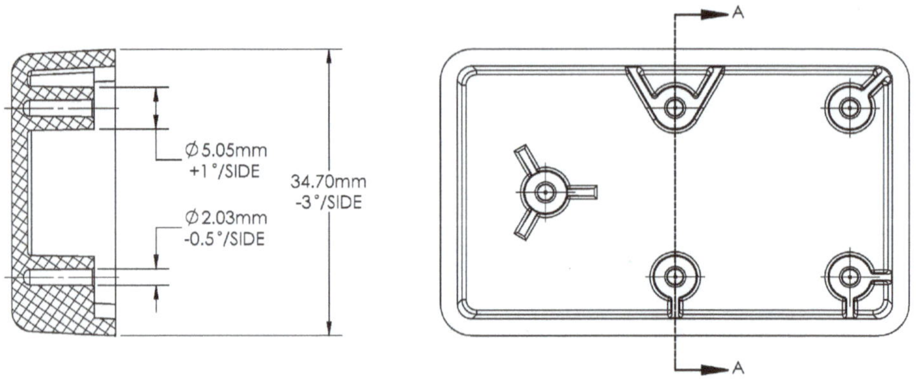

Figure 7.7 A simple example showing the +Draft from the dimensioned surface the part gets larger, and −Draft from the dimensioned surface gets smaller.

difficult task but it is laborious and dull. I tend to document, even if by hand sketching, each fit analyzed and the results for each fit. I keep these in my project file and have them for future reference. On some projects, especially high volume multiple tool and/or multiple cavity projects, I will be more formal and prepare a well-documented account of each fit analyzed and make this document part of the project archive (Figs. 7.8 and 7.9).

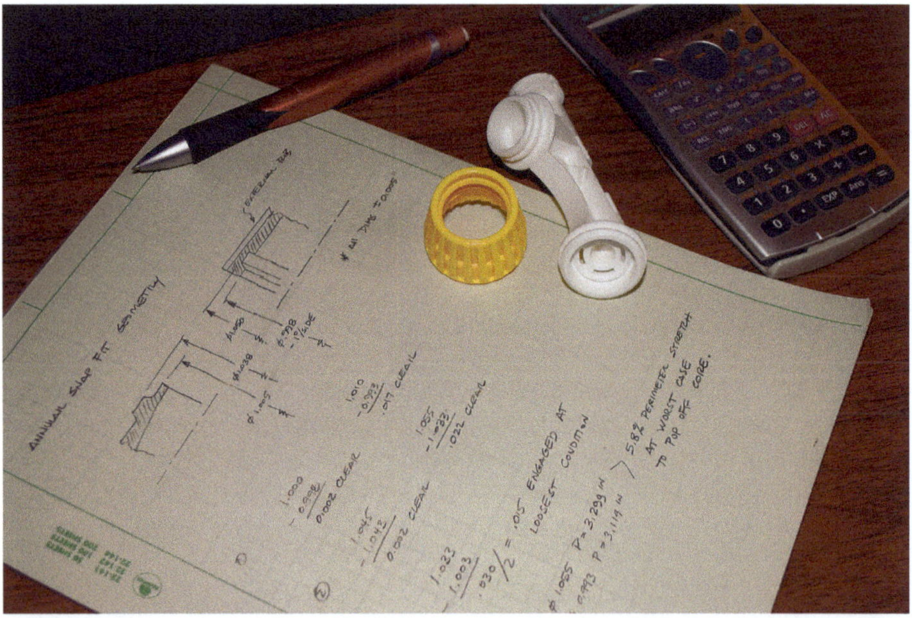

Figure 7.8 On most projects a simple handwritten analysis is all that is required. Show the dimensions and the mating parts, then do the maximum and minimum material condition math and get the results. On some analysis, with seals or snaps, there are additional stretch calculations to be reviewed at minimum and maximum material conditions as well.

Part:	Upper Eductor
Part Number:	2964663
Revision:	A
Revision Date:	17 OCT 2002
Material:	Polypropylene
Part Volume:	4007.8 mm^3
Color:	Natural

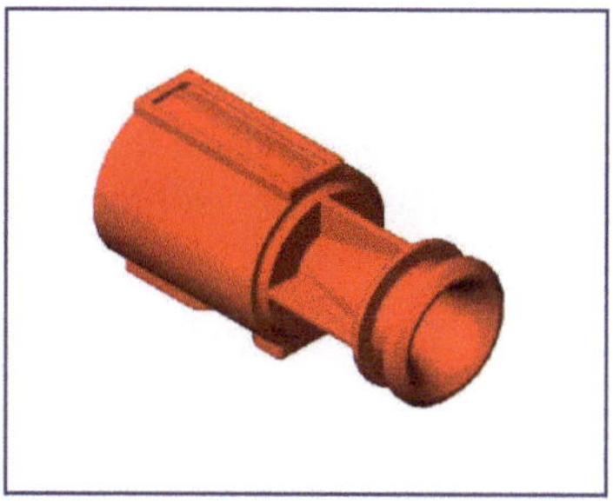

Upper Eductor part

Considerations for fit to other mating parts:

1) Fit to Body part
   a) O-ring on Upper Eductor to fit and seal to bore on Body – over entire translation range.
      i) Interference detection on 8546200 Dispenser Assembly Rev3.sldasm shows no interference with solid models.
      ii) Tolerance Analysis: UE o-ring groove ∅11.2±0.1mm
         AS568-111 o-ring ∅10.77±0.13mm I.D.
         % Stretch ((11.2/10.77)-1) = 4.0% yields 4.8% CS Reduction
         O-ring CS w 4.8% reduction ∅2.494mm
         Bore on Body ∅16.0±0.1mm – zero draft over translation range
         Gland opening = (16.0-11.2)/2 = 2.40
         % Squeeze = ((2.494/2.4)-1)= 3.9% squeeze
         (1) NOTE: This loose fit is intentional for allowing Body to be steel safe for first trials. Actual dimension for bore in Body should be around ∅15.6±0.1mm to yield 13% squeeze. We want this seal to be light to avoid drag during operation of the device. This seal is a vacuum seal only seeing atmospheric pressure (outside in) during operation (carrier stream flowing). It does not see carrier stream pressure at any time.
   b) Flat key areas on top and bottom are to fit Body with running clearance,
      i) Interference detection on 8546200 Dispenser Assembly Rev3.sldasm shows no interference with solid models.

ii) Tolerance Analysis Top:	Upper Eductor width	9.7±0.1mm	
	Body	10.0±0.1mm	
	Fit	0.3±0.2mm	Clear at all conditions.
iii) Tolerance Analysis Bottom:	Upper Eductor width	7.7±0.1mm	
	Body	8.0±0.1mm	
	Fit	0.3±0.2mm	Clear at all conditions.
iv) Tolerance Analysis – top to bottom:	UE	19.7±0.1mm	
	Body	20.0±0.1mm	
	Fit	0.3±0.2mm	Clear at all conditions

Figure 7.9 On some projects, formal documented part assembly dimensional analysis will be required, showing pertinent information regarding the part and the mating features with adjacent parts within the assembly—this is an image showing the first page of a four page report as an example.

7.5.2 Draft analysis—3D CAD analysis

Mentioned earlier, many CAD programs today offer some fairly useful simulation and design checking features; one of these is draft analysis. Given that the 3D part file is created and fully defined, including split lines at the parting line for all surfaces and that drafting has been added to all surfaces, the draft analysis tool in the CAD system can show any surfaces where the designer might have omitted draft or perhaps even applied a taper in the wrong direction (it happens). Using the draft analysis tool is simple and takes only a few minutes to apply and get results; hence it should be used on every part before calling the part complete (Fig. 7.10).

7.5.3 Corner radii and wall transitions

Hopefully the part design was created using all of the information relayed and taught in Chapter 6 of this book, but sometimes a part design may come from an outside source or from an older iteration of a similar part or assembly. Hence, it is always wise to review a part from the basic rules perspective, looking for uniform walls, smooth and gentile wall transitions, flowing from thick to thin sections, and so on. This includes the use of fillet and corner radii to assist flow and to help reduce the possibility of induced shear and stress concentration possibilities.

7.5.4 Gate location and mold fill analysis—CAD simulation

Specific to injection-molded parts, there are numerous CAD simulation tools available for determining optimum gate position and to analyze the mold cavity filling sequence (Fig. 7.11). MoldFlow® is perhaps the best known and most widely used; however, Sigmasoft® and SolidWorks Plastics® are both similar analytical tools. From a part design vantage point the designer can use these tools to verify the fill assumptions made during the design process. Often a short round rib or slightly thicker wall in a particular position will help fill the part in a more uniform fashion. This knowledge can be held and used after initial mold trials if there is some process issue or postmold warp issues. I tend to be conservative when using these tools. I run the analysis and discuss it with the tool designer and process vendor, but refrain from making changes to the design until after initial mold trials, when I can see how closely the analytical solution matches parts from the mold, in the correct material, and with the process intended for production. However, if the analysis shows some drastically bad knit line or air entrapment, or worst yet a nonfill potential, then I will review the design and look for geometry changes or gate position changes to remedy the particular situation. The use of experienced processors or toolmakers as a resource when considering part design changes to avoid these nonfill or air entrapment conditions is always recommended. These potential processing challenges show up on more parts and projects than you would think and many experienced processors have fought them and beat them before. Take advantage of this experience before the tool is constructed!

A common procedure that is used to verify the filling simulation can be completed once the mold is fabricated and operational. I will typically attempt to attend the initial mold trials for any new mold for a part, which I have designed or have responsibility for through development. During these mold trials, I will wait until the process technician has gone through their process experiments and has settled in on their desired process for the part. I will then ask them to make me a series of less-than-full "short shots" sequentially dialing down the part volume, ideally 10% at each step. The resulting series of not-full part segments shows us the exact flow of materials into the part void within the mold

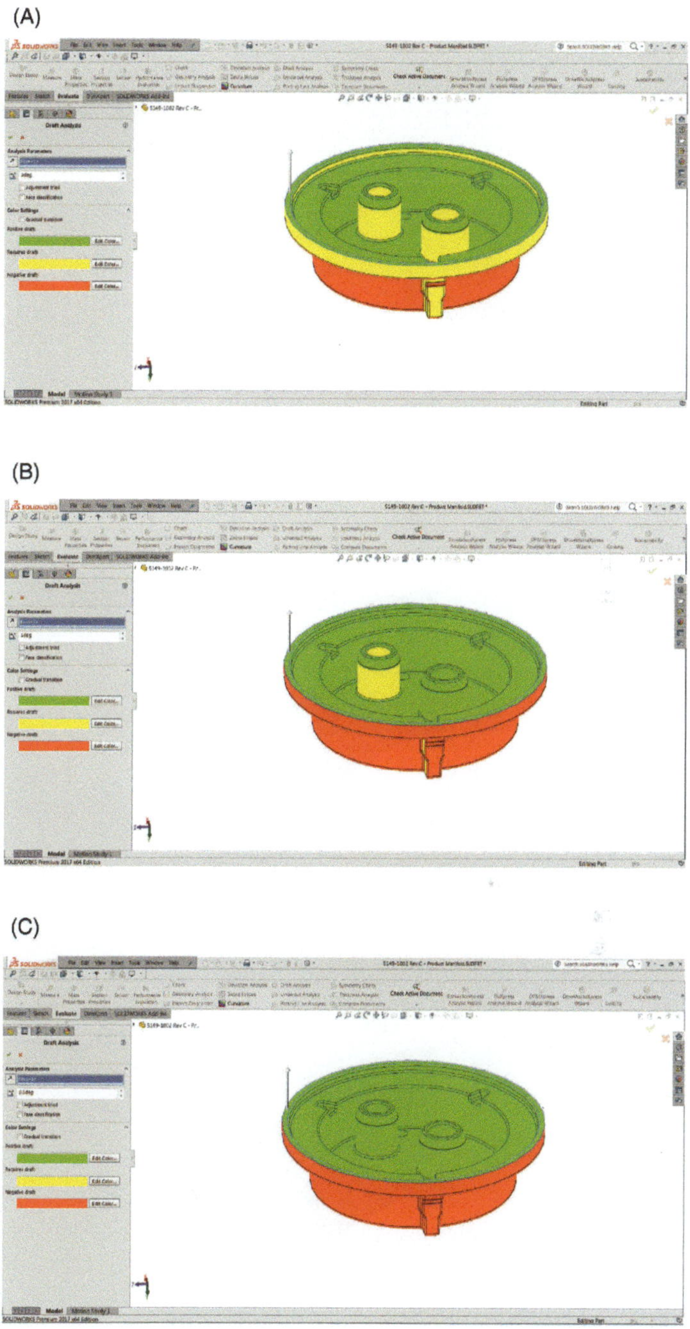

Figure 7.10 CAD generated draft analysis from the suite of tools within SolidWorks. Draft analysis set to check for a minimum of (A) three degree draft, (B) one degree draft, and (C) 0.5 degree draft.

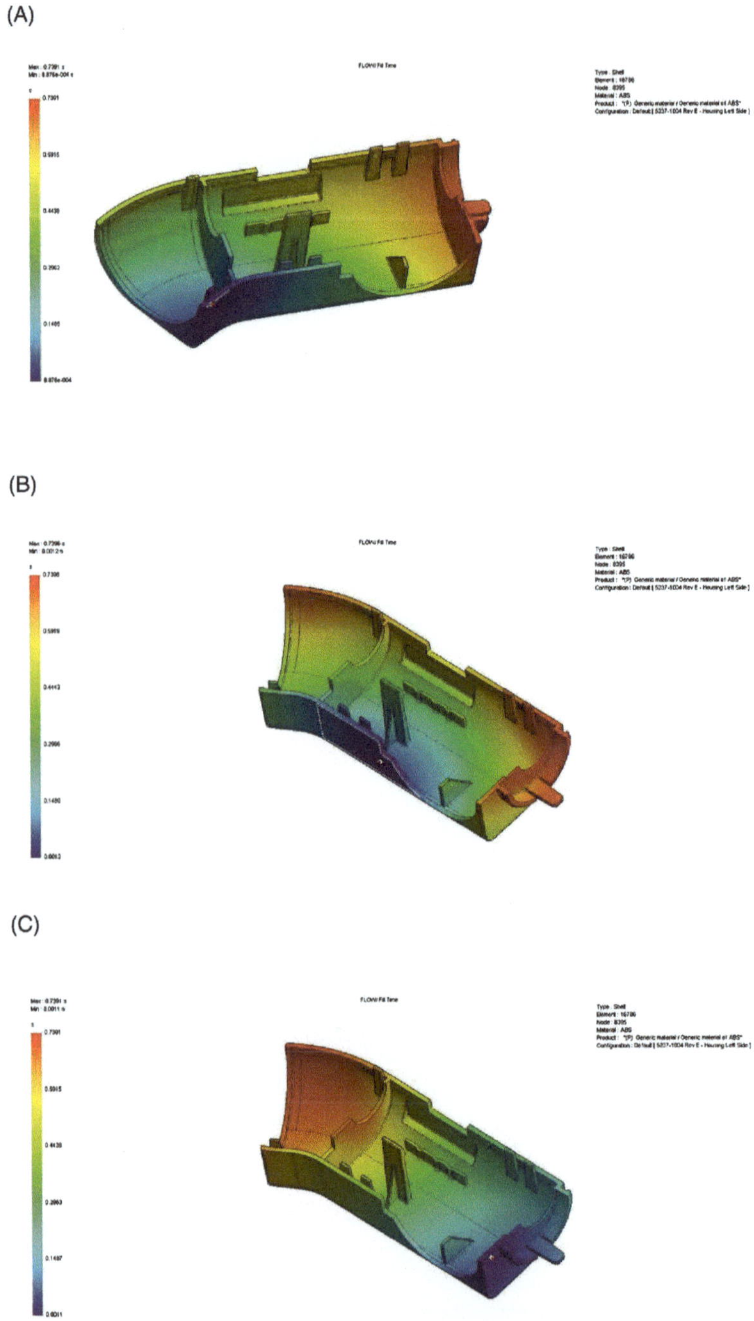

Figure 7.11 Mold filling analysis can be used by the design team for initial review of gate location and flow (polymer orientation) consideration. As with everything in plastics, balance is the goal; get the part filled as quick and as uniformly as possible. (A) Gate at extreme end, (B) gate at central point, and (C) gate at other extreme end.

in a sequence, which should match the fill sequence predicted by the simulation software. If it does then all is good and we have confidence in our assumptions used for the part. If it does not, then we need to revisit and ensure that there are no hidden risks—specifically looking for undesired knit lines, unwanted high residual stress areas, and so on, which may have been predicted incorrectly also. This checking method is often dismissed as overkill, however, experience shows it is far better to do the homework up front than to risk some failure once the part or product is in the stream of commerce.

7.5.5 Performance loading FEA stress analysis—CAD simulation

Part stress analysis can be analyzed by using tools contained within several of today's CAD programs and standalone FEA analytical program tools. Again, injection-molded parts with bending features, like cantilevered snap arms, can be analyzed by applying a deflection (the snap over bead height) and looking for maximum stress within the feature (Fig. 7.12). Adding or removing material, modifying corner radii, making the feature wider or taller, all can have an influence and through iterative trials, the design can be optimized to keep the maximum bending or torsional stress within the capability for the material to be used. In addition, impact loads can be applied to simulate drop strength for a part or assembly; however, the assembly gets considered as a unitary rigid body in most analysis tools; hence if parts can move within the assembly during the impact, the analysis will be misleading at best.

In addition, use caution when applying standard material properties, sometimes already resident within the analysis software, as these are taken directly from the material standard data specifications. Make sure; no, *make double sure*, that your material properties are those applicable at the *desired operating temperature* for the part being analyzed—remember that polymer properties vary greatly with temperature. As with all analysis, garbage in will result in garbage out, so pay attention and if possible, perform some reality check on the results.

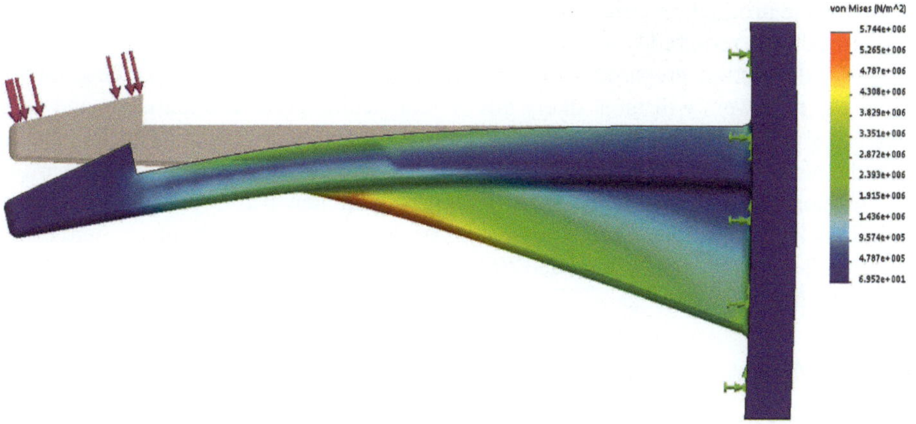

Figure 7.12 A representative analysis of a snap feature (broken out from the original part for simplification of the analysis). This analysis was produced with the simulation tools from within SolidWorks®.

7.6 Review and update FMEA

Once the detailed design is complete and before the final design review session, I like to recommend the team revisit the FMEA and go through it line by line. Sometimes it is illustrative to add a line below each attribute that has changes from the initial FMEA; in this manner the original information is retained and the new information, based upon the remedial changes made during the final detailed design process, can be recorded in this new line to show the improvement (or not) from the initial FMEA results. I like to add a notes column as well and record in the notes, on the new information line, just what changes or alterations were made. This method of keeping the original data and recording the new data with good notes can sometimes be invaluable later in the production cycle if some failure should begin to manifest itself. In which case, if remedial actions are taken to correct the failure, it might be wise to note these on the FMEA data sheet and perhaps even review any attributes affected by the remedial change and update the analysis accordingly.

Archive the final FMEA for future reference. The FMEA data sheet is a useful document when the next project team is tasked with the next generation product, or for continuous improvement once a product or part is in use in the stream of commerce and has some field acceptance and feedback from users.

7.7 Prepare for patent review

Information, both written descriptions and graphic illustrations should be produced to accurately describe the functions and features of the new part or product to be used to perform a review of the concept and product from an intellectual property (IP) standpoint. The review will be both from an infringement avoidance perspective and from an IP protection standpoint. This review may be just an update from an earlier provisional application made during the concept stage or it may be the first real review for the product. In either event, full specification and detailed description of all the attributes is required.

The information prepared will need to be complete and extremely descriptive, since the reviewers will most likely not be technical experts. Comparisons and case examples of similar or existing competitive solutions are always good to include so that the new and novel elements can be compared and contrasted to the existing art. A fair portion of the IP review will be differentiating the new part or product from other solutions in practice or protected by patent or other IP currently in place; hence any differentiations that can be included from the design team will assist the review team and help the process along.

7.7.1 Written description of the function of the product or part

Typically the written description takes the form of an executive brief, as a high-level overview style of document, carefully including every discreet detail need for differentiation, but not getting too far into the weeds with noncritical features or functions.

The graphic illustrations will tell the story for the most part, the written portion just sets the scenes for the reviewers.

7.7.2 Illustrations showing all the functional features of the part

The illustrations used for the reviewers needs to lead them through the story. I typically start with an overall isometric or perspective view so they can get a high-level feel for the entire assembly or part. Then I begin to show various side views, top view, bottom view, and so on. Finally I get into the detail, showing section views or exploded view with enlarged detail to show specific features or operating positions or conditions, and so on. I always attempt to make my written description follow the graphic illustrations, so that the detailed explanation of some specific function "A" comes before the detailed description of another feature "B", both in the written document and in the graphic illustrations (Fig. 7.13).

7.7.3 Comparative analysis

Any comparative analysis for existing products, which compete from a functional vantage point in solving the same "need" should be recorded as part of the information for review. This information may be in written form, in illustrations, or both. The importance of clearly distinguishing the novelty of the new product or part from those existing products or existing information taught in current or historic patents cannot be overstated. The examiner will use everything at his/her disposal to prove that the new product or part is indeed not novel; it is their job to thoroughly determine that the claims within the application are indeed novel. The comparative analysis materials created and provided by the design and development team are often the best source of information for the preparation of the formal application.

7.7.4 Record of references

Record any patents or other materials that might have been reviewed during the development process for use as cited references on any patent application which might be made. This is good housekeeping, which will assist in preparation of an application and will provide evidence of the research completed prior to the application.

7.8 Final design review

Yes, one final review with all of the stakeholders to ensure we have everything correct and ready for implementation.

7.8.1 Gather the stakeholders and prepare for review

As with the preliminary solution design review discussed in Chapter 6 of this book, the final design review should include parties from every functional group that will have any stake or hand in bringing the product through the development stage and

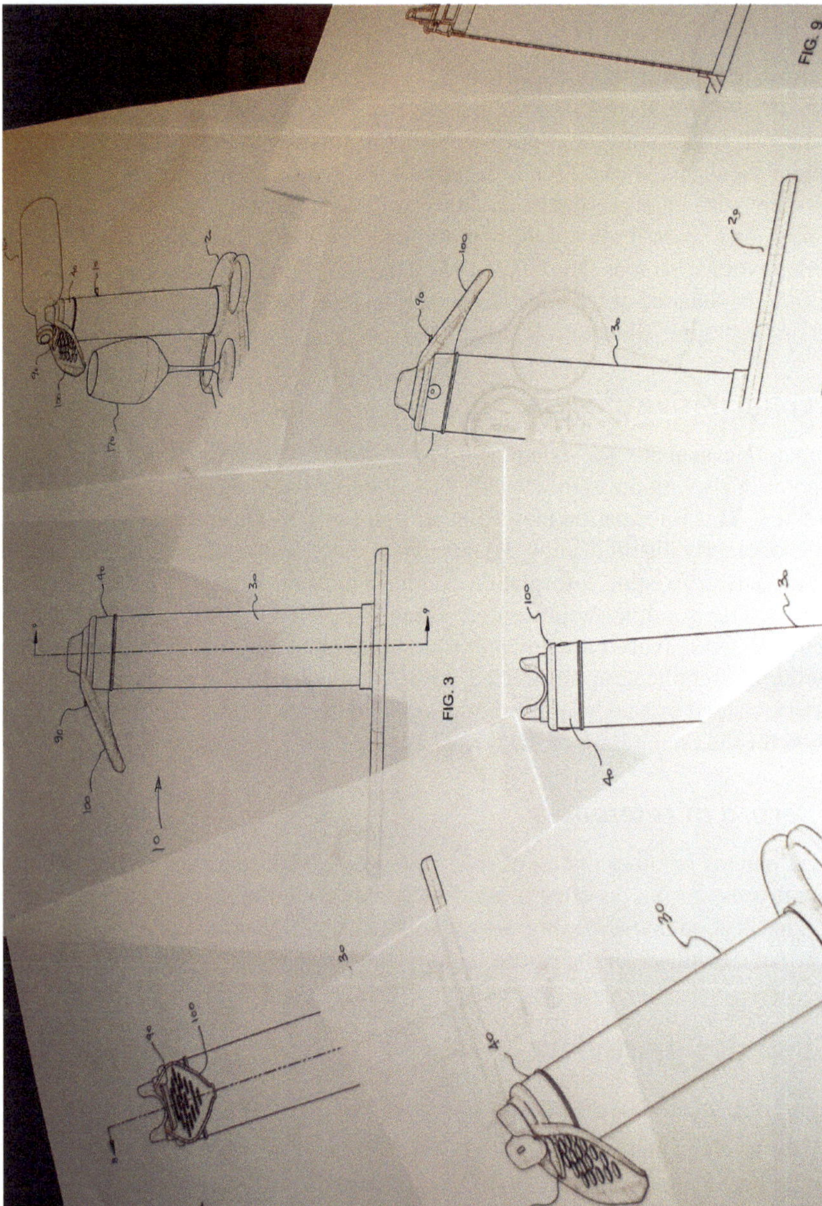

Figure 7.13 A representative sample of illustrations produced directly from the CAD layout data used for provisional patent filing purposes.

into production and sale in the stream of commerce. This will include all of those listed previously plus, perhaps, a few additional interested contributors. The list may include, but is not limited to, the following:

- design and engineering,
- management,
- sales and marketing,
- strategic sourcing, and
- manufacturing operations.

The additional interested contributors may be prospective vendors that have been prequalified by the strategic sourcing team as prime candidates in the supply chain for the project, including tooling sources, material suppliers, processing sources, critical component suppliers, and the like.

Good practice would dictate providing the stakeholders with some information prior to the meeting. In some cases an executive brief is sent to the design review team along with the agenda, a week or so prior to the meeting. The idea is to get the team thinking about the part or product ahead of time, wet their appetite, so that when they arrive at the meeting they are ready to participate and do not need to use the first hour simply learning what they could have read beforehand.

Also, it is a good practice to have parts, assemblies, prototypes, and even competitive products available for the design review participants to touch and handle during discussions. Marketing may wish to bring any proposed graphics or package art, anything directly related to the part or product in review. The object is to stir the intellect and the emotion in the room to have complete focus on completing a thorough review.

7.8.2 Full scale design review

This is the last opportunity to fully question and dissect the design prior to going out for estimates and implementation; hence it is important to be thorough and complete. A good working agenda, which leaves ample time for the questions and discussions for each topic is recommended to keep the group focused and to ensure no topic is missed or forgotten. The success of the project and product rests on this assembled group to bring and use their collective experience and wisdom to ensure that the product and project are ready for implementation.

7.8.2.1 Review against specifications

The design first and foremost needs to meet or exceed the functional specifications prepared and agreed upon at the beginning of the project. The formal function specification document should be the roadmap that is followed, item by item, for the specification review portion of the meeting. Many times the design and engineering team will produce a score sheet showing each functional specification and the resulting test result values (when available) as proof that each has been completed, and can be expected to be validated in a production part or product.

7.8.2.2 Review against budget and timeline

The project budget and timeline have not been addressed formally within this chapter, but every project will have a budget and will have a timeline; my experience is that the budget will always be too low and the timeline too short, but that is another discussion. Although formal ready for production estimates have not been gathered at this stage, informal estimate most certainly have been, as the strategic sourcing team were doing their research on potential vendors or as in-house operations groups were briefed about certain parts or assemblies. Just plain old project experience by the engineering team or by the strategic sourcing team will allow reasonably accurate assessments of costs and timelines for the development stage as well. Does the design provide the cost structure to hit the target retail price point? Will the capital investment have a reasonable payback period? Will the design require heavy investment in marketing resources to aid in creating demand? All of these sorts of questions should come up in this portion of the design review meeting.

7.8.2.3 Review against marketing's wish list

Does the proposed final design deliver all of marketing's desired features and benefits? How does the design satisfy the "need" relative to the user studies or feedback? Is the part or product intuitive in its application and use? Does it provide an excellent user experience? All questions that should be addressed in this portion of the design review meeting.

7.8.2.4 Review FMEA results

The revised and refined FMEA data should be reviewed by the group, with special attention given to those attributes still scoring in the upper range for the product. Is there anything further that can be done to reduce the high RMA scores prior to implementation?

7.8.2.5 Review any patent or IP research or opinions

If initial search results are complete or if there is a legal opinion rendered regarding the part or product from either an infringement standpoint or from a potential for protection stand point, that information should be reviewed. If there is any possibility of infringement, even a remote possibility, the review group should use extreme caution in moving the project forward without further investigation. Marketing and management will need to assess the risk versus reward of progressing, with any potential infringement litigation possible. Better off to kill a project that brings risk than to risk the entire company's fortune and future. In many cases, licensing agreements can be arranged to solve these potential issues, providing a path forward for the project and providing some benefit to the IP holder, which the part or product may be considered to infringe upon. Full review of the IP situation by the design review team and management is extremely important and should not be overlooked or discounted.

7.8.2.6 Document the design review

Good practice would dictate appointing one person, or perhaps two, to be record keeper for the design review meeting and complete minutes of the meeting should be distributed to all participants, including all data discussed during the meeting. The design and engineering team is typically the group within the organization responsible for archiving the development work once the project goes into production. They should consider the design review minutes, notes and final report a valuable resource for archiving. The review notes will prove invaluable in the future should problems arise or should the part or product be considered for development into a next generation product or device. Knowing and understanding the results of the initial development is a certain good starting place for any new development initiative.

7.9 Preparing for implementation

Once the final design review is complete and documented, each team within the project group will need to begin preparation for implementation. Transferring documentation, allocating funding, filing applications, negotiating final terms with vendors, you name it. With all systems go, every project subteam will be busy.

7.9.1 Design team

The design team will need to ensure that all technical data files and drawing specifications are updated per any last minute changes from the design review. Once updated then the information needs to be posted to whatever library resource is used by the team, for example, many of my clients use DropBox or OneDrive for a secure file sharing means among team members. All project data needs to be securely archived as well by the design team.

7.9.2 Marketing and sales

The team from marketing and sales will need to go back and review the information from the final design review and ensure that they have everything in order for the marketing roll-out, especially if there were any functional, aesthetic, or visual changes since the last design review. They should clearly understand the timeline for implementation, including qualification and validation. If they are planning to exhibit at upcoming trade shows or to begin any marketing campaign they should keep the remainder of the project team up-to-date as to their schedule and commitments. Some projects use the part qualification milestone as the delimiting marker for any public announcement or roll-out activities, preventing the possibility of a delivery delay. If marketing can't talk about it until qualification is successfully completed, then product should be available on time.

7.9.3 Operations and manufacturing

Perhaps the busiest project subteam will be the operations team, as they now kick into high gear and begin to requisition all needed capital and equipment to put the product or part into a production mode. Coupled with the strategic sourcing team, operations will be responsible for choosing resources, distributing the required information, tracking the progress of every tool in development and preparation for in-house or subcontract integration into a final working assembly process. Weekly meetings with each contributing or contracted resource will assist in keeping project momentum and good communication between the various disciplines.

7.9.4 Quality assurance

Like their counterparts in operations, the quality assurance team will be in high gear soon, hence they will be getting ready to assist with each outside vendor and in-house manufacturing group. An overall plan, encompassing the raw materials, the primary manufactured parts (injection molded, blow molded, stamped, thermoformed, or other processed part), the subassemblies and final assemblies, right down till final pack out need to be considered in the overall quality plan that will be implemented incrementally as each of the pieces and parts comes on-line.

Resources for further study

S. Levy, J.H. DuBois, Plastics Product Design Engineering Handbook, second ed., Chapman & Hall, New York, 1984.

S.E.M. Selke, J.D. Culter, R.J. Hernandez, Plastics Packaging—Properties, Processing, Applications, and Regulations, second ed., Carl Hanser Verlag, Munich, 2004.

J. Rotheiser, Joining of Plastics—Handbook for Designers and Engineers, third ed., Carl Hanser Verlag, Munich, 2009.

G.L. Beall, Rotational Molding—Design, Materials, Tooling, and Processing, first ed., Carl Hanser Verlag, Munich, 1998.

E.R. Larson, Thermoplastic Material Selection—A Practical Guide, first ed., Elsevier, Oxford, 2015.

R.A. Malloy, Plastic Part Design for Injection Molding: An Introduction, second ed., Carl Hanser Verlag, Munich, 2010.

P.A. Tres, Designing Plastic Parts for Assembly, seventh ed., Carl Hanser Verlag, Munich, 2014.

G. Boothroyd, P. Dewhurst, Product Design for Assembly, Boothroyd Dewhurst Inc, Wakefield, RI, 1989.

Web links of note

Great Design information from DuPont can be found here:

http://www.dupont.com/content/dam/dupont/products-and-services/plastics-polymers-and-resins/thermoplastics/documents/General%20Design%20Principles/General%20Design%20Principles%20Mod%201.pdf.

Sabic provides many tools for designers here (you may need to register):
http://www.sabic-ip.com/webtools/toolsHomePage.faces.

The UL Prospector site is a valuable tool for any designer reviewing materials for any application (you will need to register):
https://plastics.ulprospector.com/.

Solvay provides a great design guide for using their Udel polysulfone materials:
http://www.solvay.com/en/binaries/Udel-PSU-Design-Guide_EN-227550.pdf.

Bayer has a nice design guide for RIM systems:
http://reactioninjectionmolding.com/wp-content/uploads/2013/09/RIM-PartMoldDesign-Guide.pdf.

Covestro offers several design related guides, including part and mold design, snap fit design, and joining techniques for plastics:
http://www.plastics.covestro.com/Library/Technology-brochures.aspx.

Section III

Implementation of the Design

8. Gathering Production Estimates 229
9. Project Review and Verification—Risk Management 245
10. Production Implementation 253
11. Production Qualification and Hand-Off 275

Introduction

Proper design is wasted without efficient, precise, and effective implementation. While the design work provided creative challenges and stimulated the inner inventor; implementation requires tenacity and perseverance to ensure that the seeds planted during the design effort can actually be nurtured and grown for a fruitful harvest.

The project fulfillment factor (PFF), which can be thought of as the project results, can be analyzed in many ways, however, one method seems to always ring true for my projects. Multiply the overall rating for the excellence of the design by the overall rating for the excellence of the implementation to gain a measure of the project fulfillment. This PFF assesses the overall project performance, start to finish. It is not a measure of the results expected in the marketplace or by the end-user, only a snap shot of the development process. However, when the metrics for the PFF are high, the product tends to do well. Alternatively, when the PFF metrics are low, well you know the story.

The method follows a format whereby the Design and the Implementation are rated from 0 to 10. The ratings are evaluated as follows: zero (0) rating for poor, five (5) for acceptable, and a ten (10) for outstanding. So in essence, great design with poor implementation equals to a zero, conversely poor design with great implementation similarly equals a zero, acceptable design with acceptable implementation would score a 25 (better than a zero by far) but looks what happens when we have outstanding design and outstanding implementation, we get a score of 100, fourfold the results of acceptable work (Figure III.1).

As we have discussed since the opening introduction in Chapter 1, designing products and parts with plastics requires adequate consideration of the *Four Pillars*: Materials, Process, Tooling, and Design. The knowledge on Materials, Processes, and Tooling were covered in *Background Knowledge* (Section I)

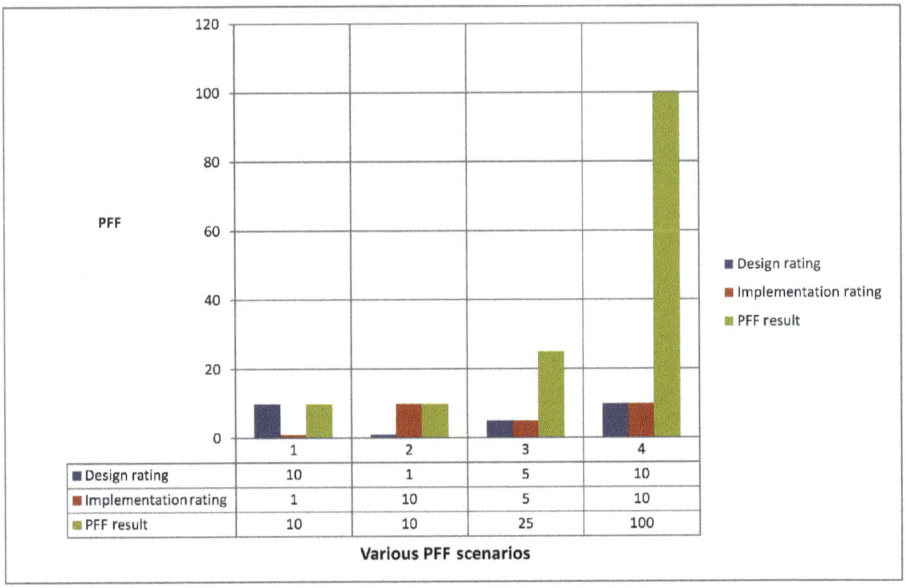

Figure III.1 The project fulfillment factor (PFF) shows one method for thinking about the relationship between Design and Implementation. The four scenarios show a range for PFF scoring—a rating of one (1) has been used in this illustration for poor so the graphic colors and magnitudes can tell the story.

of this book. Design, the primary focus of the text, was covered in *Designing with Plastics* (Section II). Now we turn our attention to the operational tasks to properly and efficiently implement the design into a production reality. Implementation, as with design, will be vastly different for each and every project; however, there are still many steps which will need to be accomplished, no matter the project type or scope. For example, tooling and parts costs will need to be gathered and evaluated, decisions on which tooling vendor or which processing vendor best suits the needs for the given part or assembly will need to be made, first time nonrecurring engineering and planning costs will certainly need to be estimated and evaluated, timelines for implementation will need to be prepared and evaluated, and total finished goods costing will need to be calculated and compared against target specifications. Any one of these seemingly mundane tasks could result in total abject failure for the project if not completed with careful thought and thorough analysis. Inadequate attention to, and analysis of, the details of each part required, including the ability of the selected processing vendor and tooling vendor to provide the qualifications required to achieve the part performance required, can lead to inferior products, no matter how good the design might have been.

Steps covered in *Implementation of the Design* (Section III) include the basic steps required to take the design from a drawing board concept into a ready for consumption product in the ubiquitous "stream of commerce" reality.

Implementation steps from precise and complete estimating and cost evaluation, through thorough project review and verification, into planned production implementation, and successful qualification of the final production product prior to production hand-off and the first sale of the product are discussed in detail. In each step, risk management is a priority; hence our discussions of each step focus on areas to watch closely during the process.

In many development projects or programs, the design, once completed, gets thrown "over the wall" to the operations team and the design group goes about finding a new "need" to prepare a new product solution (fun work). Since early in Chapter 1, I have been espousing an integrated project team, involving material suppliers, tooling vendors, and processing vendors (all implementation and operations types) early in the design process. Now that the project moves into an implementation mode, all of those involved early in the project will be prepared to move efficiently forward; however, just as the various operations team players (process professionals, tool makers, materials suppliers, etc.) responsible for implementation were vital to the design effort, I suggest that the design team needs to be a major part of the implementation process as well. The design intent and the desired end-user experience demands that the design team stay involved and assist to ensure the desired criteria are protected and met, or exceeded, in the final production unit.

Products with good design input and with excellent implementation will typically do well in the market; these tend to find themselves at the higher end of the price range and sell well because they ooze with value. Look at the range of products produced by companies, such as Apple, Mercedes, Ping, and the like; you won't find any of these anywhere but at the top of the price range, yet they are among the most popular in each of their respective categories. Why? Simple, they provide value; it is worth spending the extra incremental funds for a product designed to do what it is supposed to do well and one that has been implemented with total focus on delivering excellent quality and an excellent user-experience.

Gathering production estimates

8

What a seemingly dull and laborious task? Who would enjoy sending out requests, waiting for responses, then reviewing and comparing them to attempt to make an intelligent choice? Most engineers and designers would pass on such a task and in many organizations they do; the task is left to the strategic sourcing or purchasing professionals. However, good estimating can be an extremely important function and can lead to project success, or failure. The first cost and nonrecurring costs are often the most visible and scrutinized with a magnifying glass, yet the piece part cost will produce more profit (or loss) over the lifespan of the product or part; however, the relative magnitude difference between the two almost always drives decisions to minimizing first costs. The key to good overall estimating is good preparation and uniform information distribution to all participating vendors. Prepare enough information that each potential vendor can understand the full scope of the project and the quality expectations for each piece of the project they are estimating. As in any endeavor, garbage in results in garbage out, in other words, bad or inaccurate estimates often are the result of incomplete information to estimate with. If your specification drawings and data files are incomplete, the estimates will be suspect or, worse yet, just plain wrong. With multiple vendors estimating, if one calls and asks a question that is the result of some missing information, call or communicate with the remaining vendors and make sure the information in question is clear. Do not provide different information to different vendors and expect competitive estimates, which are truly accurate and truly competitive. At the end of the process remember, the vendor will be part of the project for perhaps the lifespan of the product, much like a joint venture, so be diligent to ensure complete transparency in all estimating activities.

8.1 First cost estimates

Initial funding is extremely visible. We are investing how much to start the project? There will be lots of exposure, be it good or bad, but the real story is in the total cost to produce whatever is being developed, for those costs go on part after part, day after day. As one story goes the engineer wanted to injection mold a particular part, but the company management could not understand why they should spend tens of thousands of dollars on tooling and opted to thermoform the part instead, saving many thousands of dollars in tooling at the start of the project. As the project went to market their price point provided a gross margin of around 50% and everyone was proud and pleased with their decisions. Marketing did a wonderful job of gaining market share and soon they were doing extremely well in the retail marketplace. Then a competitor entered the market, lowering prices to gain market share. The upshot was that the competitor

Designing Successful Products with Plastics. http://dx.doi.org/10.1016/B978-0-323-44501-6.00008-8

had a much lower per unit base factory cost; they had invested in production tooling that could efficiently produce a similar device. Now the company was faced with larger volumes and lower margins, doing much more work and earning less revenue. Looking back, the cost difference between the thermoformed part they were producing and the injection-molded part the design team wanted when multiplied by the tens of thousands of parts produced would have funded the first cost for the injection-mold tooling many times over; and, the base factory cost with the injection molded solution could have competed handsomely with the new competitor, disallowing them the ability to gain market share based on a price war. First costs are the entry fee, but be careful to analyze the total cost before making decisions.

8.1.1 Capital tooling

As mentioned earlier, be sure to provide the best possible information for your potential vendors to use during the estimating process, deliver the exact information to each potential vendor, and ask for the same estimate delivery date. In this manner, each potential vendor has the same opportunity to become the preferred vendor for this particular project. However, do your homework; do not rely on the estimate alone when determining the best vendor for a particular mold or tool. Some shops are really good with fine detail and excel with small intricate parts and complex things, other shops might be better equipped to handle larger and heavier molds, and tools. Be sure to research the potential vendors to understand that their strengths will support the part or assembly which they are being considered to manufacture. In addition, a tooling vendor that has some experience with the chosen materials is also a plus; runner system design and gate design can, and does, vary from material to material, hence it is good to find a tooling vendor that has successfully built molds and tools for the specific materials to be used. If you have an idea of the processing vendor, or if you will be processing it in house, it is good to find a tooling source that the processing team is familiar with or, better yet, a vendor that they recommend. A close working relationship between the tooling vendor and the processing team will be beneficial to the project during mold trials and debug, as well as for recurring tool maintenance once in production. The goal is to have the tooling vendor, the processor, quality assurance (QA), and product design engineering all on the same team during mold or tooling development and especially during initial tool processing trials and debug (Fig. 8.1).

Sometimes the processing vendor will be producing the capital tooling as well, or they may be brokering it through, accepting responsibility for the final end-part or product. This is often a favorable arrangement, as single source responsibility simplifies each and every decision, and negotiating point. However, when dealing with this situation be careful to fully clarify the status of the capital asset, as to ownership, rights of removal, and so on. A common estimating and proposal means is to refer to the capital tooling as something else, for example, "… custom tooling development engineering services" and the like. Any time the tool is not fully identified, as to its exact physical identification, for example " … a single cavity injection mold, constructed from P-20 steel for use in producing part number XXX-YY from a Polypropylene

Figure 8.1 The injection mold shown, considered a capital asset, is an unscrewing mold that produces four large diameter jar closures designed to affix a dispensing apparatus. *Source*: Courtesy of Quality Molding, Inc., Somerset, WI.

Copolymer resin" the legal ownership of the tooling can become questionable. Now there are several ways of working with a contract vendor and sometimes the processor may accept this single source responsibility but will expect some loyalty in return. It is not uncommon in these arrangements for there to be some minimum production term specified, for example, "... the tooling will remain captive at ABC facilities during the first 500,000 purchased, delivered and paid parts produced from said tooling, after which time the tooling can, at the request of XYZ company, and upon fully satisfied payment of any and all open orders from said tooling, be removed from ABC facilities by XYZ company." The caution herein is not to direct one method of vendor contract over another but to inform and to be clear about the tooling, as to its legal ownership and the rights of both the customer and the supplier.

Selection of a capital tooling vendor is an important step in the overall project. The vendor must have the capabilities to design the capital tooling to produce the specified part to the tolerances and performance criteria supplied for the life of the program. Then, they need to have the capability to manufacture and fabricate the tooling to the exacting specifications required within the tooling design. To do this they need two critical components, the correct equipment required for such tooling and most importantly experienced and knowledgeable personnel to use the equipment to build the tooling. One thing that I look for in a capital tooling vendor is a range of experience within the shop. For example, I am particularly impressed when I find a shop with an internal apprentice program (there aren't many) because that tells me that the experienced toolmakers are working with and training the younger or less experienced technicians and machinists. This arrangement seems

Figure 8.2 Fixtures for holding the work or transporting the work or both, as shown in this figure will need to be specified and designed as a part of implementation.
Source: Photo from frank11—stock.adobe.com.

to always lead to a workplace where there is good comradery between the shop personnel and everyone seems to stay focused on the overall quality of the finished tooling. Their name will be associated with that tool and when it is in production if it needs constant repair or has processing issues on a regular basis, it will be bad. Hence, while estimated cost and timeline are important criteria, sometimes a simple visit to the tooling vendor will provide insight which is invaluable when making a decision.

8.1.2 Fixtures for production and quality assurance

Fixtures to assist during production, or to test during production, or for use in QA activities can range from extremely complex and expensive to relatively simple and straightforward (Fig. 8.2).

Some fixtures may be considered just a part of production overhead and may have a short lifespan; others however, can be quite expensive and would be considered capital investments. The design team will certainly know and understand the relative magnitude of fixturing that will be required as they go through the design process. Whether the design team is tasked with the design of these fixtures, or if it is the responsibility of some other internal group, or even if it is to be outsourced, the design team has the responsibility to plan for and advise that capital funding

Figure 8.3 Custom parts handling fixtures or containers are required for most multipart assembly operations and will need to be specified and designed for implementation.
Source: Photo from olly—stock.adobe.com.

will be needed within the budget for the project. In addition, whether an internal activity or an outsourced activity, estimates will need to be gathered based on the information that the design team provides; hence the design team should, as with the capital tooling, prepare and provide the most complete specification possible for each fixture required.

8.1.3 Material handling

Often custom parts are produced in the processing department which may be at the other end of the factory from assembly, filling, or final packaging operations; or, the processing may be done totally off-site at another factory location or subcontract vendor. In process handling and packaging becomes a puzzle and the design team is in the best position to recommend both means and methods to provide an efficient solution (Fig. 8.3).

The solution may be a capital item, for example, reusable container specifically designed for the part or assembly in question that are durable and will be used for many years. Other times the solution may be a noncapital solution using a disposable handling means or a recyclable means that can be used for other operational benefits. Obviously, every part manufactured in every plant or factory will be different and will be made in differing quantities and will need to be handled in different ways, so there is no one correct method or guideline to follow here. The point to be made is however, that the design engineering group, during the design of the part or assembly, understands whether special handling might be required once in a production mode and has the responsibility to plan accordingly. At minimum, the topic should be brought up for budgetary consideration and the operations team should be aware that there will be some handling puzzle to be solved.

Case Example

On one particular project, the injection-molded housing component was an extremely aesthetic component for the overall final product. As such, it needed to be molded well and then handled carefully to avoid any damage after molding and prior to final assembly operations. For this product, the injection-molded parts were produced in one factory and the final assembly operations were performed at a different facility in a neighboring town. The original plan was to have custom corrugated container with dividers used to hold the housing molding, fitting 2 layers of 24 each into each container. The corrugated containers would then be returned to the molding factory and reused. However, this bulk pack method allowed the parts to attract dust and to get scratched while vibrating against the corrugated dividers within the container. Changes were necessary and the team was brought together to evaluate how to better protect the moldings.

The team kicked around many different ideas for fixes to assist the bulk pack methods, however, they all added expense and all produced extra waste (bubble wrap, foam liners, etc.). The team then set about thinking through the problem from a fresh perspective, looking not just at the immediate failure/reject issue but looking at overall quality and efficiency. The result was a new approach altogether. The molding factory was supplied the final shelf ready point-of-sale carton for the finished assembly (single unit) that the injection-molded housing in question was to be used in and each housing molding was placed into a polybag and into the point-of-sale carton—as an individual unit. Yes, the molding facility had to invest more time and was paid additional funds for this extra investment of time. However, overall finished unit costs were reduced.

The assembly factory was restructured such that the assembly line turned back upon itself; it finished one row of tables over from where the line started. This allowed the molded housing to be removed from the carton and placed onto the assembly line to begin the assembly process. The polybag was placed back into the carton and the carton was shifted to the adjacent end-of-line table where the next finished unit was covered with the polybag and then inserted into the carton, then taped shut for final palletizing for transportation to the distribution center. Final carton fabrication was no longer needed at this assembly factory, saving floor space and warehouse space. Return freight for the reusable bulk pack-out containers was eliminated and the original scratched or damaged housings problem was eliminated. Overall quality was enhanced and costs were reduced; hence, it pays to "think first" about every single step in the process.

In gathering estimates for these special handling requirements, the design team would prepare just as they have done for the tooling and fixtures. That is, prepare enough information that the potential vendor(s) can fully understand the requirements and the application. Again, just as with the tooling vendors, it is prudent to get an understanding of each potential vendor's strengths, so the selection decision is based on the vendor's abilities to perform, as well as meet the budgetary requirements.

8.2 Nonrecurring cost estimates

Nonrecurring project tasks, much like the initial design and engineering investment, are investments which need to happen at the beginning of the production phase of the project but once completed will not need to be repeated. There will be some continuing product and process improvement activity but generally that activity is categorized under the overhead of continuing production operations. Most often there will be production support staff and manufacturing engineering staff that will perform the design and implementation of these sorts of tasks, however, the initial design team will know, during the design efforts, if special fixtures will be needed for assembly, filling, testing, labeling, and so on, and should communicate this to the manufacturing operations team as early as is feasible. Often this occurs during the design review sessions where participants from all the functional groups within the organization are together scrutinizing the design, each from their own particular vantage point.

Nonrecurring costs are first cost expenses to the project but are not considered capital. That is, once completed there is no "thing" to point to and say that it is used in the production of the product. Rather, much like the design efforts invested earlier in the project, these nonrecurring efforts are expensed to the project as the first costs of bringing the product to a production ready state. These might include preparation of special software for product tracking in production or for lot identification, preparation of certain workspaces to support a certain environmental need (low humidity, low lighting, clean room environments, etc.), and other similar one-time expenses that are not capital in nature.

8.2.1 Design and development of assembly fixtures

These can range from simple nests to hold a part or assembly in a certain orientation for the operator or machine, to extremely complex automatic mechanisms that perform several operations simultaneously. Hence, while the design effort is considered a nonrecurring cost, the actual fixture may be considered capital, obviously depending upon the cost and projected lifespan of the fixture. The design and debug time invested prior to implementation in production is typically estimated as a nonrecurring investment.

Design of the assembly fixtures can be assisted by the part design itself and in many instances, the part designer will include features or fixturing locations to assist. Hence, allowing the design team to participate and convey their design intent to the manufacturing engineering team seems like a reasonable thing to do. Had the manufacturing operations team been included as part of the initial development team, as recommended throughout this book, then the information regarding the design intent would have been conveyed early in the process and the design of the fixturing could have been accomplished in a parallel fashion allowing a savings on the overall project timeline.

8.2.2 Design and development of test fixtures

Test fixtures fall into two basic categories, fixtures to qualify the part or assembly and fixtures to be used as in-process performance testing devices. The fixtures used for qualifying may only be used during initial qualification and perhaps never again. Often these are retained and used if changes are made which would necessitate requalifying the part or assembly. Changes to materials may also cause the need to requalify a part or assembly which may necessitate the use of the initial qualification fixture later in the lifespan of the part or assembly. In-process test fixtures, may be used on every part or assembly manufactured, or they may be used at some frequency to provide a level of satisfaction that all materials and processes are within the applicable thresholds. As with assembly fixtures, these can be relatively simple or they can be deceivingly complex and capital intensive.

In most every instance possible, I look for ways to accomplish multiple things within each fixturing operation. For example, when fixturing a part for an assembly step, I would look for some automatic way to verify the assembly step was completed correctly. This can include some sensing means that might give the operator a simple green or red indication of successful completion, or it could be a bit more complex and interlock the fixture so the assembly cannot be removed until the operation is completed to specification. In some instances, when assembling a sealed set of components, I will look for ways to verify a good seal right there during assembly, thereby limiting the opportunity for nonsealed assemblies to go any further down the line. Always attempt to verify each step as early and as often as possible, especially if this can happen without further movement or without further dwell time. Be sure to challenge any fixture designer or vendor to include some verification means with each and every operation and to include these verification means in the nonrecurring estimates for the fixture.

8.2.3 Design and development of quality assurance/control fixtures

Inspection fixtures differ from in-process performance in that the inspection fixtures are typically used or applied to incoming parts from the primary manufacture or processing, or to final goods finished and ready to send off to a customer or to distribution. These fixtures are similar in many respects in that they still must be designed and made, therefore adequate estimating should be considered for the budgetary process. This is not to say that some fixtures for QA would not be used for periodic in-process checks, indeed they should be used at every stage of the process where a potential fault can be caught prior to adding any additional investment into the assembly. In fact, in some QA protocols the inspection fixture is actually used by the assembly operator to self-inspect and when an out of compliance event occurs, the operator will call in the QA personnel to take responsibility for the event and the remedial action (if any).

8.2.4 Qualification testing by independent laboratories

Some finished goods or assemblies may require independent certification to published standards; this will be part of the functional specification for the part or product if it is required. If this is the case for the product or finished goods under development, the cost of the qualification testing and certification must be estimated. Securing estimates from and working with independent labs can consume valuable project schedule time, hence start gathering these estimates just as soon as the detailed design review is complete and there is a listing of the specifications that the finished product must meet or exceed. This qualification testing is usually completed with production quality units at the start of production and does not need to be repeated at any given frequency. However, if there are any changes to the design, the materials, or the process used to manufacture the product, a reevaluation or requalification process might be in order, depending upon the magnitude of the change and the potential risk to final product performance as a result of the change.

8.2.5 Regulatory compliance testing by independent labs

Independent lab testing for regulatory compliance is similar to independent lab qualification testing in many ways, however, they do differ. Regulatory compliance testing can sometimes be done internally but must be verified by a third part outside lab. In addition, the regulatory compliance testing and certification is an ongoing, repeatable event, sometimes occurring on a quarterly or annual basis. It is not uncommon for the regulation to include language requiring production audits to occur at some frequency and sometimes unannounced.

8.3 Production unit cost estimates

Per unit cost estimating and analysis is most often the domain of the strategic sourcing or purchasing groups within an organization, however, from a project management perspective the gathering of initial production per unit costs should be done with a heavy design and engineering influence. Selection of materials, processes, design geometry and features, all lead up to and influence per unit costing; and the design team has been considering these costs throughout the design effort. Hence, the design group has keen insight as to what the per-unit costs should be and can often spot estimate inconsistencies when they arise. Be sure to understand that per unit cost estimates are just that, the total cost that makes up one unit. Not just material costs, or processing costs, or assembly costs, but freight costs, scrap costs, replacement unit cost, sales samples cost, and so on. Thorough estimating will include costs that have not been realized yet but will arise once in production.

8.3.1 Piece part costs

Perhaps the easiest per unit costs to gather and to analyze, piece part costs comprise the cost of each piece that is used in the final assembly or product. Many organizations make the mistake of tallying the piece part costs within an assembly, adding in some labor factor and calling this total their base factory cost. Pair this with their anticipated price point and the margins look great. Be careful not to fall into this false pretense as other costs will soon change the picture. But to be sure, the piece part costs are the most visible and are probably one of the largest cost components (percentage wise) in the final total per unit cost. As such, it is prudent to gather the piece part costs at various price break points and to review and analyze the number of inventory turns where the greatest efficiencies can be achieved.

Piece part costs can be influenced by a number of factors; raw material cost, freight costs for the raw materials incoming (which will be volume dependent), equipment overhead costs, raw materials lost in set-up, time lost in set-up, scrap materials during the process, extreme part specifications, and so on. As a design team, the goal is to provide a design that can maximize the benefits of the material, can be processed with ease, can allow for regrind and reprocessing of all scrap (runners, etc.) and has specifications that do not demand precise processes or handling. Not all of these can be achieved on every development project but the goal should remain to achieve as many as possible. One colleague used to advise the team "… good quality parts, low-cost parts, or quick turn times—choose any two …" as a means of reinforcing that if high-quality parts are required, then they need to be willing to pay for them or wait for them.

Case Example

One part which I had the pleasure of troubleshooting was used in a handheld battery operated ambulatory diagnostic medical instrument and was injection molded from polycarbonate. It was both a structural part and an optical lens part. It held a pair of LCD display modules in position and was affixed to the main instrument circuit board, electrically connected to the two LCD displays. The structural portion had four cantilevered snap arms which telescoped through slots on the printed circuit board and latching to the back of the board to squeeze the LCD display against a resilient conductive polymer matrix (commonly referred to as zebra strip) which in turn contacted a series of conductive pads on the circuit board allowing the circuit board to drive the display output. The mechanical specifications for this structural part were not too bad and the part could be processed fairly easily, but the lenses caused some issues. Black flecks began to show up in the lens area and since the part specifications called for optical clarity and no foreign matter to be visible in the lens area, this began to produce a fair number of rejects. In addition, the part specifications specifically called for virgin resin with 0% regrind allowed. Not surprisingly, the initial piece parts costing did not survive; and, after the first round of black fleck induced rejects, the piece part costs were adjusted to accommodate the part specifications. Yes, the part cost was now more than double the original estimates.

Analysis of the rejects and discussions with the material supplier yielded some interesting facts. The black flecks seemed to be prevalent in most all of the virgin resin. We had assumed it was burning in the barrel/screw or other degradation happening with the process. Also, the functional and aesthetic performance influence of the black flecks was not nearly the failure mode anticipated by the specification requirements. In fact, on a majority of the rejected parts, the black flecks were present but they were not in areas of the display where they would interfere with the text and graphics being displayed. In addition, when the black flecks were present, there were typically less than six or seven flecks on the part, and when assembled over the dark gray LCD background they were hard to identify, unless you knew exactly what you were looking at. Now to be true, in some instances they were large and very obtrusive; but the instances where they were truly offensive were a small percentage of the total rejects reviewed.

In this case, the part specifications were meant to keep the product quality in line with customer expectations, but the specification was extreme and was interpreted far too extremely. With help from the material supplier (whom had seen this kind of issue numerous times previously) the specifications were rewritten to specifically call for no more than six black flecks in the lens area and with no black fleck to be larger than 0.020 in. in a circumscribed diameter. This increased the time invested in quality checking efforts but reduced the number of rejects by about 80% and resulted in a much less severe cost increase for the part.

Hence, piece part costing can be influenced by design specifications in both good and in negative ways. Anytime costs are changing it makes good sense to see what can be done to attack the root cause for the cost increase. Often some remedial actions can help to ease the pain.

8.3.2 Freight costs

Often overlooked, freight costs affect the final product cost more than most imagine. Every ounce of raw material, whether in the product or in the packaging, or used in the process during manufacture, needs to be delivered from its point of origin to the factory. Sometimes the materials are moved two, three, or four times as they make their way into the final product and out the back door (yes to be shipped again) in route to your customer. Some of these costs we can control, most we cannot; however, be sure to consider them when selecting vendors and when organizing the production process flow. Being cognizant of the freight costs and including freight costing as a real portion of the final per unit cost calculations provides a much more realistic view of the attainable margins for the product.

8.3.3 Assembly or labor input costs

Depending on the product, labor costs can be the largest single cost in the final per unit cost both by absolute cost magnitude and as a percentage of total cost. The design team certainly understands this and has done whatever possible to reduce the labor input required to a minimum, however, the labor is still required and must be estimated accurately. While design has an influence on the time required for assembly, other influences can and do enter into the equation when estimating the labor costs on a per unit basis, including set-up time, clean-up time, scrap fallout, workmanship standards, adequate and complete instruction of the workers, and motivation of the workers to provide high-quality workmanship at a fair production capacity rate.

8.3.4 Quality costs

When planning for production it is easy to forget or to lessen the actual cost of the quality assurance (QA) efforts that will be required once in a production mode. Be careful to assess and estimate the QA and Quality Control load that will be required based upon the type and complexity of product being built, filled, or assembled. Adequate QA/Quality Control is not a luxury, it is a necessity and ignoring the need will be a large mistake. Plan for extreme quality to start, if it proves to be excessive, then back down. As mentioned earlier, when possible self-inspect at each operation or in each fixture to the extent possible. Early detection of an out of specification condition will allow early action to save the added time and costs invested in the nonconforming portion that would normally make its way down the production path where more parts and labor is added sequentially.

8.3.5 Packaging costs

Packaging costs can be near zero or can be the major cost of production (women's cosmetic products for instance where the flashy package can sometimes cost 100 times the active ingredients within the package). Be sure to evaluate and include estimates for not only product packaging for retail display or for point of sale, but also bulk packaging for the cartons of units that must be transported from the factory to and through distribution to the retail or end-user. Cartons, gaylords, pallets, and containers, it all adds up; and, the cost of each per unit part or assembly shares a small fraction of these costs. Be sure to include some factor to account for the various packaging materials and components required to get the product to market.

8.3.6 Financing costs

One cost often overlooked by the design team is the cost of money. The funding required to implement the tooling and remaining first cost items, as well as the first turn of inventory to be sold on the open market will amount to a large sum of funds. These funds could alternatively be invested in some other investment vehicle and generate revenue for the company. Otherwise, the company may need to borrow the funds to implement the project. In either case the funds used to carry some overhead cost, either lost revenue or the direct cost of interest paid to borrow the funds. These costs need to be included in any evaluation of the final project costing.

8.4 Total cost evaluation

How are decisions really made? I mean, when you get down to the final decision regarding which material to use, which process and process vendor to use, which packaging to use, how is the decision really made? Is the sourcing gig really fixed and the vendor that sends the CEO on the best ski trip or beach trip wins? Is it the material supplier with the best looking booth attendants at the annual trade show? Or, is it really based on performance and costing? In my small world, performance and costing are indeed paramount; however, I have seen clients make decisions for reasons that only the gods themselves might know; no earthly being could ever figure them out or discern the rationale behind the decisions.

For me, the total cost of ownership has always been a benchmark that is hard to argue with for comparative analysis. In making purchasing decisions for a project it simplifies to the total cost evaluation (TCE). Meaning is quite simple that we look at all the costs and roll them into a first cost of development and a per-unit cost in production. What does this provide? It provides a means for analyzing one production system or method against another. Not one vendor against another, but one system against another. So, for example, material A is 1.6 times more expensive than material B, hence we choose material B for our project. However, when looking at the processing cost, material A processes twice as fast as material B and 100% of the scrap can be reground and fed back into the feedstock (after adequate drying). Now, given this processing information which material should we choose? Well, if we understood that material A was nonrecyclable and material B was the leading recycled material on the market and that our end-user was the home owner or consumer who felt that recycling was our duty, now what? This decision is becoming more and more complex. What value do we place on the recycling ability; it means little to our operations costs. However, marketing says it will help to sell more products to environmentally conscious consumers (Fig. 8.4).

This holistic view of total cost is messy but extremely effective, when one system is compared against another. Yes, values need to be assigned to subjective attributes like consumer preference, but as long as the valuation means are fair between competing systems, the final analysis and decision will be valid. This applies to the evaluation of first costs as well. Everyone works hard to find the lowest and fastest possible tooling source and is championed for doing so. However, once the cheap and quick tool makes it to production and is down for 4 or 5 weeks of remedial work at a repair cost of near the first cost, does anyone really question the initial decisions. Again, I am reminded of one of my old colleague's advice regarding fabrication of capital tooling, "...high quality, low cost, or quick turn time, choose any two ..." Once committed

TCE–system one of three

Cost component	Part A	Part B	Part C	Part D
Raw material	$ 0.0153	$ 0.0624	$ 0.1006	$ 0.0987
Process	$ 0.0220	$ 0.1823	$ 0.1108	$ 0.1654
Secondary operations	$ —	$ 0.1172	$ —	$ —
Finishing	$ —	$ —	$ —	$ 0.2200
QC	$ 0.0300	$ 0.0400	$ 0.0300	$ 0.0450
Packaging	$ 0.0050	$ 0.0080	$ 0.0050	$ 0.0300
Transport	$ 0.0150	$ 0.0190	$ 0.0150	$ 0.0400
Incoming QC	$ 0.0200	$ 0.0200	$ 0.0200	$ 0.0500
Warehouse	$ 0.0190	$ 0.0190	$ 0.0190	$ 0.0190
Rejects/loss	$ 0.0200	$ 0.0623	$ 0.0350	$ 0.0850
Piece part subtotal	$ 0.1463	$ 0.5302	$ 0.3354	$ 0.7531
Req'd per finished unit	3	1	2	1
Parts cost/unit	$ 2.39			

Cost assembly	Step 1	Step 2	Step 3	
Line set-up	$ 0.0500	$ 0.0500	$ 0.0500	
Operation	$ 0.1324	$ 0.2270	$ 0.1080	
Visual check	$ 0.0200	$ 0.0400	$ —	
Transport	$ 0.0150	$ 0.0150	$ 0.0300	
Packaging	$ —	$ —	$ 0.0750	
Warehouse	$ 0.0050	$ 0.0050	$ 0.0300	
Rejects/loss	$ 0.0200	$ 0.0010	$ 0.0500	
QC	$ —	$ —	$ 0.1200	
Assembly subtotal	$ 0.2424	$ 0.3380	$ 0.4630	
Ass'y cost/unit	$ 1.0434			
Total (w/o overhead)	$ 3.4364			

Figure 8.4 The total cost evaluation (TCE) method is not rocket science, it is just plain and simple list and analyze each and every cost for each competing prospective production system as a balanced comparison to be used as a basis for a decision between competing systems.

and the path is set, it is hard to turn back; hence, make the decision based on the total cost projections. If you know there will be remedial changes and costs, then load them into the estimates up front and prepare accordingly.

TCE is not an easy task and you will make mistakes and errors in assigning values. However, the odds will favor you if you do your homework. Don't let expediency or one particular "excellent opportunity" distract or sway you away from looking at the total picture. I assure you the total picture is much clearer for the long-term than the "excellent opportunity" that attempts to take your eye off the goal. Don't pay any attention to the shiny metal object in the room (it's just electroplated plastic anyway).

8.5 Preparation for project verification and implementation

Once the estimates are gathered and analyzed for accuracy, and after using TCE to select the best sourcing system, the information needs to be prepared for presentation to the project team. Typically the costing information will be one of the major presentations during the project review and verification phase (Chapter 9). It is important to present the information in a form as it will be needed or used; for example, the estimates should be broken into first costs, or investments, and per unit production costs. TCE analysis systems should be presented and reviewed so that every team member can understand the competing systems and understand the decisions leading to the final selections. In this manner, as mentioned in previous discussions, the team as a whole buys-in to the selection and everyone shares the responsibility when implementation time comes.

Resources for further study

S. Levy, J.H. DuBois, Plastics Product Design Engineering Handbook, second ed., Chapman & Hall, New York, 1984.

S.E.M. Selke, J.D. Culter, R.J. Hernandez, Plastics Packaging—Properties, Processing, Applications, and Regulations, second ed., Carl Hanser Verlag, Munich, 2004.

J. Rotheiser, Joining of Plastics—Handbook for Designers and Engineers, third ed., Carl Hanser Verlag, Munich, 2009.

G.L. Beall, Rotational Molding—Design, Materials, Tooling, and Processing, first ed., Carl Hanser Verlag, Munich, 1998.

E.R. Larson, Thermoplastic Material Selection—A Practical Guide, first ed., Elsevier, Oxford, 2015.

R.A. Malloy, Plastic Part Design for Injection Molding: An Introduction, second ed., Carl Hanser Verlag, Munich, 2010.

P.A. Tres, Designing Plastic Parts for Assembly, seventh ed., Carl Hanser Verlag, Munich, 2014.

Web links of note

Great Design information from DuPont can be found here:

http://www.dupont.com/content/dam/dupont/products-and-services/plastics-polymers-and-resins/thermoplastics/documents/General%20Design%20Principles/General%20Design%20Principles%20Mod%201.pdf.

Sabic provides many tools for designers here (you may need to register):

http://www.sabic-ip.com/webtools/toolsHomePage.faces.

The UL Prospector site is a valuable tool for any designer reviewing materials for any application (you will need to register):

https://plastics.ulprospector.com/.

Solvay provides a great design guide for using their Udel polysulfone materials:
http://www.solvay.com/en/binaries/Udel-PSU-Design-Guide_EN-227550.pdf.
Bayer has a nice design guide for RIM systems:
http://reactioninjectionmolding.com/wp-content/uploads/2013/09/RIM-PartMoldDesign-Guide.pdf.
Covestro offers several design related guides, including part and mold design, snap fit design, and joining techniques for plastics:
http://www.plastics.covestro.com/Library/Technology-brochures.aspx.

Project review and verification—risk management

9

What is product development really about—it's about creating the next best widget, right? Well, that is at least a romantic's view. In reality, it is creating the next product or service that can generate the revenue to see the company through the next few years without putting the company in any further risk. If you think about everything we have discussed thus far in this book, I mean really give it some thought; it is really all about creating opportunity while limiting risk. There it is, said out in the open, this is not a creative lifestyle at all, it's an ultra-conservative lifestyle. Wow, my balloon is now fully deflated, here I thought of myself as a starving artist type; well at least the starving part was close to realistic.

Full review and thorough verification of the design before sending the first purchase order out for any capital tooling is only natural. Spend a week or two doing the due diligence before several hundreds of thousands of dollars or more is committed. Remember, once you step off that 30th floor balcony and begin the earthbound trip it is extremely hard to turn back.

The following few topics are a mere listing of procedural steps encountered on most development projects. Use these as a potential reminder list so that, as they say, no stone is left unturned.

9.1 Gather the verification team

As with our design reviews during conceptual development and detailed development, we need to identify the stakeholders at this stage of the program and make sure that each is represented well for the verification review discussions and consensus. Remember this is the final review prior to committing valuable company resources and funds for the project.

9.1.1 Design and engineering team

Members of the design and engineering team need to be available to bring the history of the trade-off decisions that were made as a part of the design process. In any validation/verification discussion there are always questions about "why" a certain feature or structural element was used and "why wasn't" this and such feature or material used or selected. The design and engineering team needs to be able to address such questions to ensure that every decision was and has been thoroughly vetted in the proposed final design.

Designing Successful Products with Plastics. http://dx.doi.org/10.1016/B978-0-323-44501-6.00009-X

9.1.2 Management team

Management team will be the recipient of the results from implementation of the proposed final design into a product. They will be the heroes if the product is successful but will also need to accept responsibility if the product fails. They have already authorized the funding and resources to get the project though development; now they will be asked to authorize potentially much larger sums for the capital required to prepare, to build, and launch the product. Management should be the most interested group in the room and should be asking the toughest questions.

9.1.3 Marketing team

Like management team, the marketing team will be accepting the responsibility to generate sales for the new product and will, like management, suffer should the product fail. Hence, marketing should have a keen interest in ensuring the product can perform as needed and can be manufactured at the price point necessary to compete in the marketplace.

9.1.4 Quality assurance team

Quality Assurance members should be prepared to support the design team with information regarding their quality plan for qualification and for production implementation. These plans, coupled with the actual design, when assessed against the functional specification should provide management with the full story of how the proposed design will satisfy the market "need" and do so without undue risk to the organization.

9.1.5 Other potential stakeholders

In addition to the highlighted stakeholders, others may be considered for inclusion in the verification review group. These can include: legal counsel, operations, manufacturing, a potentially critical material supplier or subcontract vendor, strategic sourcing, and others within the organization that might have valuable input or who could be impacted, either positively or negatively, by the implementation of the design.

9.2 Verification of the design

The verification review team, comprising the members as described earlier, will be the responsible group for final approval of the design prior to implementation; they will put the puzzle together, so to speak, for the final move to commercialization of the product. In a perfect world, the verification team will require the design team to prepare a presentation outlining the proposed design, the budget, the QA plans, the functional specification, and all other relevant information. The presentation should be circulated to everyone in the group prior to the review meeting, allowing everyone

Figure 9.1 The verification team reviews the information gathered thus far in the project and then set out to solve the puzzle and create the optimum implementation plan. *Source*: Photo from iStock/.

on the verification team to become familiar, with the information and prepare for the review meeting adequately. Each member will need to come to the review meeting with their particular puzzle piece ready to mix and match with every other team members piece, attempting to find the optimum implementation plan (Fig. 9.1).

9.2.1 Review the design against the written functional specification

The functional specification is the primary reference document against which the design will be verified. The proposed design should be reviewed and tested against each attribute or functional requirement within the specification. Some groups use a grading system wherein the proposed design is scored on a 1–10 basis against each functional requirement. For example, a score of 1 might represent the lowest condition, such as "does not meet the requirement at all" a score of 10 might represent "exceeds the requirement" and a score of 5 might be "meets the minimum necessary for the requirement," and so on. The proposed design should be tested against every attribute or requirement within the written specification, do not skip a single requirement. Any testing results and observations compiled during the prototype testing or analysis should be available to be reviewed by the verification review team.

9.2.2 Review the design against the conceptual and final design review notes

Notes from both the conceptual design review and the final detailed design review will often highlight features and functions of concern to those review groups. It is wise to go through those and ensure that whatever concerns were discussed in those reviews were indeed addressed and do not still pose any concern. The design team should be prepared to highlight any changes implemented as a result of the discussions from both the conceptual and the detailed design review sessions.

9.2.3 Review the design against the budget and costing targets

Unlike the previous design review, the group should now have detailed cost estimates for both the capital investment required to implement the project and the per-unit costing once in a production context. Given this fairly precise financial information, management can determine the potential margins based on the proposed price structure from marketing. Any surprises in per-unit costing or in capital funds required will certainly be topics for review and discussion by the group. Design and engineering should be prepared to support decisions or features that increase cost or capital requirements over or above the preliminary budgets earlier in the project. Conversely, design and engineering should be prepared to show the group any proposed additional features or capital additions which might bring value to the program either at present or in the near future. Sometimes these proposed additional features or capital improvements come in the form of statements such as "once in production and once sales are meeting expectations, we can foresee increasing capacity and lowering per unit costs by implementing a new piece of equipment and process which will require a capital expenditure in the neighborhood of …" or similar. Management will often be pleased to see staged implementation where the capital investments can be gated according to results in the market. For many capital requirements this is not possible, we must build the mold or tooling before we can mold or process the part. However, we can sometimes plan the staged implementation; for example, build a single cavity mold in a mold base designed for two cavities, or perhaps even four, thereby limiting the upfront capital requirements but allow room for future capacity increase and cost reduction.

9.2.4 Review FMEA results for the design

During project verification it is prudent to review the failure modes and effects analysis (FMEA) results with the group. This is as much about proving that the risks have truly been reviewed and to the greatest extent possible reduced, as it is to refresh the groups awareness that failures do still happen and that we must be prepared for them. Quality assurance members will want to illustrate any procedures or system checks developed to be implemented as a direct result of any FMEA attribute or condition. The design team will want to illustrate any design changes or remedial actions implemented to correct any attribute or condition from the FMEA. All of this should build confidence in the project and provide the due diligence comfort needed by the group to verify and approve the project.

9.2.5 *Review any intellectual property analysis or opinion on the design*

Any potential infringement issues should be resolved prior to this verification review meeting, however, they should be discussed in group; especially highlighting areas where there is a need to be careful with respect to some other intellectual property. This is as much to educate the design and engineering staff as it is to illustrate to marketing what not to highlight in their promotional materials and advertising. Any potential licensing deals to assist with any infringement issues should be reviewed and discussed. The group should be advised whether the presence of any of these agreements should be public knowledge, or should be kept company confidential.

Any proprietary information that will not be part of any patent application should be identified and the group should be clear that the information considered proprietary is not to be used or disclosed, in any form, for any reason. This proprietary information may indeed not be technical information at all, rather it may be the market rollout plans, or the market research results, or other nontechnical, nonpatentable information which could be of great benefit to a rival or competitive organization.

In addition, any patent or trademark applications which have been made or are in process should be discussed so that everyone in the group understands the importance of each. Marketing should be promoting any product on which a patent application has been filed as a "patent pending" product. Likewise, all trademarks should be identified as such on all promotional or advertising copy.

9.3 Agreement to move to production implementation planning

Once all pertinent data and issues have been reviewed, the group must come to a consensus to proceed to production implementation. If the group cannot do this, then the issues preventing the agreement need to be identified and the responsible parties need to be identified and tasked with resolving the issues or finding the requisite information needed to gain full verification and agreement to approve and move forward. Once agreement and approval to proceed is achieved, the group begins the next phase of work preparing for production. Getting our ducks in a row, or structuring the functional groups for effective transition to implementation is the goal.

9.3.1 *Review production estimates and timeline*

Now that the project review is complete, the group should review the preliminary plans for said implementation, including the timeline estimates and overall implementation schedule. While representatives from most functional groups are still together in the review session, with all of the project details are fresh in their minds, it is an excellent time to discuss how to arrange the implementation schedule for maximum efficiency and effectiveness. This is the time to identify the critical milestones in the timeline which will drive the schedule and provide the decision points for critical marketing roll-out campaigns, and so on.

9.3.2 Review the selected vendors

An open and frank discussion regarding the selected vendors for the capital tooling and for the production processing should be part of any full project validation review. In many instances these vendors may hold one or more critical components and their ability to both meet quality plans, and to meet capacity requirements may be crucial to the product line and to the company. Don't let a vendor who is at capacity in one industry become the weak link in the supply of components for this project, even if they have the cheapest price, or offer the best terms. If the processes are to be done in-house for production, make sure your production equipment is up to the program and ensure the production group has the experience and level of staffing necessary to handle the anticipated program. Get these vendor issues out in the open and discuss them BEFORE giving them the orders and suffering through the consequences. Do your homework!

9.3.3 Assign implementation leaders and reporting protocol

Since production implementation requires efforts from most every functional group or team within the organization, a leader or group of leaders should be identified and tasked with organization and control of the implementation process. In addition, some formal reporting protocol, both from the functional teams to the leaders and vice versa should be determined and agreed among the various functional groups represented. This is just good project management housekeeping policy.

9.3.4 Determine the implementation status meeting frequency

Before calling for adjournment, the assembled group should determine and set the time, place, and date for the next meeting, as well as what an appropriate meeting frequency should be. On most production implementations for plastic parts and products, there is a huge lead time for tooling, molds, and so on, at the beginning of the implementation process. Then as process trials begin and part samples begin to be available for qualification and marketing samples, and so on, there is a flurry of activity. As such, many times we will see status meetings held monthly at the beginning of the implementation project which will then shift to weekly meetings as the samples start to arrive and debugging begins. Additionally the group can determine and define the reporting frequency and responsibility. Keeping all stakeholders up-to-date on a regular basis will help to ensure that everyone can see the project dynamics and will have the information they need to understand when and how their contribution is required. Good communications leads to well-run projects—enough said!

9.4 Preparation for transitioning into production

With the review complete and with agreement by all parties that the project has a green light to transition into production, the various functional teams must prepare for their respective tasks. Each functional team will need to allocate resources, staff, and funding for their portion of the project. Every member of every team will need to be in synch and be ready to support the project well.

Resources for further study

S. Levy, J.H. DuBois, Plastics Product Design Engineering Handbook, second ed., Chapman & Hall, New York, 1984.

S.E.M. Selke, J.D. Culter, R.J. Hernandez, Plastics Packaging—Properties, Processing, Applications, and Regulations, second ed., Carl Hanser Verlag, Munich, 2004.

J. Rotheiser, Joining of Plastics—Handbook for Designers and Engineers, third ed., Carl Hanser Verlag, Munich, 2009.

G.L. Beall, Rotational Molding—Design, Materials, Tooling, and Processing, first ed., Carl Hanser Verlag, Munich, 1998.

E.R. Larson, Thermoplastic Material Selection—A Practical Guide, first ed., Elsevier, Oxford, 2015.

R.A. Malloy, Plastic Part Design for Injection Molding: An Introduction, second ed., Carl Hanser Verlag, Munich, 2010.

P.A. Tres, Designing Plastic Parts for Assembly, seventh ed., Carl Hanser Verlag, Munich, 2014.

Web links of note

Great Design information from DuPont can be found here:

http://www.dupont.com/content/dam/dupont/products-and-services/plastics-polymers-and-resins/thermoplastics/documents/General%20Design%20Principles/General%20Design%20Principles%20Mod%201.pdf.

Sabic provides many tools for designers here (you may need to register):

http://www.sabic-ip.com/webtools/toolsHomePage.faces.

The UL Prospector site is a valuable tool for any designer reviewing materials for any application (you will need to register):

https://plastics.ulprospector.com/.

Solvay provides a great design guide for using their Udel polysulfone materials:

http://www.solvay.com/en/binaries/Udel-PSU-Design-Guide_EN-227550.pdf.

Bayer has a nice design guide for RIM systems:

http://reactioninjectionmolding.com/wp-content/uploads/2013/09/RIM-PartMoldDesign-Guide.pdf.

Covestro offers several design related guides, including part and mold design, snap fit design, and joining techniques for plastics:

http://www.plastics.covestro.com/Library/Technology-brochures.aspx.

Production implementation

10

The focus throughout the design and development stages of a project are aimed at squeezing the most out of the materials, the process, the tooling, and the design, in an effort to get the greatest benefit for the end-user, at the most efficient use of the available resources. Now the project is shifting to a purely implementation mode, causing the focus to shift as well; shifting to being far more risk adverse. During the implementation stage of the project there are abundant opportunities for issues to arise; only vigilant and disciplined management can reduce these risks. In many organizations, the minute the project moves to implementation, the design team signs off and goes away to find some other creative endeavor. The rigorous details of implementation do not provide the exciting creative outlet desired by most in the design field; however, the good project manager will insist the design team remain engaged to assist during implementation, especially during initial manufacturing trials for the first parts, first pilot assembly trials, and during testing and debug stages. Just as we condone involvement of the manufacturer, processor, tooling designers, and so on, during the early stages of the design process; we also condone design team involvement until the product or part is fully qualified and in a routine production mode.

The production implementation phase encompasses many different resources accomplishing a multitude of dissimilar tasks, from reviewing estimates to placing capital orders, to checking the legal language in agreements, to reviewing tooling designs, to planning quality assurance for both part and assembly production, to running first trials of capital tooling and molds, to initial assembly fitting and testing, to figuring out packaging for parts and assemblies needing transport during manufacturing, and the list goes on. Our review of production implementation will begin, with the nuances of placing capital orders and will walk with the project chronologically through the final tooling trials and process validation trials.

10.1 Placing capital tooling orders

Capital tooling orders, often placed by the strategic sourcing or purchasing personnel within an organization, should be reviewed and approved by the design team; at least on a technical merit basis. The purchase order is a legal document used in commerce and, as such can define the rights and obligations of the agreement between the purchaser and the supplier. What should be a simple matter of listing all of the details regarding the capital item to be purchased is often reduced to some statement like "… injection-mold tooling to produce XYZ, as per the estimate dated (day) of (month) of the year (year)." Instead, the order could list all of the details

Designing Successful Products with Plastics. http://dx.doi.org/10.1016/B978-0-323-44501-6.00010-6

discussed and agreed upon during vendor selection. The list could include items are as follows:

- Accurate part number, revision identification, and part name/description that the tooling is to produce when complete and in production.
- Materials of construction for the tool (type of tool steel, hardness, surface treatments, etc.).
- Standards to which the tool is to be built (ANSI, ISO, etc.).
- Materials, which the tool will be built to accommodate and run in production—be as specific as possible.
- Specific tool features required (runner system, gates, parting line locks, oil line, hydraulics, etc.) and what preliminary design information (tool design layout, mold filling analysis, etc.) must be approved by the design team before tool fabrication can commence.
- Tool marking for identification of the tool owner, the part name and/or number, the date the tool was manufactured, the shrink rate used for the cavity sizing, and the like.
- Requirements for first article inspection (FAI) or qualification piece parts (quantity, materials, colors, etc.).
- Capital tool ownership statement and any provisions for insuring the asset during development and into production, should production be performed by the same vendor.
- Statement of responsibility for routine tooling maintenance and bearer of costs for same—most often the vendor in whose shop the capital tooling is resident will bear the responsibility and costs for routine tooling maintenance. However, it is good to nail it down in writing on the PO to be sure.
- Statement of ancillary equipment, which may be part of the capital tooling but is separable when not in use—for example, a multichannel temperature controller that is purchased and used for an injection mold with a hot runner multizone system. Ensure that this controller is part of the capital equipment, is visibly marked as such, and must always remain with the capital tooling; and, that it is not be used with any other capital equipment owned by any other customer of said vendor.

Be extremely careful when placing an order and referencing a proposal or estimate from the selected vendor on the purchase order. There may be some actionable legal information that gets tied to the purchase agreement by reference. For instance, some estimates will include language like "...engineering service for the development of production capable tooling to produce XYZ ..." that basically says they are proposing to sell you a service, not a capital asset. In this particular case, the asset may still in fact be the legal property of the vendor who provided the estimate; hence when disagreement arises and it becomes time to select an alternate vendor, the tooling cannot be removed from that original vendor. Be clear in the formal written purchase order or agreement that the purchase order or agreement is the single and final agreement, and that any and all attachments are for reference only.

In some cases, the manufacturer will negotiate some term or time limit of exclusive manufacturing rights, in exchange for some intellectual property tied to their particular tooling or process capabilities. These specifics need to be clearly identified and annotated as to exactly to the date or to the production number of parts produced and should provide some mechanism to allow for early termination by either party. In many of these arrangements, it is often wise to seek counsel from attorneys practicing in manufacturing agreements specifically.

10.2 Review the preliminary tooling design layout

On any given project, whether injection molding, blow molding, rotational molding, or thermoforming is used for the process, a preliminary tool layout drawing is typically produced and reviewed, with the processing vendor or department, and with the design team. As mentioned earlier, this should be specified on the purchase document and on the 2D part specification drawing. Sometimes this will be a full blown tool layout file, other times it may just be the part 2D print marked up to show the mold features and their respective locations. In any event, the design team should be anxious to review this information PRIOR to any actual construction of the tool or mold. These tool features will become permanent features for the life of the part, hence if there is any question regarding the feature, it is better to discuss it and make a change before the first steel block is ordered (Fig. 10.1).

10.2.1 Parting line

On most parts, the parting line will be exactly as designed and the tooling designers understand where this feature has to be. However, in some cases the parting line may need to step up or down to accommodate some feature or side action that the part geometry dictates. With these cases, the tool designer will have some liberty as to where the feature steps and what that geometry will be in the tool. If the step is occurring on an appearance surface, or a functional surface, the design team will want to be extremely sure that the witness lines, the reversed draft surface, and so on, will not detract from appearance or functionality of the finished part. Usually, discussion with the tool designer regarding the functional use or appearance needs for the area can result in a design that is acceptable to both. Hopefully, if the tool designer has been up to date, from being involved in the design reviews, and so on, as recommended, they will already know and anticipate these requirements.

10.2.2 Gates

Gate style, number of gates, and gate locations will be recommended by the tool designer, unless they have previously been specified by the design team. In reviewing a proposed tool design where the gates are specified, the design team will be looking at the position of the gate(s) as they relate to appearance and/or functional surfaces; the position of the gate(s) as they relate to holes or other geometry, which may place a knit line in a bad location; the position of the gate(s) as they relate to features where flow will dictate polymer orientation, such as living hinges; and the style of gate(s) that will assist in efficient processing of the tool. Gate sizes and styles are material dependent and some materials have more critical gate requirements than others; hence, make sure the gate selections are in line, with the recommendations from the material supplier for that particular material and grade. Visualization through the use of mold filling analysis software can be extremely useful when having gate location discussion during this preliminary tool design process.

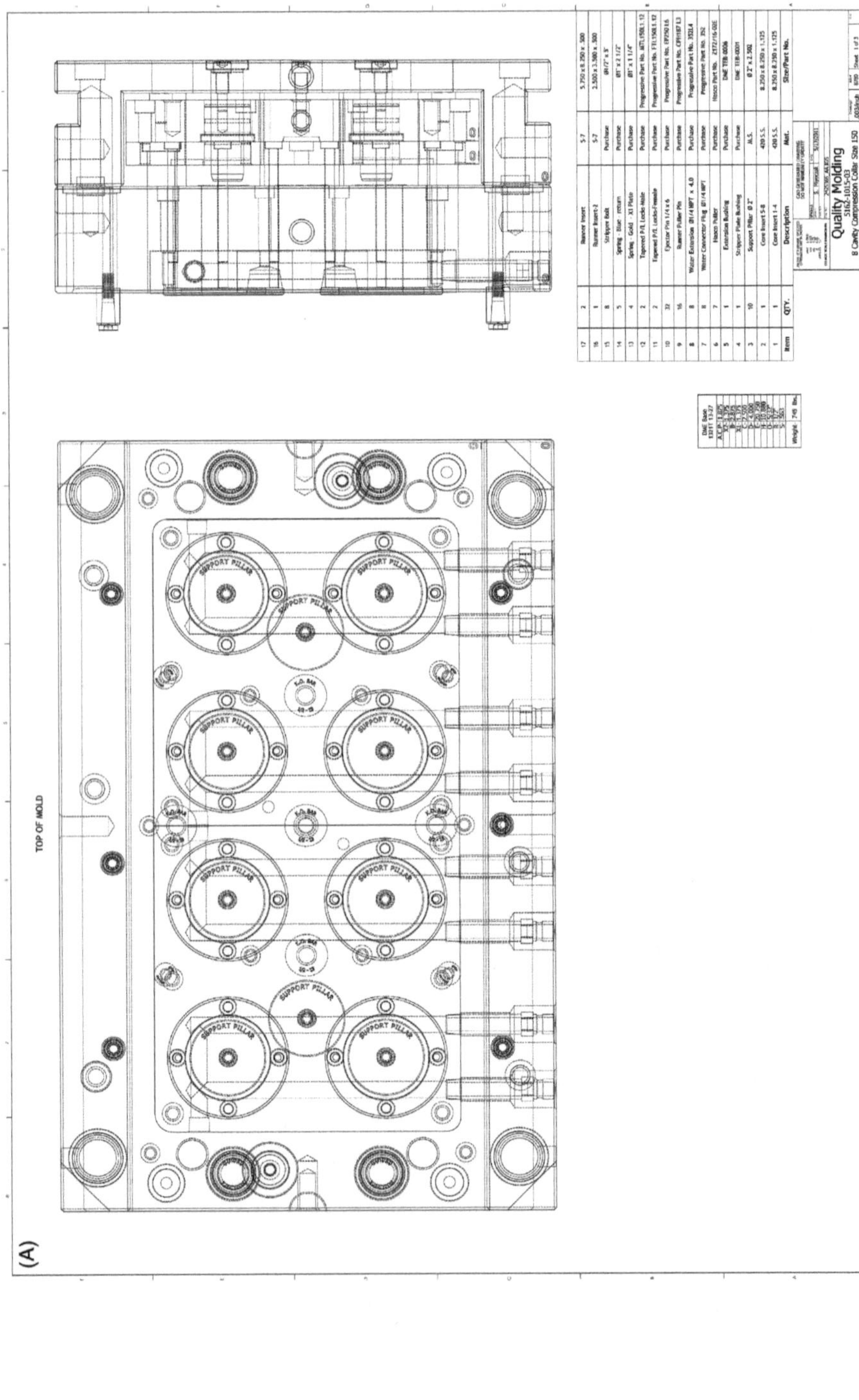

Figure 10.1 (A–C) These three illustrations are three sheets from an example of a preliminary mold layout drawing used to relay the mold design information from the mold designer to the product design team. In these drawings the design team can see the gate location, the ejector system and locations, the venting arrangement, the cooling lines, special movements, and so on.

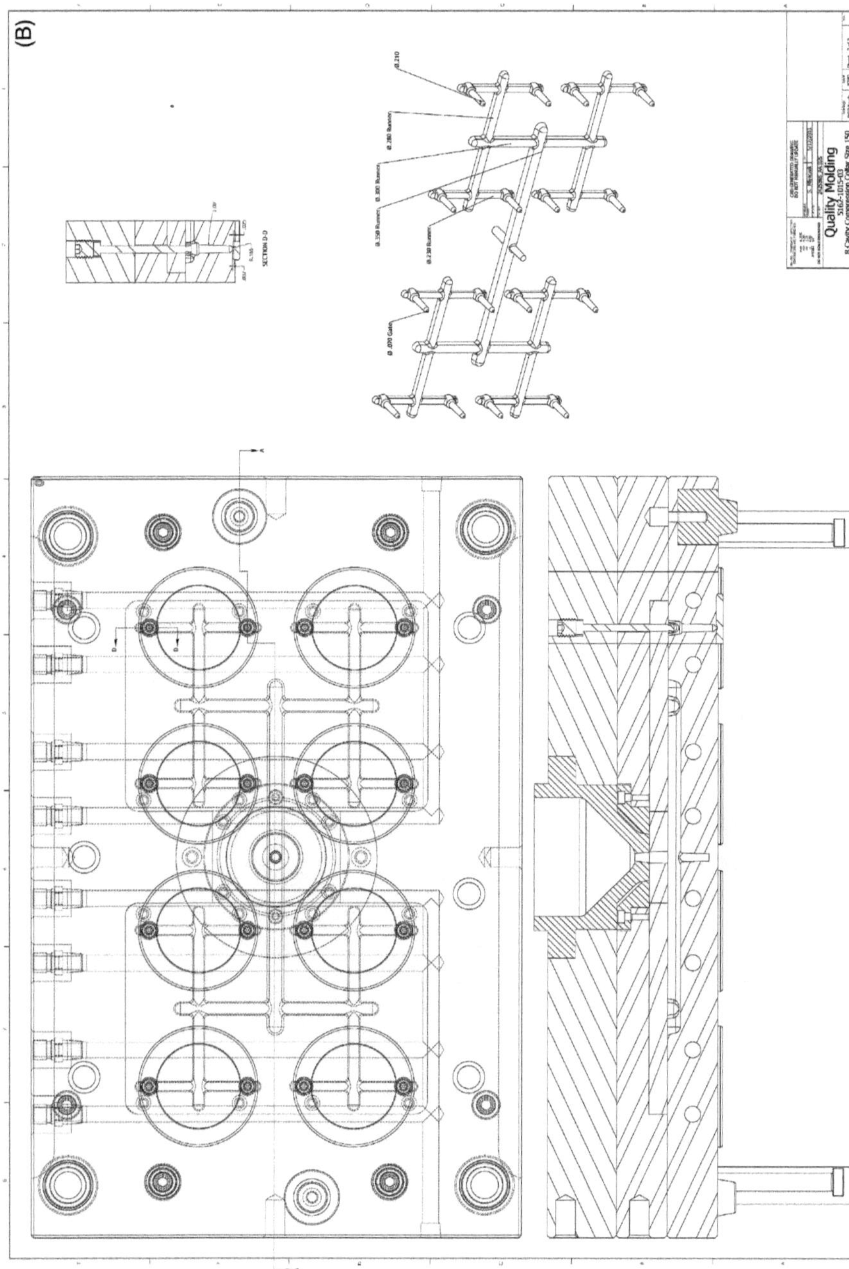

Figure 10.1 *(cont.)*

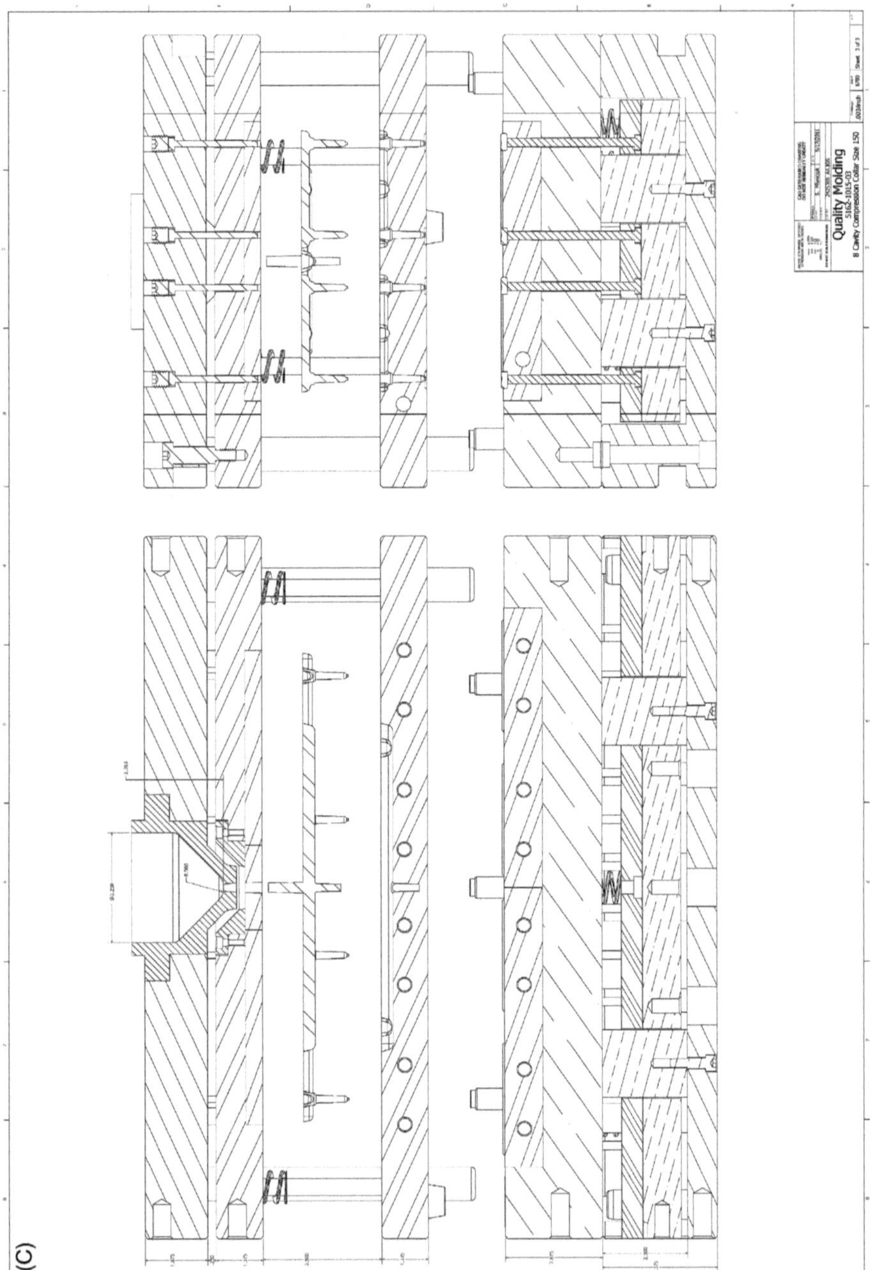

Figure 10.1 *(cont.)*

10.2.3 Knit line locations

As mentioned earlier, when reviewing the gate locations, the areas where knit lines will form will be somewhat apparent. If these suspected knit line locations happen to fall in an area that will be subject to some loading or stress, the area should be evaluated to ensure that the knit lines will not become a failure risk. If so, then an alternate gate location might be suggested. Sometimes, mold filling or flow analysis reports will be provided with the tool preliminary design as a reference and knit line predicted locations are often included. This information can be useful if available, hence it makes sense to request it when placing the order for the tool.

10.2.4 Venting

Venting is not often specified on any preliminary tool drawing or layout. However, if the design team feels that there may be any area of the part where venting may be an issue, it is certainly prudent to include some request for specification regarding the venting means to be included within the preliminary tool layout information. Hopefully, through collaborative work earlier in the design process the tool designer will be already informed of such concerns and should be motivated to include any specific tool design features that would be used or included to help remedy the concern. For injection-molded parts, review of any mold filling analysis results will assist in showing the areas where additional venting might be needed. Often the mold filling analysis is a key component of the mold design review process for injection-molded parts.

10.2.5 Runner systems

Depending upon the part design geometry and/or the production capacity required, the mold or tooling may be designed to have one of many different filling and runner systems. If the tool in question is a multiple cavity tool the preliminary tool layout should specify exactly what this system is to be. For example, given a four cavity injection mold for simple part geometry, the tool may be specified for example, "… a four cavity injection mold, with a balanced parting line runner system, using edge tab gates at each cavity." In reviewing the tool layout, this may not have much effect, unless the tool will be processed in-house and the operations team wants to limit secondary operations. In this case they may direct the tool designer that they would prefer sub gates or some automatic trimming style gates. Making a change like this during the preliminary review requires a phone call or an email at this point in the tool development. Once the tool is complete and shows up for production, the change would be far more difficult and expensive to implement.

If the tool or mold is to have a more complex filling system, like a three-plate mold, or if is to have a hot manifold style filling system, then the preliminary mold layout will either show the mechanical detail of such or will callout the specifics regarding the hot manifold system and what type of temperature controller is required to control the hot manifold system in production. It might also be wise to review the materials being used in these kinds of filling situations as not all materials are friendly in these

Figure 10.2 A runner system with tab gates in an injection mold.
Source: Photo from iStock.com/robynmac.

arrangements. Just be sure the material suppliers understand the filling system and have them verify that their materials will process well in these filling systems. Don't rely on the tooling source that does not spend time daily running materials through the tooling they produce (Fig. 10.2).

10.2.6 Slides or lifters

If slides, side actions, or lifters are to be used in the tool, the preliminary tool layout should make note and should propose the positions and the means by which these will operate (cam, hydraulics, pick-out, etc.). As with the review of the parting line mentioned above, if the slide mating area to the part is either an appearance surface or a functional surface, then more precise tool design information may be required to ensure the desired appearance or function. In addition, it should be understood if the slides will carry cooling or heating lines at this preliminary layout step. If the slides need to be interchangeable, this requirement should also be considered and clearly understood by the tool designer at this preliminary layout stage.

10.2.7 Cooling (or heating)

The processing of plastics materials is essentially always a thermal event; hence heating and cooling are almost always incorporated into the process in one form or another. At this preliminary tool design review, it is wise to understand just what form of cooling (or heating) is to be used, how will the connections between the tool and the cooling (heating) source be made, and how will the cooling (heating) be controlled. In addition, if any special cooling or heating means or methods are to be employed, these should be noted and reviewed as well (bubblers, heat exchangers, conformal baffling, etc.). This specification will be more important for review by the processing team who will be responsible for using the tool to manufacture the part to specification. Adequate thermal control will provide the required processing window to allow good cycle control, resulting in good part-to-part (cycle-to-cycle) consistency.

10.2.8 Tool materials

Anticipated total life cycle capacity required from the tool will, to a large extent, determine the materials used in construction of the tool. In addition, certain materials to be processed by the tool will determine the tool material, for example, PVC will most always require stainless steel as it tends to be a corrosive material during processing. Sometimes functional aspects will determine tool material, for example, an optical lens part that requires a highly polished tool to produce the clear optical lens surface would define a material that could be polished and that would hold the polish over the life of the tool. If there are any special features that the design team knows will be an issue, for example, a long thin core pin, some discussion may be in order regarding the cooling capacity for the particular feature and alternate materials may be part of that discussion. For the thin long pin a material with a higher heat transfer coefficient may be desired, something like beryllium copper, for example.

10.2.8.1 Coring

Cores, or features within a mold which act to remove material to achieve uniform walls, or which are used to produce hollow round hole features for fasteners, or for fluid flow, are areas which need to be evaluated in the tool design. These features are often slender and thin, and will tend to heat up as the mold cycles, as their cross-section does not allow enough heat transfer to the cavity material at the base of the feature. In addition, these features are often blind, that is they end without sealing-off to the other half of the mold. This blind condition can sometimes lead to gas trapping during the mold filling cycle, resulting in either surface imperfections, nonfilled areas, or burning of the part and mold. The burning is from gas compression causing the gas to heat up, further adding heat to the core material. Anytime these core features can be tied to any external wall, the process will benefit; either through better venting, or additional cross-sectional area for heat transfer, or both. When reviewing the preliminary mold design, be on the lookout for these thin slender features; sometimes we don't see them in the detailed part designs but when the tool is designed we can see how the tool steel must get thin to produce a certain feature on the part. Always discuss these thin tool steel features with the toolmaker to see if they can be eliminated either though redesign of the mold or of the part. Alternatively, work with the toolmaker/designer to provide enhanced cooling in the area where these thin features are located. Perhaps make these as inserts, which can be higher conductivity materials to assist. Review these areas with the processor as well, for the processor has probably fought with similar conditions in other plastic parts, which they have molded over the years.

Case Example

On one project we had a long slender round core, which was removed from the part by a side action slide, cam driven by the mold opening movement. The parts were molded in a polypropylene resin and the end of the slender round core sealed off against another core with a face seal arrangement. Our design called for zero flash at this seal off location because this was a fluidic device and the open area of the round feature was used to determine a flow rate for the end product. The mold was designed, the mold design was checked, the processor approved the mold design, and the mold maker went to work and fabricated the mold. During mold trials the parts measured dead-on and they were deemed good parts. The first trial parts went to performance testing and the seals worked as designed, the flow rates were within the targets and in general the parts performed as intended. During assembly for testing it was noticed that there was an anomaly with the assembly because the part could be assembled two different ways (180 degree upside down) and in one orientation the fluidics functioned differently than in the other. Hence a design change was made to provide a feature to ensure proper orientation by the assembly operator. All was good, right up until ….production started.

Now, the processor was tuning his cycle to get optimum cycle times and had to make some adjustment for the new orientation feature. Once the mold was up to temperature and running smoothly, QA measured parts and inspected for flash, and everything passed the metrics and production commenced. However, the long slender pin was dragging material out when it was pulled from the part. Not too much, but enough to cause the ID formed by the long slender pin to be roughened, sort of wavy with small tears circumferentially along the bore. This was not seen in any of the QA inspections or in-process checks, but when the performance testing was underway the finished products were failing the fluidic performance testing.

Initial parts tested from the initial process trials did not exhibit this core removal surface tearing/pulling phenomenon and they functioned just fine, but add in the surface imperfections and the flow through the part was now more turbulent and killed the fluidic action required for functional use of the part and product.

Remedial actions included increasing the cooling in the slide behind this long slender core pin, and changing the core pin materials to a material with higher conductivity. This higher thermal conductivity material was softer then the material in the original design and the seal off design in the mold need to be changed to ensure a zero flash condition. An accommodating surface with some degree of resilience was added in the seal off zone to assist and the routine maintenance for the mold called for this core pin to be inspected and the fit tightened on a much more frequent schedule.

The remedial actions implemented allowed the mold to be useful and the product went into full production with little problems thereafter, but extremely small details can make a large difference.

10.2.9 Seal-off details

Similar to the parting line discussion earlier in this section, seal-off areas need to be reviewed if the area involved will be either an appearance surface or a functional surface. Typically the seal off areas, where tool features from one side of the mold pass through the part and need to seal against the other side of the mold, are internal to the part and typically are not an aesthetic concern, but this is not always the case. There will always be a witness line where the seal off occurs, unless it is a core pin face-to-face seal off. These witness lines are never obtrusive but they are always there and sometime with tool age these sealing surfaces will wear and flash will begin to appear. If the presence of flash will be detrimental (as in a fluidic orifice part or feature) be sure to note the maximum flash allowed in the seal-off area.

If flash at a seal-off could impact part functionality, then think about how the seal-off is constructed and determine where flash will first occur. For example, if you are sealing off a round core pin against a flat faced core (the part would have a hole through an exterior wall) then the flash would occur between the end of the core pin and the flat faced core, effectively closing the hole at the perimeter. If this hole were used to flow a fluid, then the reduced cross-sectional area (made smaller by the presence of flash) might have an impact on flow performance. On the other hand if the core pin telescoped into a hole on the flat faced core, then the flash will follow the core pin and produce a raised ring on the inside of the part (which may get sheared off when the part is ejected off the flat faced core)—with no reduction in cross-sectional area. Remember, there is often more than one design option for the mold, just as there is for the part, so, pay attention when reviewing the preliminary mold design. Asking for a core to telescope at mold design is far simpler than reconstructing the mold after the initial mold trials (and far less expensive and time consuming).

10.2.10 Ejection system(s)

The style of ejection system and the location of the ejection surfaces should be indicated on the preliminary tool layout. Review to ensure that these locations do not interfere with an appearance surface or a functional surface on the part. Understand that the ejection components are moving elements in the mold and can change overtime with wear or mold maintenance. There will usually always be indications and marks in the part where the ejector components are located. The processing technicians will want to understand that there is adequate surface area allocated for the ejection force required to avoid long-cycle times waiting for the part to freeze sufficiently before ejection (avoiding push through). Ejection plate operation can be different from region to region. Make sure your processing team understands what kind of ejection system is being fabricated and let them advise what system they wish to use for their equipment.

10.3 Quality planning for production (again management of risk)

Given that each company or organization is different, the quality planning for each will be different as well. Some will have the quality plan complete well before capital tooling is ordered but for many the dwell time awaiting completion and first piece samples from capital tooling is the ideal time to prepare the quality plan. Typically the quality plan is prepared and delivered by the quality assurance team; however, the design team should be a major contributor, at least during the technical development of the plan.

10.3.1 Assist in adequate specification of quality parameters

The design team understands better than anyone in the organization the requirements of the part or product at this stage of the development process and are in the best position to help determine the parameters or attributes, which when checked or measured will give an indication, relative to final use or function, of acceptability. Likewise, from a performance standpoint, the design team understands the requirements best at this stage, and can recommend and discuss the performance criteria, which might be testable or measureable for the best indication of final product capability. Hence, during the creation of the plan, the design team should be in a position to assist when developing the specifications.

10.3.2 Review and recommend functional in-process checks

Once a production process plan is in place, the design team can participate by providing a review relative to its design intention. During this review, it will be apparent (or it should be) where a functional in-process measurement or check would be appropriate. If not already in the quality plan specifications, it can be added and if it is in the specification, then values for acceptability can be assessed, evaluated, and made final. It is important to remember that not all specifications will be a measurement, with an exact go/no-go value to asses against; indeed some will be qualitative and inferential, hence adequate comparison evidence is required to allow multiple inspection or production personnel to arrive at similar judgmental decisions.

10.3.3 Review and recommend performance tests

Performance testing, whether in-process or as part of a final inspection process, can be tricky to develop. Some elements of performance can be tested without a great degree of difficulty, other elements require much more effort and some elements simply cannot be tested without damaging the part or product. The design of the testing methods requires innovative thinking and needs to consider the time cost of the test relative to quality of the information obtained. Sometimes it is better to test from some inferential vantage point or by using some technique that is not exactly the intended application use. For example, we will often leak test a vessel using pressurized air, rather than water or some other fluid. In this manner, the part or assembly does not require drying prior to final packaging. However, if the vessel does not leak a gas, then it will most certainly perform with a liquid. Alternatively, some tests cannot be accomplished without an actual in-use situation; otherwise the information gained provides no indication about the real performance of the part or product.

Often the FMEA discussions and results will point to areas of a design that are prime candidates for performance testing as a means of mitigating the potential failure risk. It is always prudent to review the FMEA results and notes when developing testing criteria.

10.3.3.1 Nondestructive examples

The leak test mentioned previously is one nondestructive test example and there are others. Each particular functional attribute of a part or product will have some element that can be tested in a nondestructive manner. However, sometimes better data can be achieved when testing to a failure point. Other examples include, pull testing on a press fit or snap fit assembly, where the pull force exceeds the normal expected loads but does not exceed the yield point for the materials at the particular cross-sectional areas being tested. In addition, these nondestructive tests can be performed at a range of environmental conditions (temperature, humidity, etc.) to simulate real world conditions. With an eye towards efficiency, the testing will almost always be based on the most severe conditions, which for plastics is usually at wide temperature extremes.

10.3.3.2 Destructive examples

Strength testing is possible to a certain degree without damaging a part or product, but gathering data to failure allows us to understand the anticipated factor of safety, which might be exhibited in a certain loading situation. For molded parts the knit line strength can sometimes be tested by bending or loading the part applying the maximum stress at the area where the knit lines are found, which is most likely the weakest portion of the overall part.

Case Example

On one project we were tasked to design a plastic vessel to act as a vacuum trap for use in surgeries where extreme volumes of fluids were used to irrigate and flush the areas being treated arthroscopically. The vessel was to be a rigid single-use disposable device, which needed to provide access to the collected fluids to allow for decontamination prior to disposal. The vessel was to be able to withstand full vacuum or at least high-vacuum situations. To allow testing we devised a pressure tank with a vacuum feedthrough, such that we could pull a vacuum on the closed vessel from outside the pressure tank, and then we could increase the pressure outside of the vessel within the pressure tank; thereby creating a pressure differential greater than the specified full vacuum or one atmosphere. Of course, this test took some time to set-up and run and it was definitely destructive, so it was used as a qualification test, not for in-process or for final inspection. However, during qualification, we used this test means, to prove that our structural ribbing design was indeed not strong enough for the conditions and we modified and retested until the failure point provided the desired factor of safety.

A common destructive test for plastics parts and products is the drop test, simulating the part or product being dropped during use. Usually, the drop is a controlled drop with some effort incorporated to ensure the impact point remains the same for each part or product tested. This can be accomplished by providing a sliding surface upon which the part or product slides until impact. The sliding surface can have grooves or features to mesh with features on the test specimen to retain the orientation and attitude of the part during the travel to impact.

Another common destructive test with injection-molded parts is a squeeze or bending test to determine the process condition at a knit or weld line. This will typically be on a round part with a gate to one side of the part causing a knit line on the side opposite the gate. If the knit line is processed correctly the part can be squeezed to attempt to open the knit line but it holds tight. If the part is marginal the knit line may open slightly during the squeeze or my split altogether. Of course, this test in inferential and material dependent, but in more shops than I can count on two hands I have seen this sort of experiential destructive testing being performed.

10.4 First tooling trials

If implementation is all about limiting risk, then the first tooling trails and the evaluation of both the tool and the parts produced from the tool are perhaps the most important risk-limiting tasks on our to-do list for implementation. This is where all of the intellectual and monetary project investments are tested for the first time and the team needs to hold true to the specification and not let ego or pride short-cut or side step any potential issue. Better to find a fault now and have the opportunity to fix it than to find it with 100,000 units in the field that will need to be recalled.

10.4.1 First process trials

First process trials will generally have two main components, first is general mechanical and thermal operation of the tool. Does the tool function as designed and can it hit the mark thermally. Second, can the tool make the part to the specification; that is, does the finished part meet all dimensional and performance specifications. Sometimes these trials happen during one session, and sometimes it takes multiple iterations to ensure that the trials are completed successfully. Once the tool seems to be functioning as designed, the process technician will then experiment, with the process variables to ascertain just how large of a processing window is available to them and where within that processing window might be the optimum process for the particular part from this particular material.

10.4.1.1 Validation of mold filling

One of the analysis that is commonly performed during final part design is a computational flow analysis of the filling and material flow into the tool cavity. Commonly referred to as a mold flow analysis, named after the software product, Autodesk Mold-Flow, used to perform the analysis. However, there are several competitive software platforms, which can provide the analysis as well. If the filling analysis was completed during the detailed design stage of the project, the information gained from the analysis was most likely used to select gate locations and to identify potential filling issues, like knit lines and air traps.

In an effort to really understand of our assumptions during the design stage were near to reality we can perform a small but important validation test. Once the processing technician has settled in on their ideal production process parameters, have them run a set of less than full parts, sequentially less full. For example, always start with a good process and a full shot, then ask the technician to back off and provide a 90% full shot, once complete, then drop to an 80 % shot and so on attempting to get a series of every smaller short shots. Now we can compare the short-shot progression with our mold filling progression and see just how closely it matches. Yes, this is an inferential test; however, if the part actually fills in a similar pattern to the predicted analysis, then the other predictions made during detailed design based on the results of the filling analysis can be trusted as well. If not, then depending upon the geometry and the criticality of the end-use application, one of two things can be advised. Either, ensure that the qualification testing incorporates satisfactory review of the other areas considered from the initial analytical results, or, rerun the filling analysis using the exact process settings used in the ideal process used to make the short-shot progression. This second option is for those of us who like to get the stars to align; by getting the computational analysis to match reality. If successful, then the assumptions can be better trusted.

10.4.2 First part inspections

Once the processor has what they determine are good parts from their selected best process for the part, the material, and the tool, the parts are then subjected to dimensional analysis. This is usually referred to as a first piece or first article inspection (FAI) and it comprises the physical inspection of every dimension on a part from every cavity of the tool. The final reported findings provide insight as to how precise the toolmaker was able to convert the CAD data into the tool cavities or forms and how well the finished part processed (Fig. 10.3).

10.4.3 First part performance testing

Like the first part inspections, first part performance testing provides insight into the precision of the tool, the quality of the process and the suitability of the material, from a performance aspect. This performance testing is comprised of two components, first is simply form and function. That is, does it fit to the mating parts correctly and is the form correct for the application. The second component is more difficult to assess and it will require the determination of features that carry a load or need to flex to certain distances for a given number of cycles, and so on. For this second component of performance testing, time in the lab is almost always required. However, the lab time is prudent to qualify the design and the material selection; and, this is the time to get the data and be certain that everything acts and reacts as desired.

Another first part analysis which must be completed is to just simply build an assembly of the parts and see how the pieces fit. This is easy and there is some small validation to be gained. However, this task gets a bit more interesting when there

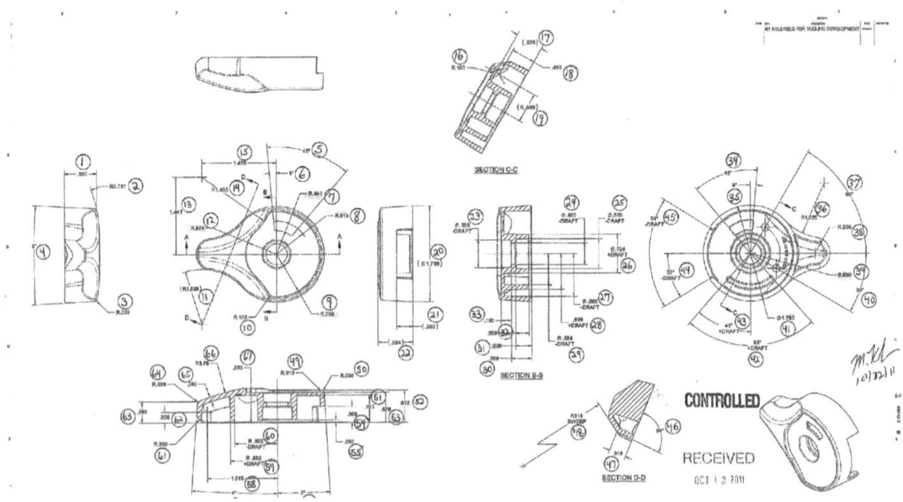

Material: ZYTEL ST801-NAT Lot#: N61L1KZ101 S.C. Inspected By: D.PAOLONI Date: 04/11/12

Comments: CAV#2 Reviewed By: [illegible] Date: 5/7/12

P/N: 5170-1003 Revision: R1 FAI Reason: ENG REQUEST Sample Date: 04/11/12

DRAWING SPECIFICATIONS ☐ INCH ☐ MM						Inspection Method	Gage Number		INSPECTION RESULTS								COMMENTS
Item	DWG Loc	Description	Spec	+ Tol	- Tol			CTP / CTF	Sample Number 1	Sample Number 2	Sample Number 3	Deviation from Center 1	Deviation from Center 2	Deviation from Center 3	Mean	SPEC STATUS	
1	E-8	WIDTH	0.6000	0.0050	0.0050	HT GAGE	420		0.6035	0.6045	0.6040	0.0035	0.0045	0.0040	0.6040	IN	RANGE AT SIDES W/OUT KO FLASH
2	E-7	RADIUS	3.7810	0.0050	0.0050	CMM	38-09		3.7760	3.7765	3.7760	-0.0050	-0.0045	-0.0050	3.7762	IN	
3	E-7	RADIUS	0.0390	0.0050	0.0050	OPTIC	38-09		0.0360	0.0365	0.0365	-0.0030	-0.0025	-0.0025	0.0363	IN	
4	D-8	ANGLE	6.0000	0.5000	0.5000	OPTIC	38-09		5.5500	5.6000	5.5500	-0.4500	-0.4000	-0.4500	5.5667	IN	
5	E-6	ANGLE	45.0000	0.5000	0.5000	OPTIC	38-09		44.5500	44.6500	44.6500	-0.4500	-0.3500	-0.3500	44.6167	IN	
6	E-8	ANGLE	5.0000	0.5000	0.5000	OPTIC	38-09		4.8500	4.9000	4.9000	-0.1500	-0.1000	-0.1000	4.8833	IN	
7	E-6	RADIUS	0.4410	0.0050	0.0050	OPTIC	38-09		0.4360	0.4365	0.4365	-0.0050	-0.0045	-0.0045	0.4363	IN	
8	D-6	RADIUS	0.6190	0.0050	0.0050	OPTIC	38-09		0.6205	0.6209	0.6204	0.0015	0.0019	0.0014	0.6206	IN	
9	C-6	RADIUS	0.7950	0.0050	0.0050	OPTIC	38-09		0.7910	0.7910	0.7910	-0.0040	-0.0040	-0.0040	0.7910	IN	
10	C-8	RADIUS	0.1000	0.0050	0.0050	OPTIC	38-09		0.0970	0.0970	0.0980	-0.0030	-0.0030	-0.0020	0.0973	IN	CONTROLLED
11	D-7	RADIUS [REF]	1.0260			OPTIC	38-09		1.0205	1.0195	1.0210				1.0203		
12	D-7	RADIUS	0.5740	0.0050	0.0050	OPTIC	38-09		0.5715	0.5720	0.5720	-0.0025	-0.0020	-0.0020	0.5718	IN	
13	D-7	TANG PT.	1.4430	0.0050	0.0050	OPTIC	38-09		1.4440	1.4440	1.4445	0.0010	0.0010	0.0015	1.4442	IN	
14	E-7	RADIUS	1.4390	0.0050	0.0050	OPTIC	38-09		1.4340	1.4345	1.4340	-0.0050	-0.0045	-0.0050	1.4342	IN	
15	E-7	TANG PT.	1.4000	0.0050	0.0050	OPTIC	38-09		1.4000	1.4000	1.4000	0.0000	0.0000	0.0000	1.4000	IN	
16	F-4	RADIUS	0.1000	0.0050	0.0050	OPTIC	38-09		0.0985	0.0980	0.0980	-0.0015	-0.0020	-0.0020	0.0982	IN	
17	F-4	DEPTH [REF]	0.5200			Drop Gage	420		0.5205	0.5210	0.5210				0.5208		W/OUT KO FLASH
18	F-3	DEPTH	0.4500	0.0050	0.0050	OPTIC	38-09		0.4500	0.4500	0.4500	0.0000	0.0000	0.0000	0.4500	IN	
19	F-4	RADIUS [REF]	0.5600	0.0050	0.0050	OPTIC	38-09		0.5640	0.5645	0.5645	0.0040	0.0045	0.0045	0.5643	IN	
20	D-5	DIAMETER [REF]	1.7900			CALIPER	131		1.7835	1.7840	1.7840				1.7838		
21	C-5	DEPTH [REF]	0.3000			Drop Gage	420		0.3010	0.3015	0.3010				0.3012		W/OUT KO FLASH
22	C-5	HEIGHT [REF]	0.6540			HT GAGE	520		0.6485	0.6480	0.6485				0.6483		
23	D-4	DIAMETER	0.5230	0.0050	0.0050	PIN	QA1		0.5220	0.5225	0.5225	-0.0010	-0.0005	-0.0005	0.5223	IN	
24	D-3	DIAMETER	0.4000	0.0050	0.0050	PIN	QA1		0.3925	0.3925	0.3925	-0.0075	-0.0075	-0.0075	0.3925	OUT	
25	D-3	DIAMETER	0.5550	0.0050	0.0050	PIN	QA1		0.5505	0.5505	0.5505	-0.0045	-0.0045	-0.0045	0.5505	IN	
26	D-3	DIAMETER MMC	0.7240	0.0050	0.0050	CALIPER	131		0.7185	0.7190	0.7185	-0.0055	-0.0050	-0.0055	0.7187	OUT	
27	D-3	RADIUS	0.8020	0.0050	0.0050	CMM	38-09		0.8000	0.8010	0.8005	-0.0020	-0.0010	-0.0015	0.8005	IN	
28	C-3	DISTANCE	0.6660	0.0050	0.0050	CMM	38-09		0.6610	0.6615	0.6610	-0.0050	-0.0045	-0.0050	0.6612	IN	
29	C-3	RADIUS	0.5840	0.0050	0.0050	CMM	38-09		0.5800	0.5803	0.5800	-0.0040	-0.0037	-0.0040	0.5801	IN	TOP OF ARC
30	C-4	DEPTH	0.3800	0.0050	0.0050	Drop Gage	420		0.3860	0.3860	0.3870	0.0060	0.0060	0.0070	0.3863	OUT	
31	C-4	DEPTH	0.3000	0.0050	0.0050	Drop Gage	420		0.3020	0.3050	0.3050	0.0020	0.0050	0.0050	0.3040	IN	
32	C-4	DEPTH	0.2500	0.0050	0.0050	Drop Gage	420		0.2510	0.2515	0.2520	0.0010	0.0015	0.0020	0.2515	IN	
33	C-4	DEPTH	0.1900	0.0050	0.0050	Drop Gage	420		0.1875	0.1875	0.1885	-0.0025	-0.0025	-0.0015	0.1878	IN	

Figure 10.3 A first article inspection (FAI) report shown, with the part specification drawing numbered to correlate with the FAI findings document. Page 1 of the findings is shown—this particular report was five pages long. For multicavity parts, the FAI would be performed for each cavity, so the reports can get quite labor intensive.

are several parts, and when these several parts are produced on tooling with multiple cavities. The best analysis would dictate building at least one assembly from every possible combination of cavities possible (because this could happen in a production context). How many test assemblies do we need to build? It could be just a few but it could be thousands, depending upon the number of parts and the number of cavities.

Sometimes we simplify and only consider the cavity combinations for those parts, which are truly "functional" or performance drivers for the product.

10.4.3.1 *Important qualification fact*

If the product being developed is a single part product, or if it is an assembly of parts, or even if it is having some combination of processing techniques and materials, one thing is certain, the product will bring benefits and risks once it is sold in the stream of commerce. The benefits are obvious; revenue is generated and begins to flow in from product sales. The risks and liabilities begin once the first unit is sold and will remain for as long as the product is sold and in use by the customer. As discussed in Chapter 11, qualification is paramount to limiting risk for any part or product placed in the stream of commerce. Be sure to do the required work and fully vet and qualify any part or product developed by you, your team, or company.

10.4.4 Documentation

First piece inspection and performance testing should be documented well. Some design engineer will be tasked with figuring something out regarding the part in a year or three or five and the information recorded when the tool and part was first built will be invaluable. The form of documentation does not matter but there are several standard forms used for the first piece dimensional inspection; and, the performance testing documentation can be as simple as a word document or a spreadsheet, so long as everything is annotated sufficiently. The reports should be filed with all pertinent project data for archiving and perhaps with the QC/QA records as well.

10.4.4.1 *Document all results and findings*

Complete dimensional data and any results from any performance testing should be recorded and stored along with sample parts from the test lot—both tested and virgin. Having first parts off the mold in the archive can be a blessing when troubles come to call later in the production life cycle. From a risk management vantage, having example parts from each material attempted and from each revision as the part progressed through the production stages is an invaluable research tool. Archiving part samples is always recommended, especially once on the other end of the product life cycle when we wished we knew what the part looking like and measured when the tool was first built.

In addition to full documentation, there are decisions which need to be made. The first piece inspection report will identify each and every dimension, which is out of specification on the part. Now what do we do with all of this information? Someone (guess who) has to review each out of specification dimension on the part and make a determination whether the tool needs to be changed to bring the dimension in specification, or, if the tolerance or the dimension on the specification drawing is modified to accommodate the real part measurements. Once the determinations are made, the tool then needs to be modified (if needed) and the specification drawings need to be updated (if needed).

10.4.4.2 Update detailed part drawings

Pending the results of the review of the first piece inspection, if necessary, the specification drawing should be updated and the revision identifier indexed to show that changes and modification have been made. In many instances, features and attributes considered out of specification on the first piece inspections are nonfunctional features or attributes. In this case, it is common to relieve the tolerance on these features to allow the measured dimensions to become in specification, thereby not requiring some tool or process modification. Alternatively, if an out of specification dimension is critical to function, and is at the edge of the tolerance zone or out of specification, some remedial action must be implemented.

10.4.4.3 Capture all process parameters

Initial production documentation should also include the recommended or ideal process for the part and many times this detail is overlooked or left to the subcontracted process vendor. Do not overlook this important step—as the part matures and as the tool and process are used overtime, things will change, the tool will wear, the cooling lines will become less efficient, the material supply chain will change, the press or equipment running the tool will change, and the list goes on. Having the baseline process used when the tooling was new is an invaluable resource for troubleshooting problems later on in the life of a plastic part.

10.4.5 Process suitability

Now that the material, the tool and the process are all determined and production of acceptable parts has been accomplished, what are the results? Are the tool and process suitable? How does the process time, part yield, and quality compare to the projected needs for the part or product? Understanding these statistics at the onset will help with planning for growth and will ensure that production targets are not unrealistic.

10.4.5.1 Capacity targets met?

Do the production tooling, material, and process have the capability to meet the specified production capacity requirements? In most cases production capacity will be greater than 1-year requirements; however, understanding the true production capacity is critical for planning; especially if positive sales growth is forecast. Knowing when to invest in additional capacity, with enough schedule time for the additional tooling, machine, or equipment sourcing and qualification, can be a most difficult task.

10.4.5.2 Process variability studies

Has the process been completely and thoroughly vetted? Is the process window as wide as possible to allow for changes in materials, to allow for wear and tool degradation overtime, to allow for use on alternate process equipment as needed? Often, a tool or mold will be set-up and run for first trials on the process equipment that is set aside for short-run process trials, not equipment allocated for true day-in and day-out production. These short-run, set-up machines, are often the newest and have the up-to-date processing controls and software, allowing the processing technician the ability to dial things in quickly and efficiently. Presto, we have a dialed in process window off the set-up equipment; then we tear it down and send it onto the production floor for real, honest to goodness production. Guess what? The production equipment is older than the set-up machinery and has thousands and thousands of hours of runtime behind it with controls from yesteryear. What will normally occur is the production technician will start with something as near to the first run set-up as possible, then tweak the process until the parts seem to be acceptable; then the production tech goes to the next level of debug and tweaks the process to squeeze out every tenth of a second possible. The result is a production cycle that is far different than the first run cycle; sometimes much better, often worse. Hopefully the quality assurance has been organized to review and capture any potential issues, which may arise from this natural process evolution situation. Just remember, once the design engineering team is finished, with the initial process trials and the part moves to production, it is a whole new ball game; hence, have the systems and controls in order to ensure the part quality required for the application.

10.5 Prepare and file patent applications

There is really no set time that is the proper time to prepare and file any potential patent applications for a new product or part, except as provided for by the statutes. That is, once a product has been offered to the marketplace the applicant has 1 year from the initial offering date to file an application. Hence, it seems that about the time initial process trials and qualifications are occurring would be a good time to have the patent process initiated as well.

There are several reasons that this particular timing is recommended. First, the design is complete and has gone through all of the vetting steps and is now on the precipice of production, no substantial changes should be forthcoming at this stage of the project. Secondly, at this point, marketing is probably preparing for the roll-out of descriptive literature and advertising information, which when published out to the marketplace will start the 1 year clock. Third, if the application can be completed and filed prior to the marketing roll-out, the product can be identified as "patent pending" which may be valuable from a marketing perspective.

Preliminary preparation is normally completed by the design engineering team and consists of a report outlining the products intended application, including some background about alternate or competing technologies used to solve the same "need". In addition, the report will detail the specific novel features or methods used by the product, which differentiate it from any other potential solution for the "need". The details will need to be specific enough to describe each feature, process step, or method to the counsel tasked with preparing and filing the application. Any patents pulled for reference during the development process should be included, as well as any published papers or technical literature referenced, which may be helpful to the patent attorneys during application preparation and prosecution.

10.6 Prepare for product qualification

Once each part, and its materials and processes have been tested, measured, and verified, the part and any other parts or components must be assembled or processed into the final finished product. These first finished product samples need to be from production materials, production processes, in production colors and if possible at all conditions of production tolerances. These will now be forwarded to the product qualification team for intensive review and analysis to ensure they are suitable for introduction and sale into the stream of commerce. Once a product enters this dangerous stream of commerce two things will immediately occur; first, revenue will begin to flow back to the company for all of the efforts and resources invested (a good thing), and second, the liability ticket is punched putting the company in the gun sights of anyone who may suffer harm from the use or design of the product (a bad thing). So thing one and thing two go hand in hand, but if the project was managed well and vetted thoroughly as recommended in the preceding chapters of this book, then thing two is mitigated.

Resources for further study

S. Levy, J.H. DuBois, Plastics Product Design Engineering Handbook, second ed., Chapman & Hall, New York, 1984.

S.E.M. Selke, J.D. Culter, R.J. Hernandez, Plastics Packaging—Properties, Processing, Applications, and Regulations, second ed., Carl Hanser Verlag, Munich, 2004.

J. Rotheiser, Joining of Plastics—Handbook for Designers and Engineers, third ed., Carl Hanser Verlag, Munich, 2009.

G.L. Beall, Rotational Molding—Design, Materials, Tooling, and Processing, first ed., Carl Hanser Verlag, Munich, 1998.

E.R. Larson, Thermoplastic Material Selection—A Practical Guide, first ed., Elsevier, Oxford, 2015.

R.A. Malloy, Plastic Part Design for Injection Molding: An Introduction, second ed., Carl Hanser Verlag, Munich, 2010.

P.A. Tres, Designing Plastic Parts for Assembly, seventh ed., Carl Hanser Verlag, Munich, 2014.

Web links of note

Great Design information from DuPont can be found here:

http://www.dupont.com/content/dam/dupont/products-and-services/plastics-polymers-and-resins/thermoplastics/documents/General%20Design%20Principles/General%20Design%20Principles%20Mod%201.pdf.

Sabic provides many tools for designers here (you may need to register):
http://www.sabic-ip.com/webtools/toolsHomePage.faces.
The UL Prospector site is a valuable tool for any designer reviewing materials for any application (you will need to register):
https://plastics.ulprospector.com/.
Solvay provides a great design guide for using their Udel polysulfone materials:
http://www.solvay.com/en/binaries/Udel-PSU-Design-Guide_EN-227550.pdf.
Bayer has a nice design guide for RIM systems:
http://reactioninjectionmolding.com/wp-content/uploads/2013/09/RIM-PartMoldDesign-Guide.pdf.
Covestro offers several design related guides, including part and mold design, snap fit design, and joining techniques for plastics:
http://www.plastics.covestro.com/Library/Technology-brochures.aspx.

Production qualification and hand-off

11

Qualification of a new design can sometimes be an extremely formal and well-documented event and other times it can be extremely informal. For most nonregulated product development projects the qualification process happens in stages throughout the development process, culminating in a finished product ready for production but perhaps without fully vetted and documented qualification. Once the product begins to take final form and become real, there is always a push to get it into production and to start the revenue faucet.

Once qualification is successfully completed, it is time for the hand-off to manufacturing. Ideally, the primary manufacturing and assembly team has been working with the design team, participating in the design review and qualification process, and have a good understanding of the product and part(s) now being transitioned into production. Whether or not this cooperative development has occurred, the design team must prepare, and transition the accumulated data and knowledge to the manufacturing and production team in a professional and complete manner.

Experience shows that in most instances, initial field failures for most any product could have been avoided if a more thorough qualification had been performed, hence the lesson is, do the homework first, then do a complete hand-off to production and reap the harvest.

11.1 Product performance verification

Performance verification is simple. Does the product meet each and every specification in the functional specification document? If it does then all is good, if not, then there must be some decision made to alter the functional specifications or to modify the product to meet the original specification. However, because the premise of performance verification is simple, it does not mean that it is easy (Fig. 11.1).

11.1.1 Specification review

Some specifications in the functional specification are easy to validate and some are not. Parameters like size, weight, color, ease of use, and so on, can all be readily assessed and quantified. Some attributes like ruggedness, UV stability, hydrolytic stability, dielectric strength, resistance to environmental stress cracking, weatherability, and so on, become more difficult and time consuming to assess. In as much detail as possible and with as much quantitative data as possible, each specification validated should be documented with the data gathered and any observation(s) which might be

Designing Successful Products with Plastics. http://dx.doi.org/10.1016/B978-0-323-44501-6.00011-8

Figure 11.1 While performance verification for most of the parts and products we design with plastics are not as critical as validation of vehicle passenger safety, the complexity of the testing and validation can sometime seem just as daunting.
Source: Photo from iStock.com/JestersCap.

useful for future product development or retesting efforts. These formal specifications along with the design teams documented notes, the quality testing data, and the final qualification report will survive and be the roadmap used when failures occur later in the product lifecycle. Hence it is important to be thorough in documenting the testing data, results, and most importantly the observation(s) of the persons doing the testing and evaluation. In addition, this documentation can be invaluable to others when and during future "next generation" product development or when "sister/brother" products are being considered and/or developed.

11.1.2 Long-term testing

Many attributes will demand testing that cannot be completed in a short time in the lab and will need to be evaluated over some longer-time period. These attributes like weatherability, UV stability, fatigue life, and so on, can be assessed with accelerated lab means; however, full understanding of the performance overtime will require a longer-range evaluation. Results from the longer-term evaluations can be used to help incrementally improve the overall quality of a product and can be used as a means to better predict behaviors of the tested attributes, if and when changes are required. In many instances, when product specimens are undergoing long-term testing, the quality assurance plan will call for new parts or products to be introduced into the long-term evaluation process whenever there are certain changes to the part or product; changes like a material or material additive change, a process change, a structural design change, a finish or coating change, and so on. As with other qualification test data and observations, it is important that the observations and results from this long-term testing to be circulated back into the design team for their review, allowing them to learn how the product is actually behaving relative to the design intent and assumptions made during the development efforts.

11.1.3 Safety review

Of the evaluations performed during validation, the safety review is perhaps the most important. This should ideally be done in a group setting or meeting with participants from outside the design, marketing, and manufacturing teams; new eyes and brains to flush out any, and all potential risks. The safety review should be all inclusive, looking at safety of the project from incoming raw materials, through production, testing, packaging, and perhaps most importantly once the part or product is in the hands of the end-user. Look at every "what-if" that can be identified and assess if there is any risk.

Risks hide in places where they are least expected, for example, the lead (Pb) poisoning risk posed to children from toys, costume jewelry and other objects where the pigment, added at even low ratios, exposed the user(s) to Pb. The design team certainly didn't assume that Pb would be present in the pigment, the manufacturer just ordered materials to match the color specification on the print, the quality assurance staff checked the color match and the physical attributes, and there was still a hazard that could be overlooked. Be vigilant; do not let some unassumed risk destroy the creative work, time, and resource investment made to develop the part or product being validated.

11.1.4 Compliance testing

For some regulated markets (medical, military, automotive, aircraft, etc.) there will be some extremely organized and discreet regulations, which will need to be adhered to during development and tested against during validation. For nonregulated markets there will still be many codes and regulations which can be used to validate or verify the part or product for the intended market application. Sometimes these are called-out within the functional specifications, other times they are not specified but can still add value to the validation process. There are many instances where a regulation written for a specific market can be used to validate a part or product for another market, so do the research and find the most applicable regulation or code for the attribute to be tested, even if it is in another market segment. Of course, compliance testing does cost time and resources, so ensure the value of testing is worth the investment, when compliance is not required by law.

11.2 Quality plan review

The quality plan prepared prior to this qualification event needs to be reviewed and followed to ensure every attribute has been qualified as per the plan. If new information is learned which should be included within the quality plan, then the plan should be formally updated to include such information. Once completed, the qualification results should become an appendix in the quality plan document; allowing future use and reference of the findings and results (Fig. 11.2).

11.2.1 Verification of quality plan actions

Testing and observing the finished product during implementation of the quality plan can be a laborious task and is often dull. We have designed the part to withstand all sorts of probable in-use conditions, we have selected materials and processes to withstand the rigors of the intended environment, now we are validating that the materials and design do what they are supposed to do. Hence, in most every test the part or product will pass—this is boring. Breaking things is much more stimulating. However, there are often subtle nuances, which can sometimes occur to a trained and watchful eye during these validation tests. Plastics do interesting things once in a while and we can only learn from these if we are paying attention.

Each verification step was thoughtfully written for some specific reason and each must be carried out exactly as specified, with the observations and results

Figure 11.2 The quality plan provides the most important framework to ensure success. *Source*: Photo from iStock.com/vinnstock.

recorded clearly. If additional steps are required or if there is some other potential failure mode that could also be verified in a particular test or step, then by all means perform the verification action, but note this in the results, then prepare a modification request for updating the quality plan to include the additions or correction for future use.

11.2.2 Qualification reporting

The qualification reporting should be a formal document, signed by those performing the validation and certifying that the validation was complete or is complete with some testing still in-progress (long-term evaluation). Recommendations for updates, additions, or changes to the quality plan to improve the qualification testing steps or to better verify or validate some particular attribute should be captured in a formal document submitted alongside the qualification report. The qualification report should become a permanent appendix within the quality plan document for the product.

11.3 Hand-off to primary manufacturing and assembly

Historically in many manufacturing concerns, the engineering and product development teams would work on their new products and when completed, they would throw them over the wall into manufacturing. Throughout this book, from the very first page, I have been espousing a cooperative, all-inclusive team approach; where manufacturing participates from the beginning of the project. If the cooperative approach has been used then the hand-off becomes a simple event. If not, then all sorts of alligators, with all sorts of unnamed diseases, are lurking to bite.

11.3.1 Transmittal of all documentation

The design review and qualification process is organized to ensure the design is ready for production and once produced will satisfy the functional specification and perform in the end-use application. Many things are learned about a particular part or assembly during the qualification and in debug prior to qualification. These learned things might include a processing set-up configuration, which allows for a better overall cycle, a means of testing a part to check knit line strength, a good way to fixture the part while performing a particular test or assembly operation, and the like. Often these simple learned facts are not part of any formal documentation procedure, but are relayed during discussions during the review process. As such, adequate and complete design review meeting notes should always be captured and distributed to attendees; these notes should also become part of the project history file that gets forwarded to the technical staff responsible once transitioned into production. This is not a difficult task to complete, but it is often overlooked and the manufacturing team goes about relearning things that have already been determined or worse yet, not relearning and not benefiting from the time and resource investment made during development.

The natural inclination at this juncture is for the design team to deliver the information and then move onto the next interesting project; however, the design team should be involved and supportive throughout the transition. Just as the manufacturing and process folks were involved early in the design process, the project can benefit from design team involvement and support during production transition and ramp up. Not to mention the design team can benefit from learning the details required during the ramp-up and hopefully be able to apply those learning's on the next project.

11.3.2 Review of all safety requirements

Design teams really have no business specifying or making prognostications about safety to the manufacturing team, however, throughout the development process the design staff has learned much regarding the product, the processes, the materials, and so forth ,with respect to the safety of those tasked with the manufacture or use of the product under development. In fact, the design team has an obligation to inform anyone downstream of any and all potential hazards known to the design team. Hazards which might be found in specifying the materials, in processing the tooling and materials, in handling the materials, in performing secondary operations, in performing assembly operations, in testing the finished product or assembly, in packaging the finished product or assembly, in storing the finished product or assembly, and so on.

Likewise, in the development of the product packaging and use instructions, the design team is obligated to present any and all potential risks to the user to be incorporated into cautions and warning statements. Many times these warnings and caution statements can be extremely hazardous to write; they become legal time bombs waiting for the first hint of any harm to bring the ambulance chasers out of the woodwork. Hence, the design team needs to provide an honest summary of any potential safety risks to the end-user, and then the legal team needs to carefully wordsmith the written and published cautions and warning statements.

11.3.3 Review all material-handling requirements

The material handling review assesses raw materials coming into the factory, lot identification and traceability to material supplier's batch, or lot identifications, storage conditions for the raw materials, storage and transportation of parts and subassemblies within the factory, and between the factories. This can be especially important to specify when using processing facilities in multiple global locations.

Unless the product will be produced in a totally vertically integrated factory, some part/assembly storage and transportation will be required between primary manufacturing and final product packaging, ready for retail. This intramanufacturing storage and transportation, sometimes many times, can be a formidable overhead cost and can be a source of loss due to damaged parts or product. When the primary manufacture is a contract or outside vendor, there is a tendency to leave the packaging and handling issues up to them; however, helping them to be more efficient and protecting the part or product simultaneously by having some discussion and cooperatively figuring out the best handling methods is a better practice. There are many examples of this working well, especially when combined with a just-in-time style of inventory management.

Reduction of total overhead cost per unit is always the goal when implementing a new project in a production context. The storage, transportation, and handling of primary manufactured parts and assemblies is a major component in the overhead cost per unit, in most instances. Effort invested in optimizing these operational events will often result in better overall cost efficiency and reduction in risk of damage to the parts or assemblies.

Examples of well thought out and implemented systems include, the use of reusable containers, which can be sent back to the primary manufacturer, sometimes in a knocked-down or lowest possible shipping volume condition. Other systems include packaging the parts or subassemblies in the final ready for retail packaging for shipment to final assembly or pack-out, where the subassembly or part(s) are removed, processed, tested, and then reinserted in the packaging for final pack-out and shipment into the distribution stream.

11.3.4 Review all tooling requirements

The tooling requirements review includes recommended tool maintenance schedules and reporting, tool capacity, tool design data files, any applicable safety considerations, and tool processing specifications from qualification. The production team should have a thorough understanding of their responsibilities and of any special requirements that the tooling might have. Also, review and make sure the processing technical and quality assurance teams understand any tooling features, which are critical to part function or performance; for example, pin shut-offs that can flash and cause defects in the part or assembly, and so on.

Specific tool features which are critical to part performance might be considered for inclusion as part of the quality plan for the part. The quality plan should, in these instances, call for frequent measurement and recording of the specified measurements as part of the continuous quality assurance requirements for the part or feature. In addition, through process trials we sometimes find specific maintenance requirements necessary to ensure part performance, for example, some resins may build up on cavity walls and need to be polished out at some frequency (every 50 K shots for example) or sometimes vents need to be polished clean at some frequency–this kind of knowledge should be documented and made part of the routine maintenance schedule for the tools during the hand-off to manufacturing.

11.3.5 Review all process requirements

Process specifications are fairly easy to document and to supply to the primary manufacturing team, however, if the part(s) or assemblies will be made in multiple locations or be produced on multiple tools even within the same factory, make sure the process specifications are complete. For example, a tool was qualified on a 120-ton, 3.5-oz. shot injection molding machine and the new primary manufacturer has decided to run the mold on a 200-ton machine, with a 6-oz. shot, that has open capacity and begins production, only now the material dwell time in the barrel is 5-shot cycles instead of two and there is a risk of material degradation. The resulting parts look the same and weigh the same but the material properties might be less than desired, certainly might be less than the tested and qualified properties from the qualification lot. Other examples exist where a tool developed for a specific material and particular piece of process equipment get modified and used for an alternate material or formulation and/or in a different piece of processing equipment. Sometimes this can work fine without issues, but when these kinds of changes are made, requalification is certainly in order.

11.3.6 Review the quality plan

Understanding the quality plan and how the plan relates to functional specification and end-use performance is crucial for the production team and the production quality assurance team. The design team must complete a thorough review with manufacturing and should remain on the transition team for several months into production to assist as issues arise. Actually, in many businesses the quality group has a permanent representative from the design team available to join them in every review and/or conformance situation for each product line.

11.3.7 Review all packaging requirements

Packaging requirements, both in-process and for final product or assembly pack out are easy to just forget and leave up to the production team. However, the design team knows and has a vested interest in protection of parts and subassemblies in-process and especially for finished goods protection during storage, shipping, handling, and industrial distribution, or retail point of sale display. The success of the product or brand will be partially dependent on the perceived quality of the product when received by the end-user(s). Hence, a thorough review of the packaging for in-process and final pack out should not be overlooked. In fact, as discussed earlier in a case example, sometimes additional efficiencies can be found.

11.4 Periodic production monitoring and review

Once a product is in production and running smooth, it is often left alone. If things are going well, leave it be; this is kind of like the old adage, don't wake the sleeping dog. However, from initial production with time, moving into a mature production process, things are learned by the processor, by the warehouse staff, by the assemblers, by the quality assurance staff, and so on. This learned information might be just a certain way to pack the parts to assist downstream or to gain a higher-packing density. It might be that grasping three parts in a certain manner allows the assembler to make one-third the motions into the carton, saving time.

A scheduled periodic review, which includes representatives from all disciplines responsible through the production process will allow the group to learn and perhaps gain insight from each other in an effort to gain more efficiency, or to eliminate waste. The design team can learn how their part design or their assembly selection (snap fit, screwed assembly, adhesive joint, etc.) has been received and what features might be added to assist or might be removed because they are unnecessary. The strategic sourcing team can learn how their outside vendors are doing and what might be done to assist internal operations. The quality assurance team might learn what tests and measurements they are continuing to perform, which may need to be altered to tighten performance or which may be loosened, lessening their time investment per unit. Hence, the day or two invested to perform this periodic review should prove profitable over the long run. In some cases once the first or second review has been completed the frequency might be reduced to an annual review, or it might be eliminated all together until some other, or new, change is implemented in the design or processes used.

11.4.1 Quality audit

The periodic production review is a proactive means to continuously improve the product and production process. A quality audit can provide valuable insight for those working in preparation for this periodic production review. In addition the quality audit can provide a metric based analysis showing why certain changes to part dimensions or tolerances should be considered, or to show why certain changes to processes should be considered. The numbers collected during routine day-to-day quality assurance tasks can be reviewed and will sometimes show how a change to a dimension on a part would be beneficial to the overall process, or how a process change might provide a more efficient cycle. However, without the audit or the long-term review of these collected metrics there can be no improvement. Hence, a quality audit, perhaps on a semiannual basis of the production plan and the quality plan from a high-level is recommended.

Resources for further study

S. Levy, J.H. DuBois, Plastics Product Design Engineering Handbook, second ed., Chapman & Hall, New York, 1984.

S.E.M. Selke, J.D. Culter, R.J. Hernandez, Plastics Packaging—Properties, Processing, Applications, and Regulations, second ed., Carl Hanser Verlag, Munich, 2004.

J. Rotheiser, Joining of Plastics—Handbook for Designers and Engineers, third ed., Carl Hanser Verlag, Munich, 2009.

G.L. Beall, Rotational Molding—Design, Materials, Tooling, and Processing, first ed., Carl Hanser Verlag, Munich, 1998.

E.R. Larson, Thermoplastic Material Selection—A Practical Guide, first ed., Elsevier, Oxford, 2015.

R.A. Malloy, Plastic Part Design for Injection Molding: An Introduction, second ed., Carl Hanser Verlag, Munich, 2010.

P.A. Tres, Designing Plastic Parts for Assembly, seventh ed., Carl Hanser Verlag, Munich, 2014.

Web links of note

Great Design information from DuPont can be found here:

http://www.dupont.com/content/dam/dupont/products-and-services/plastics-polymers-and-resins/thermoplastics/documents/General%20Design%20Principles/General%20Design%20Principles%20Mod%201.pdf.

Sabic provides many tools for designers here (you may need to register):

http://www.sabic-ip.com/webtools/toolsHomePage.faces.

The UL Prospector site is a valuable tool for any designer reviewing materials for any application (you will need to register):

https://plastics.ulprospector.com/.

Solvay provides a great design guide for using their Udel polysulfone materials:

http://www.solvay.com/en/binaries/Udel-PSU-Design-Guide_EN-227550.pdf.

Bayer has a nice design guide for RIM systems:

http://reactioninjectionmolding.com/wp-content/uploads/2013/09/RIM-PartMoldDesign-Guide.pdf.

Covestro offers several design related guides, including part and mold design, snap fit design, and joining techniques for plastics:

http://www.plastics.covestro.com/Library/Technology-brochures.aspx.

Section IV

Lessons

12. Creating Value through Good Design Practices 287

Introduction

This final section is really nothing more than a reminder of all of the lessons that have been presented throughout the book. When I was in high school, one of my instructors taught us that there were three things that needed to be done in any written work. Tell the reader what you are going to tell them, then tell them, then tell them what you told them. Hence this section is the "tell them what you told them" section; the wrap, or conclusions portion. Not bold or sexy, but a concise recap of the major lessons that any reader should want to come away remembering.

Creating value through good design practices

12

The basis for this book was to demonstrate the practical methods and processes utilized successfully for the design and development of products and parts manufactured from the material class known as plastics. The story told was derived not from academic pursuit, but from hands on, one step forward and two steps back, real world product development experience; the so called, school of hard knocks. This book was created and organized in much that same manner as any part or product made from plastics might be; that is, we first gather and review the background information, then we dive into design solutions, we vet those solutions and fully detail and specify them, then and only then do we implement the design into a production part or product. The structure of this book leads the novice plastics designer through the process just as they would if they were creating a solution for a "need" for some specific set of "users" to be manufactured from plastics.

The first section of this book dealt with the *Background Knowledge* needed by anyone tasked with designing a part or product from plastics. It highlighted the *Four Pillars* of plastics part or product design; Materials, Processes, Tooling, and Design. In fact, Materials, Processes, and Tooling were discussed in detail with each having its own chapter within Section I. *Design*, the main topic of this book, became the sole focus for the entire Section II of the book, and was covered in three distinct steps, *Conceptual Design, Preliminary Design,* and *Final Detailed Design.* Risk management, always present in any new product development program or project, became the focus for Section III of the book covering *Implementation of the Design.* Thorough review of the design, careful estimating, exhaustive vendor selection and vetting, organized project control methods, and the like, all are used to implement the design into a successful production item or product and are part of the topics covered in Section III of the book.

So which is it that determines good product or part design with plastics? Is it good design practices, following all of the design rules religiously? Is it good project management, involving all the stakeholders early in the project? Is it the artistic creativity of the design team, or the years of experience within the operations/implementation team? From my perspective, as an old guy who has designed thousands of plastics parts, some that were wildly successful, some that never made it to implementation, and some that were outright failures, I say it is all of the above.

12.1 The *Four Pillars*—complete integration is a requirement

Good design practices require the complete integration of the *Four Pillars* at each step and with every decision made during the plastic product or part design and development process. Leave out or ignore one of the pillars and the design will

Designing Successful Products with Plastics. http://dx.doi.org/10.1016/B978-0-323-44501-6.00012-X

suffer. Bad things happen to those who ignore this basic rule of plastics part design. I know, I've done it and felt the shame when later in life I realized just how simple it would have been to just do my homework and pay attention to this one basic rule. It is really not difficult, during every decision just ask yourself, how does this decision affect each of the *Four Pillars*; positive, negative, or no affect, it matters not, just the process of thinking through the question will probably bring some relationship to mind that will cause you to check and perfect one other feature or attribute on the part design.

12.1.1 Which of the pillars takes precedent?

Does any pillar really supersede the others? In my world the simple answer is no. However, if we absolutely had to categorize them, or assign relative importance, then I would opine that of the four, Design carries the highest level of importance. Design ties all *Four Pillars* together and great design utilizes the other three pillars to their fullest capacity. Design creates the specifications from which the other three pillars are extracted and defined relative to the "need" and "user" for that the solution is required. However, as important as the design pillar is, it is useless without the remaining three pillars. Great design without an excellent material, or an excellent process, or for which there is inadequate tooling, is of no value; so any plastics part or product design MUST be derived with ample consideration of all *Four Pillars*.

12.1.2 Risks if we ignore one or more of the pillars

It is fairly common to find products or parts in production for which one or more of the pillars were not considered carefully during the design and development of the part. As an independent consultant, I see this often and in many cases, I am hired to fix problems which should have never occurred (and for that I am eternally grateful). In fact, there is an enormous amount of work that I have been blessed with simply because the designer did not even remotely consider the *Four Pillars*. The scenario goes like this, I get a call from a custom injection-molding house, "…Hey Mark, can you take a look at this new project we are estimating? It is the most beautiful design style that we have ever seen, absolutely gorgeous; however, there is one problem, we cannot mold any of these parts, there is trapped steel undercuts everywhere, wall thicknesses that we just can't fill, they are asking for a material that is special order, and the list goes on. You really need to get in here and talk with this customer before they go much further…." So now, the customer has the design of their dreams with nobody able to make it—now what? We meet with them and advise them of their options and attempt to rescue the project with the least amount of additional schedule time and cost possible. Still, had the original design been completed with even some consideration of the *Four Pillars*, the project would have been in a far better place.

12.1.3 Inclusion of stakeholders from each of the pillars in early discussions and reviews

Inclusion of anyone who has experience with the materials to be used, the process to be used, or the tooling to be used, is almost a no brainer in my world. Why wouldn't you ask the material supplier, "Hey, have you ever used this material in this kind of application before?" and if so, then the question becomes, "well, how did it perform, what issues were there, would you recommend it for our application that is similar?" Why wouldn't you ask the processing team or vendor, "Hey, have you ever molded a part like this from this particular material, with these discreet functional requirements or attributes?" and if so, then the follow on questions are, "Great, how did it work, what problems did you need to overcome, do you think our part design can be molded easily over a wide processing range?" and so on. Why wouldn't you ask the toolmaker, "Hey, what do you think of our part design? Ever make a tool for a part like this, from this particular material?" and if so, the follow up questions are, "well then, how has that tooling (that is similar to what we need) performed? Has it needed much maintenance? How many good parts have been produced from that tool?" and so on.

It is so easy to ask questions of experienced practitioners, yet many designers don't. Don't ask me why, because I don't get it, but I know when I ask questions, the toolmakers, or molders, or material suppliers are always ready to provide answers based on their experience. Most are eager to assist and feel a reward from participating in the design discussions. This builds team loyalty to the project and I have found that early inclusion of the stakeholders has helped me shorten timelines and keep projects within budget. If these early involvement practitioners are internal then they have a great head start on the project when it goes to implementation. If they are outside resources, bidding or estimating to win the project, then again they have a head start and some edge on their competition as a result of participating early. I have found that these early involvement vendors will almost always underbid any other competitive vendor and it is easy to see why. Their knowledge base is far advanced from the early involvement and they can be far more precise with their estimating. This fact of life benefits the project as much as it benefits the outside vendor.

12.1.4 Foundation of the solution is the customer and the need

If our *Four Pillars* support the final "Optimum" part or product, then what supports the *Four Pillars*? The foundation for the *Four Pillars* is the "need" for which we are seeking the solution and the "user" or customer that has this "need". Without a "need" and a "user" there is no call for a solution and the *Four Pillars* are of little use. Always remember to "think first" placing the "need" and "user" as the basis for any consideration of any of the *Four Pillars* during the design and development process. Preparing solution without keeping the "need" and "user" at the foundation will lead to an unfocused product design and a finished product, which may find itself without any real commercial market appeal.

12.2 Basis for design decisions rests on the functional specification

Once the conceptual design pool has been reduced to an ideal design to go into the preliminary detailed design phase, the functional specification is formulated (most often from the conceptual specification), finalized, and documented. Any and all decisions during design must be based upon specifications and the functional requirements for the product or part, which are found in the functional specification document. Every single decision must be compared against the functional specification; if the decision does not satisfy the functional specification then it needs to be discarded. Many times, a decision seems innocent, like "...well, just this one time we will make this change to help our toolmaker stay on schedule..." however, if the decision causes some violation of the functional specifications, then you will, at some point, suffer. Little deviations and omissions, come back later in the project, once all the hardened tool steel molds are completed and raise their heads up to let you know—hey dummy remember that decision you made 2 months ago, well it wasn't a good one. Every designer that I know has scars from these sorts of decisions, so you are not alone, but you have been warned.

12.3 Verification and qualification is mandatory

Thorough review of each decision, of each design feature, of the material selection, of the process selection, of every choice during the design is mandatory. Speed to market is wasted when the product or part does not meet requirements because thorough vetting of the design was not implemented as a priority. The opportunity costs of a month of lost sales because we did our homework and vetted the product completely are a visible investment in the brand, or product loyalty, or company reputation. Alternatively, a product recall, or worse yet a failure in the field that causes some damage or harm, would cost orders of magnitude more. Sure we have product liability insurances to protect us from those who might misuse our products, but in my world the best insurance is to do the job right from the start. Make the time and treasury investment to qualify the part or product PRIOR to putting it into the stream of commerce.

12.4 Final part or product safety is our responsibility

Parts in final geometry, from the final tooling, in the final process, and in the final materials, must be tested in real world conditions to validate the design PRIOR to placing the part or product into the stream of commerce. All the design simulation tools in the world are no substitute for real world testing, analysis, and validation of the finished goods before they are sold. No product liability insurance policy will prevent harm, as well as good design and thorough finished product testing and validation can.

This mentality of thorough testing against functional specifications of final part geometry, in final materials, from final processes, benefits both the manufacturer and the end-user, and limiting risks and potential harm. Remember when high levels of lead (Pb) were found in the promotional toys given out for free as part of a children's meal promotion? Well, just think, if these promotional toy parts were tested and verified prior to being placed in the stream of commerce? They certainly would have never been approved and some large number of children would not have been exposed the lead (Pb) unknowingly. Risk management by doing the verification homework is easy and it is ethically and morally correct; yes there is schedule cost and treasury cost but these are small compared to potential risk to the brand and the company reputation. No designer would ever want to find out that some part or product that they have created would be cause for harm to any human, child, pet, or any other living thing.

12.5 Big picture thinking

Design consumes a fraction of the total time to market and a fraction of the total nonrecurring costs preproduction, yet the design determines, not only final product costs, but final product quality, reliability, and suitability for the application; in short the success of the product or part is almost entirely determined by design, and this before the first capital dollar is spent or first cut is made in producing the first tool. So why then do we always attempt to be so frugal with the design budget and timeline (Fig. 12.1)?

Figure 12.1 Big thinker—I will wager he is pondering why most companies don't invest more on upfront design efforts on all of their plastics product design projects.
Source: Photo from iStock.com/ToniFlap.

For most projects it is simple: everybody understands that the capital tooling will take 16–20 weeks or more, so the sooner we can get the tooling ordered, the sooner we can have parts to show our prospective customers. Hurry up you designers, you are keeping us from investing tons of our treasury funds in capital tooling so we can go and make a name for ourselves with this new product. Don't worry so much about the small details, we can fix those once we are selling product and have a revenue stream to fund all of this effort. If you are new to product design, these statements may seem outrageous, but if you are a seasoned practitioner, I know you have heard these sentiments, over and over again.

Great design process implementation does take time and does cost nonrecurring funds; however, the investment in great design will pay dividends for years. What's more, the incremental difference in investment between rushed speed to market design efforts and thorough well vetted design efforts is amazingly small. The bean counters will always be speaking about lost market opportunities, when they should be asking for excellent user experience and quality.

However, as creative types, designers need to be careful to know when to say when. I am not talking about Friday night at the pub; rather I am speaking to their creative efforts. As bad as a rushed design schedule can be, an extended design schedule for the wrong reasons can be worse. Don't continue solving the problem; once there is a solution, which satisfies the functional requirements, and once the design review team has given the solution the green light, don't look back. The propensity to keep innovating is great and is what has made many companies grow far beyond their competition. On a project by project level, however, we need to know when to stop inventing and start making product.

In similar circumstances, I can recall more projects than I wish, where once we were through the design review and ready to implement, someone (usually the business owner, boss, or some higher ranking individual in the business) would say, "… well in my dreams last night I saw the solution to all of our problems, all we need to do is change this and such to …" and then we would be asked to investigate, just to make sure that this new inspiration isn't the best thing since sliced bread. There is something deep in our egos that causes us to need to own some part of a new product; that was my contribution, I came up with the portion that was claimable in the patent, it never would have been successful without my ideas, and so on. If the design has gone through the design process as taught herein and has passed the metrics of the design review, then politely tell the boss, or whoever has had the great inspiration, that the design is frozen and that we will gladly record the idea for the next generation product iteration. Be polite and thank them profusely for the marvelous insight, but firmly tell them that the idea will be so great when it is implemented on the second edition of the current design now heading to production.

12.6 Thanks and do your part!

For those of you who have read this book, I am truly grateful. It was a ton of effort and I hope you gained some useable knowledge that will benefit you, your career, and your projects. As with everything, it is time to ask for the sale, to close the deal, to gain the commitment, so here it comes; my simple request to you as a designer is to do your part. What I am asking is, for you to create "optimum" parts from plastics (and all materials for that matter) every chance you get. Life is too short for troubles and angst caused by poor design. In addition, I ask for your help in spreading the word; find a newbie product or part designer and take them under your wings, show them what "optimum" design is all about, teach them about the *Four Pillars*, help them understand why good design can be so rewarding.

Good design—pass it on!

Mark MacLean-Blevins

Resources for further study

S. Levy, J.H. DuBois, Plastics Product Design Engineering Handbook, second ed., Chapman & Hall, New York, 1984.

S.E.M. Selke, J.D. Culter, R.J. Hernandez, Plastics Packaging—Properties, Processing, Applications, and Regulations, second ed., Carl Hanser Verlag, Munich, 2004.

J. Rotheiser, Joining of Plastics—Handbook for Designers and Engineers, third ed., Carl Hanser Verlag, Munich, 2009.

G.L. Beall, Rotational Molding—Design, Materials, Tooling, and Processing, first ed., Carl Hanser Verlag, Munich, 1998.

E.R. Larson, Thermoplastic Material Selection—A Practical Guide, first ed., Elsevier, Oxford, 2015.

R.A. Malloy, Plastic Part Design for Injection Molding: An Introduction, second ed., Carl Hanser Verlag, Munich, 2010.

P.A. Tres, Designing Plastic Parts for Assembly, seventh ed., Carl Hanser Verlag, Munich, 2014.

Web links of note

Great Design information from DuPont can be found here:

http://www.dupont.com/content/dam/dupont/products-and-services/plastics-polymers-and-resins/thermoplastics/documents/General%20Design%20Principles/General%20Design%20Principles%20Mod%201.pdf.

Sabic provides many tools for designers here (you may need to register):

http://www.sabic-ip.com/webtools/toolsHomePage.faces.

The UL Prospector site is a valuable tool for any designer reviewing materials for any application (you will need to register):

https://plastics.ulprospector.com/.

Solvay provides a great design guide for using their Udel polysulfone materials:

http://www.solvay.com/en/binaries/Udel-PSU-Design-Guide_EN-227550.pdf.

Bayer has a nice design guide for RIM systems:

http://reactioninjectionmolding.com/wp-content/uploads/2013/09/RIM-PartMoldDesign-Guide.pdf.

Covestro offers several design related guides, including part and mold design, snap fit design, and joining techniques for plastics:

http://www.plastics.covestro.com/Library/Technology-brochures.aspx.

Index

A

Action items, departing conceptual stage, 123
  design team, 123
  legal, 124
  manufacturing operations/strategic sourcing, 124
  marketing and sales, 123
Additives
  for aesthetic properties, 45
  manufacturing, 64
Agreement, 249
  assign implementation leaders and reporting protocol, 250
  determine implementation status meeting frequency, 250
  review production estimates and timeline, 249
  review selected vendors, 250
Amorphous materials, 19, 20
  properties, comparison with semicrystalline materials, 21
Annealing, 182
Annotations and descriptive notes, concept with, 114
Annular snap ABS injection-molded parts and assembly, 155
Annular undercut, 153
Appearance properties, 43
  color/special effects, 43
  secondary processes for, 43
  surface finish, 43
  texture, 43
Assembly layout, 3D CAD master checking, 201
  recommended checking method, 202
Assembly operations, 127, 129
Assembly time/physical parts, add features to reduce, 134
ASTM D149, 36
ASTM D257, 35
ASTM D638, 27, 29
ASTM D785, for plastics, 31
ASTM D523 Gardner gloss, 39
ASTM D790 test standard, 30
ASTM test method, 27
Attention, 140, 220, 225
Auditory properties, 44
Autodesk MoldFlow, 266

B

Background Knowledge, 1, 287
Big thinker, 291
  picture thinking, 291
Bill of materials (BOM) files, 182, 184, 197
  updation, 197
Blow molding, 68–70, 79, 82, 84
  extrusion blow molding (EBM), 68, 69
  injection blow molding (IBM), 69
  stretch blow molding (SBM), 70

C

Capacity-related tooling decisions, 89
Capacity target, 270
Capital asset, 231
Capital funding, 232
Capital investments, 232
Capital tooling, 230, 292
Captive tooling, 95
  customer base, 95
  vendor, 231
Capture all process parameters, 270
Cavity ID locations, 94
Cavity pressures, 80
Chemical stresses, 133
CO_2 emissions, 6
Commercialization, 118, 246
Communications, 250
Company management, 229
Comparative analysis, 115
Competitive products, 110
Compliance testing, 277
Computer aided design (CAD)
  layout process, 140, 142, 192
    3D, 197
    design, 139, 140
  program, 137

Concept specification, 120
Easy-Clean bird feeding apparatus development project, 107
Conceptual design, 116, 127, 128
balanced review of concepts against need, 118–119
concepts infringe for existing IP, 119
economic realities for need, 119
IP protection for solution for need, 119
solution having market appeal, 119
success chances, 118
value offering, 118
failure, due to proposed concepts, 120
documentation, 120
review, 118
Conceptual development, 120
Concurrent design engineering, 14
Consumption product, 226
Continuous process improvement, 283
Coring, 261
ejection system(s), 263
seal-off details, 263
Corrosive materials, 81
Cost estimates, 229
capital tooling, 230–231
vendor selection, 231
delivery date estimation, 230
fixtures for production/quality assurance, 232
injection mold, 231
material handling, 233–235
quality assurance (QA), 230
runner system design and gate design, 230
use, vendors potential, 230
Custom parts handling fixtures, 233
Custom tooling development engineering services, 230
Cut sheet thermoforming process
process steps required, graphic illustration of, 61

D

Degree of remedial changes, 197
Design decisions, 127
and functional specification, 290
Design effort, 2
Designer's domain, 93
Design for assembly (DFA) methods, 134
Design for Six Sigma (DFSS), 108, 109
Design from aesthetic perspective, 140
auditory considerations, 141
master external style CAD surface file, 140
surface finish, and color considerations, 140
surface touch/feel considerations, 140
weight distribution considerations, 141
Design from disassembly perspective, 159
Design from end-of-life perspective, 160
Design from end-use application, 132
additional structure to allow for changes in properties, 132
geometry to support anticipated misuse, 133
physical structure, adequate, for anticipated loads, 132
Design from labeling/packaging perspective, 142
structure required to
facilitate labeling process, 142
facilitate packaging method to be used, 142
Design from material flow perspective, 157
flow knit line locations *vs.* structural features, 159
part fill, 157
flow analysis run within SolidWorks Plastics, 158
residual stresses from molding, 158
venting cavity, 159
Design from part assembly perspective, 134
design for assembly (DFA) methods, 134
features to reduce assembly time/other physical parts, 134
sealing or joining features, 137
snap features, 139
Design from part/product safety perspective, 133
Design from tooling perspective, 142
cavitation considerations, 145
hot runner systems, 147
runner systems, 145
three-plate mold, 145
gate and flow considerations, 143
gate location, 143
type of gate, 144
venting considerations, 145
parting line, starting line, 142

undercuts, 147
lifters, 148
side-action slides, 148
slide-by sealing, 148
strip-off or "pop-off" features, 151
Design practitioners, 127
Design process, 13, 101, 117, 227, 292
Design project, 101, 102, 182, 185
Design review, 217
full scale design, 219
stakeholders gathering and review preparation, 217
Design rules review, 160
blow-molded parts, 169
bottom push-up features, 171
corner radii, 173
draft, 173
neck area transitions, 170
pinch-off areas, 171
ribs and tack-off features, 174
injection-molded parts, 160
bosses, 164
corner radii, 168
draft, 166
intersecting ribs, 161
uniform wall thickness, 161
Design solution, 130
Design team, 225
Design verification, 188
Desired criteria, 225
Destructive examples, 265
case example, 265
Detailed part design phase, 198
DFSS. *See* Design for Six Sigma (DFSS)
Dielectric strength, 275
Document, results and findings, 269
Durability, 132

E

EBM. *See* Extrusion blow molding (EBM)
Ejection systems, 129, 156
knock out ejector pins, 156
stripper plate, 156
Electrical insulators, 36
EMI/RFI shielding, 9, 40
End milling machine
black Delrin (acetal) part, machining of, 59
End-use environmental conditions, 132
End-users, 118
emotions, 130
satisfaction, 130
Energy conservation, 6
Environmental stress cracking (ESC), 10
ESC. *See* Environmental stress cracking (ESC)
Experienced practitioners, 289
External core features, 155
Extrusion, 79
Extrusion blow molding (EBM), 68, 69, 82, 83
mold cavities, 83

F

Failure modes and effects analysis (FMEA), 190
preliminary design, 190
review and update, 216
score sheet, 193
scoring/theoretical rating
for likelihood, 191
for severity, 191
FDM. *See* Fused deposition modeling (FDM)
Feature design, detailed, 198
breakout features, 198
case example, 200
3D CAD layout, for detailed part design, 198
2D specification-drawing document, 198
processing and tooling source, 200
representation, of removing individual part file from layout, 199
review and opinion, 200
rules, for materials and processes, 200
Final part/product safety, responsibility, 290
First capital costs, 229
evaluation of, 241
funding, 240
for injection mold tooling, 229
nonrecurring costs, 235
First process trials, 266
First tooling trials, 266
Flexible sheet manufacturing, 4
Flexural modulus, 30
Flexural strength, 31
Flexural stress *vs.* flexural strain, 30

Four pillars-complete integration, 287
foundation of solution, customer and need, 289
inclusion of stakeholders, from each of pillars in, 289
pillars take precedent, 288
risks in ignore one/more of pillars, 288
Functional specifications, 120–123
functional categories, 122
type of value, 122
value, 122
verification method and timing, 123
Funding initial, 229
Fused deposition modeling (FDM), 64, 65, 89, 182
additive manufacturing methods, 81
3D printing technology, 59
parts traditionally used for, 183

G

Gaining background knowledge, 2
Geometry, 128, 129
optimization, 116
to support anticipated misuse, 133
Gloss surfaces, 39
Glow wire ignition test, 37
Graphical tools, 128
Graphic language, 5

H

Hand-off to primary manufacturing and assembly, 279
review
material-handling requirements, 280
packaging requirements, 282
process requirements, 281
quality plan, 282
safety requirements, 280
tooling requirements, 281
transmittal of all documentation, 279
Hand-off to production, 275
Hazards, 280
High-capacity tooling, 88
High-cavitation molds, 91
High-level CAD illustrations, of trigger sprayer rapid fill concept, 115
Holistic visualization, 127, 129
Housekeeping policy, 250
Hydrolytic stability, 103, 275

I

Ideation, 101
session room, created during session, 111
IEC test, 37
Implementation
of design, 287
preparation, 221
Important qualification fact, 269
documentation, 269
Independent lab testing, for regulatory compliance, 237
Industrial design (ID), 5
firm, 139
Industrial designer, 5
Information goods, 108
Initial computer aided design (CAD) layout in 3D, 130
Initial solution sketching, 113–116
comparative analysis *versus* competitive products/concepts, 115
intellectual property (IP) and patents, initial review, 116
pencil sketches/high-level CAD visualization renderings, 113–114
written form descriptions, 113
Injection blow molding (IBM), 69, 82, 83
Injection molding, 19, 66, 79, 80, 81, 83, 86, 88, 95, 231
basic steps, 67
housing, 234
molding demands, 79
preform parts, 70
single high-pressure process, 79
Injection stretch blow molding (ISB), 82
Intellectual property (IP), 116
standpoint, 216
Internal features, 155
Intuitive instructions, 130
IP. *See* Intellectual property (IP)

K

Knowledge, 1, 99, 281. *See also* Working knowledge

L

LCD display modules, 238
Lead (Pb), 290
Lessons, 1, 285
Longer-term evaluations, 276

Long-term testing, 276
Low pressure gas connection, original design of, 138
Luggage roller assemblies, in manufacture, 133
Lustran 248 ABS, 27, 48

M

Machined from stock parts, 181
  annealing, 182
  stress concentration features, 181
Machining processes
  waste from, 58
Management, 248
Market appeal, 119
Marketing strategy, 114
Market opportunities, 292
Material degradation, 281
Material removal processes, 8
  computer numerical controlled machining, 8
  laser cutting, 8
  water jet cutting, 8
Material selection, conceptual stage, 116–117
Melt viscosity, 39
Mental exercises, 127
Modulus of elasticity, 27
Mold cavity, 127
Mold shrinkage values, 92
Mold trials, 262
Multicavity mold, 82, 92
  plastics mold, 81, 82

N

Needs
  listing, in reference patents for proposed bird feeding apparatus, 105
  for product, part, or assembly, 101–109
  qualification, 104
  specification, 106–107
  user for potential design project, 102
New design, qualification of, 275
New product development, 109
"Next generation" product development, 275
Nondestructive examples, 265
Non-Newtonian behavior, 40
Nonrecurring costs, 235
  estimation of, 229, 235
  assembly fixtures, design/development of, 235
  qualification testing, by independent laboratories, 237
  quality assurance/control fixtures, design/development of, 236
  regulatory compliance testing, by independent labs, 237
  test fixtures, design/development of, 236
Nonrecurring project tasks, 235

O

Operations team players, 227
"Optimum" design, 292
Optimum solution, 109
Organic growth, 119
Organizational tool, 108
O-ring support part moldings, in black PVC translucent amber PSU, 104
Overhead costs
  continuing production operations, 235
  equipment, 238

P

Packaging requirements, 282
Part design checking, 209
  corner radii and wall transitions, 212
  draft analysis, 211
    CAD generated draft analysis, 213
  gate location and mold fill analysis, 212, 214
  handwritten analysis, 210
  part assembly dimensional analysis, 211
  performance loading FEA stress analysis, 215
    representative analysis of snap feature, 215
  tolerance analysis, 209
Part design specification, specifying tool, 92–93
Part drawings, and specifications, 203
  cavity number and mold numbers, 209
  2D part drawings, with specifications, 204
    with notes and finished injection-molded part, 205
  example, of +Draft and –Draft, 210
  individual 3D data file, 204
  reference plaque, 207
  sample color chips, 206

Patent
pending, 249
protection, 106
review, 216
representative sample of illustrations, 218
PEEK, 90
Performance verification, 275, 276
Periodic production monitoring and review, 282
quality audit, 283
Per-unit costs, 237
estimating, 237
in production, 241
PET water bottle, 8
PFF. *See* Project fulfillment factor (PFF)
Physical properties, 132
Placing capital tooling orders, 253
Planned production implementation, 226
Planning costs, 225
Plastic part cost evaluation, 238
Plastic parts design, 3, 4
designer, 5, 11
four pillars, 11–15
application, 11
concurrent design engineering, 14
design, 13
materials, 12
processes, 12
quality considerations, 15
stakeholders involvement, 13, 14
tooling, 12
processes, 6, 12
quality assurance, 8
replacement material, 6
vs. product design, 4–6
Plastics, 7. *See also specific entries*
apocalypse, four horsemen of, 10
ban on, 6
design opportunities, 9
drawbacks, 7
in frictional contact, 45
optimum parts, 3, 4, 11, 13
product, 1, 99, 225
tactile properties of, 44
Plastics materials, 6, 9
aesthetic properties, 19
appearance properties, 43
benefits, 7
change in state, 8
classes
thermoplastic materials, 19
thermosetting materials, 19
economic efficiency, 7
economic properties
manufacturing efficiencies, 46
total energy input discussion, 47
weight reduction effects, 47
final unit in-use benefits, 48
part handling benefits, 47
part shipping benefits, 47
electrical properties
arc resistance, 36
dielectric constant, 36
dielectric strength, 36
dissipation factor, 36
high amp arc ignition (HAI), 36
surface resistivity, 35
volume resistivity, 35
flammability properties, 36
burning rate, 37
glow wire flammability index, 37
glow wire ignition temperature, 37
UL flammability rating, 37
formulations, 40
additives for lubrication/reduced friction effects, 41
aesthetic properties, 42
fillers for electrical/thermal conductivity, 40
flexibility/enhanced elongation, 42
sterilization processes, 41
stiffness/strength, 40
UV stability, 42
impact properties, 31–33
dart impact strength, 32
gardner impact strength, 33
notched impact strength, 31
unnotched impact strength, 32
mechanical properties, 27–31
coefficient of friction, 31
flexural modulus, 30
flexural strength, 31
flexural text machine, 30
hardness, 31
nominal tensile strain at break, 29
stress *vs.* strain curve, 28
tensile elongation, 28

tensile modulus, 27
tensile strength, 28
tensile test, 29
need, 11
optical properties
Gardner gloss, 39
Haze value, 39
transmittance, 39
packaging industry, use in, 6
physical properties, 19
processing benefits, 8
process-specific properties, 39
apparent viscosity, 40
melt viscosity, 39
recycle, 6
selection, 6–10, 19, 48–49
by aesthetic properties, 42–45
by economic properties, 45–48
by material properties, 21–42
for product development, 6–10
selection, risk management, 49
suitability of, 10
supply source, 14
thermal properties, 33–34
coefficient of linear thermal expansion (CLTE), 34
continuous uses, 33
deflection temperature under load (DTUL), 33
glass transition temperature, 34
specific heat, 34
thermal conductivity, 34
Vicat softening temperature, 34
tooling benefits, 8
Plastics, process selection, 51, 66, 75
by capacity (parts per year), 58–71
high-volume processes, 66–72
blow molding, 68–70
injection molding, 66–68
profile extrusion, 71
thermoforming, 71
low-volume processes, 58–64
fused deposition modeling (FDM), 64
machining, 58
reaction injection molding (RIM), 63
rotational molding, 60–62
thermoforming, 58–60
vacuum casting, 63
comparative attributes, 52
by considering secondary processes, 72–75
annealing, 74
assembly and joining, 72–73
welding, 73
electroplating, 74
machining, 73–74
painting, 73
plasma treatment, 75
by cost economics, 72
NRE and tooling, 72
production per-unit costing, 72
by functional requirements, 51–56
end-use, 56
strength considerations, 56
surface properties, 56
by other criteria, 75
part size considerations, 75
Plastics tooling, 79, 81, 83
capacity dependent, 88–89
manufacturing location dependent, 90
material dependent, 90
part design dependent, 91
process dependent, 79–86
high-pressure processes, 79–84
blow molding, 82–83
injection molding, 80–81
profile extrusion, 84
low-pressure processes, 86
resin casting/vacuum casting, 86
rotational molding, 86
thermoforming, 86
sourcing of, 93–95
contract based sourcing, 95
external sourcing, 95
internal sourcing, 93
Polymers, 7, 12
chain, 20
orientation, 132
clarity and transparency properties, 45
properties, 7
Polypropylene copolymer resin, 230
Polysulfone (PSU), 26, 103
Polyurethane paint system, 63
Polyvinylchloride (PVC), 81
Preliminary bill, of materials, 182
identify and list each part, 183
proposed process, 183
quantity required, 183
tentative material, 183

Preliminary design solution, review of, 186
critique concepts and design, 188
end-users feedback, 188
alternate materials and/or processes, 189
against budget and timeline, 189
compliance issues, hazards, and risks, 189
cost and lifespan targets, 189
plan B, 190
with respect to end-user/marketing "wants", 189
against written specification, 188
gather stakeholders, 186
engineering team, 186
marketing/sales, 188
material suppliers, 188
part manufacturing, 187
product assembly, 187
quality assurance, 188
tool designers, 186
Preliminary detailed design stage, action items upon departing, 192
design team, 192
legal, 194
manufacturing operations and/or strategic sourcing, 194
marketing/sales team, 192
quality assurance, 194
Preparation
file patent applications, 271
to move to production, 250
product qualification, 272
Pressure forming, 58–60
Price range, 225
Process capabilities, 76
Process machinery, 112
Process production capacity
comparisons of, 57
Process requirements, 281
Process selection, conceptual stage, 116
Process suitability, 270
Process variability studies, 270
Product begins, 275
Product design, 4, 287
Product designer, 3, 4, 11
Product development
plastics selection for, 6–10
project, 101
specifications, 106
Production estimates
cost estimates, 229
capital tooling, 230–231
delivery date estimation, 230
fixtures for production/quality assurance, 232
injection mold, 231
material handling, 233–235
quality assurance (QA), 230
runner system design and gate design, 230
selection of capital tooling vendor, 231
use, vendors potential, 230
nonrecurring costs, 229
assembly fixtures, design/development of, 235
qualification testing, by independent laboratories, 237
quality assurance/control fixtures, design/development of, 236
regulatory compliance testing, by independent labs, 237
test fixtures, design/development of, 236
project verification/implementation, 243
total cost evaluation, 241–242
unit cost estimates, 237
assembly/labor input costs, 240
financing costs, 240
freight costs, 239
packaging costs, 240
piece part costs, 238–239
quality costs, 240
Production injection-mold tooling, 63
Production qualification, 275
Product lifespan, 6
Product performance, 107, 155
verification, 275
Product safety, 120
Products design, 225
Profile extrusion, 180
corner radii, 181
die, 85
profile interior detail, 181
wall thickness, 180
Project fulfillment factor (PFF), 225
Project housekeeping, 120
Project management, 250

Project review, 249
Project validation, 250
Prototyping, 99
  during active design, 183
    prototyping cautions, 185
    visualization benefits, 185
  testing, 247
  tooling, 81, 86
P-20 steel, 230

Q

QFD. *See* Quality function deployment (QFD)
Qualification test data, 276
Quality assurance (QA), 194, 230, 282
Quality function deployment (QFD), 108–109
Quality plan, 278, 282
  for production (again management of risk), 263
    assist in adequate specification, of quality parameters, 264
  review and recommendation, 278
    functional in-process checks, 264
    performance tests, 264
    qualification reporting, 279
    verification of quality plan actions, 278

R

Reaction injection molding (RIM), 63, 64
Recyclable bits, 130
Residual stresses, 158
Resin casting, 86
Retail packaging, 281
Review preliminary tooling design layout, 255
  cooling (or heating), 261
  gates, 255
  knit line locations, 259
  parting line, 255
  runner systems, 259
  slides or lifters, 260
  tool materials, 261
  venting, 259
RIM. *See* Reaction injection molding (RIM)
Risk management, 49
Risk mitigation, for development, 281
RMA values, 191, 193
Rockwell hardness number, 31
Room temperature vulcanizing (RTV), 63
Rotational molding, 60–62, 86, 128
  higher volume applications, 60
  molded parts, 178
    closely spaced features or parallel walls, 179
    corner radii, 179
    draft, 179
    ribs, 179
    undercuts, 179
  schematic view, 62
  tool, 87
  vs. thermoforming, 62

S

Safety requirements, 280
Safety review, 277
Sand cast aluminum shell, 86
SBM. *See* Stretch blow molding (SBM)
Sealing/joining features, 137
Sealing surfaces, 137
Secondary operations, 127, 129
Semicrystalline materials, 19, 20
  chemical resistance of, 19
  in solid polymer, 20
Short-run plastics tooling, 86
Snaps, 139
  features, 139
    annular snap features, 155
    configurations, 139
  retention, 155
Society of Plastic Engineers ANTEC conference, 182
Solution ideation, 109–112
  alternatives, 112
  expansive creation, of solution possibilities, 109–111
  weed-out process, 111
Specification review, 275
Stakeholders, 118
Standard machine shop practices, 58
Standard test ASTM C351, 34
Sterilization processes, 41
Strength testing, 265
Stress *vs.* strain curve plot, 28

Stretch blow molding (SBM), 66, 84
high-capacity injection molding, 89
Structural capability, 132

T

Teambuilding, 110
Teflon, 41
Tensile elongation, 29
Tensile modulus, 27, 28
Tensile strain, 29
Tensile strength, 28, 31
Tensile testing
machine, 29
sample, 29
Test fixtures, 236
Thermoformed, 8, 12, 19, 58–60, 71, 175
corner radii, 178
cubitainers manufacturing, case study, 71
die, 88
draft, 175
ribs, 178
rotational molding, 89
sheet thickness *vs.* draw depth, 175
tooling, 88
in-line, 89
trimming/routing secondary operations, 178
undercuts, 178
Thermoplastic materials, 19
categories
amorphous, 19
semicrystalline, 19
Thermosetting materials, 19
Thinking processes, 127, 128
first process, 128
Three-dimensional (3D) computer aided design (CAD) layout, 130
Tool designers, 3
Tooling for plastics. *See* Plastics tooling
Tooling quality, 90
for part quality, 91
Tooling vendors, 225, 235
Toolmakers, 5
Tool quality
part quality, relationship to, 91
Total cost evaluation (TCE) method, 241
Trademark applications, 249
Transitioned into production, 275
Trimming operation, 178
Twin sheet forming, 58–60

U

UL-94
classification rating, 38
flammability rating test, 37
UL Prospector database, 27
UL test, 36
Unit cost estimates, 237
assembly/labor input costs, 240
financing costs, 240
freight costs, 239
packaging costs, 240
piece part costs, 238–239
quality costs, 240
Update detailed part drawings, 270
Users
classes, 106
for product, part, or assembly, 101–109
UV stability fillers, 42

V

Vacuum
casting, 63
forming, 58–60
Validation, of mold filling, 266
first part inspections, 267
first part performance testing, 267
Verification
of design, 246
review design against
budget and costing targets, 248
conceptual and final design review notes, 248
written functional specification, 247
review FMEA results for design, 248
review intellectual property analysis/opinion on design, 249
qualification, mandatory, 290
team, 245, 247
design and engineering team, 245
management team, 246

marketing team, 246
potential stakeholders, 246
quality assurance team, 246
reviews the information, 247
Visualization, 127, 128
Voice of Customer analysis, 108

W

Weatherability, 275
Welding, 73
hot air welding, 73
hot plate welding, 73
requirements, 73
solvent welding, 73
spin welding, 73
ultrasonic welding, 73
vibration welding, 73
Working knowledge, 1
Work transporting, fixtures for holding, 232

FSC
www.fsc.org
MIX
Papier | Fördert gute Waldnutzung
FSC® C083411

Kolibri 360 GmbH
Ferdinand-Jühlke-Straße 7
99095 Erfurt, Deutschland
produktsicherheit@kolibri360.de